T0176669

ORGANIC SYNTHESES

ORGANIC SYNTHESES

AN ANNUAL PUBLICATION OF SATISFACTORY METHODS FOR THE PREPARATION OF ORGANIC CHEMICALS

VOLUME 87
2010

PETER WIPF
VOLUME EDITOR

BOARD OF EDITORS

The procedures in this text are intended for use only by persons with prior training in the field of organic chemistry. In the checking and editing of these procedures, every effort has been made to identify potentially hazardous steps and to eliminate as much as possible the handling of potentially dangerous materials; safety precautions have been inserted where appropriate. If performed with the materials and equipment specified, in careful accordance with the instructions and methods in this text, the Editors believe the procedures to be very useful tools. However, these procedures must be conducted at one's own risk. Organic Syntheses, Inc., its Editors, who act as checkers, and its Board of Directors do not warrant or guarantee the safety of individuals using these procedures and hereby disclaim any liability for any injuries or damages claimed to have resulted from or related in any way to the procedures herein.

ORGANIC SYNTHESES

Out of print.
†*Deceased.*

*Out of print.
†Deceased.

*Out of print.
†Deceased.

*Out of print.
†Deceased.

NOTICE

Beginning with Volume 84, the Editors of *Organic Syntheses* initiated a new publication protocol, which is intended to shorten the time between submission of a procedure and its appearance as a publication. Immediately upon completion of the successful checking process, procedures are assigned volume and page numbers and are then posted on the Organic Syntheses website (www.orgsyn.org). The accumulated procedures from a single volume are assembled once a year and submitted for publication. The annual volume is published by John Wiley and Sons, Inc., and includes an index. The hard cover edition is available for purchase through the publisher. Incorporation of graphical abstracts into the Table of Contents began with Volume 77. Annual volumes 70–74, 75–79 and 80–84 have been incorporated into five-year versions of the collective volumes of *Organic Syntheses.* Collective Volumes IX, X and XI are available for purchase in the traditional hard cover format from the publishers. The Editors hope that the new Collective Volume series, appearing twice as frequently as the previous decennial volumes, will provide a permanent and timely edition of the procedures for personal and institutional libraries. The Editors welcome comments and suggestions from users concerning the new editions.

Organic Syntheses, Inc., joined the age of electronic publication in 2001 with the release of its web site (www.orgsyn.org). Organic Syntheses, Inc., fully funded the creation of the free website in a partnership with CambridgeSoft Corporation and Data-Trace Publishing Company. The site is accessible to most internet browsers using Macintosh and Windows operating systems and may be used with or without a ChemDraw plugin. Because of continually evolving system requirements, users should review software compatibility at the website prior to use. John Wiley & Sons, Inc., and Accelrys, Inc., partnered with Organic Syntheses, Inc., to develop the new database (www.mrw.interscience.wiley.com/osdb) that is available for license with internet solutions from John Wiley & Sons, Inc. and intranet solutions from Accelrys, Inc.

Both the commercial database and the free website contain all annual and collective volumes and indices of *Organic Syntheses*. Chemists can draw structural queries and combine structural or reaction transformation queries with full-text and bibliographic search terms, such as chemical name, reagents, molecular formula, apparatus, or even hazard warnings or phrases. The preparations are categorized into reaction types, allowing search by category. The contents of individual or collective volumes can be browsed by lists of titles, submitters' names, and volume and page references, with or without structural information.

The commercial database at www.mrw.interscience.wiley.com/osdb also enables the user to choose his/her preferred chemical drawing package, or to utilize several freely available plug-ins for entering queries. The user is also able to cut and paste existing structures and reactions directly into the structure search query or their preferred chemistry editor, streamlining workflow. Additionally, this database contains links to the full text of primary literature references via CrossRef, ChemPort, Medline, and ISI Web of Science. Links to local holdings for institutions using open url technology can also be enabled. The database user can limit his/her search to, or order the search results by, such factors as reaction type, percentage yield, temperature, and publication date, and can create a customized table of reactions for comparison. Connections to other Wiley references are currently made via text search, with cross-product structure and reaction searching to be added in the near future. Incorporations of new preparations will occur as new material becomes available.

INFORMATION FOR AUTHORS OF PROCEDURES

Organic Syntheses welcomes and encourages submissions of experimental procedures that lead to compounds of wide interest or that illustrate important new developments in methodology. Proposals for *Organic Syntheses* procedures will be considered by the Editorial Board upon receipt of an outline proposal as described below. A full procedure will then be invited for those proposals determined to be of sufficient interest. These full procedures will be evaluated by the Editorial Board, and if approved, assigned to a member of the Board for checking. In order for a procedure to be accepted for publication, each reaction must be successfully repeated in the laboratory of a member of the Editorial Board at least twice, with similar yields (generally ±5%) and selectivity to that reported by the submitters.

Organic Syntheses Proposals

A cover sheet should be included providing full contact information for the principal author and including a scheme outlining the proposed reactions (an *Organic Syntheses* Proposal Cover Sheet can be downloaded at orgsyn.org). Attach an outline proposal describing the utility of the methodology and/or the usefulness of the product. Identify and reference the best current alternatives. For each step, indicate the proposed scale, yield, method of isolation and purification, and how the purity of the product is determined. Describe any unusual apparatus or techniques required, and any special hazards associated with the procedure. Identify the source of starting materials. Enclose copies of relevant publications (attach pdf files if an electronic submission is used).

Submit proposals by mail or as e-mail attachments to:

Professor Charles K. Zercher
Associate Editor, Organic Syntheses
Department of Chemistry
University of New Hampshire
23 Academic Way, Parsons Hall
Durham, NH 03824

For electronic submissions: *org.syn@unh.edu*

Submission of Procedures

Authors invited by the Editorial Board to submit full procedures should prepare their manuscripts in accord with the Instructions to Authors which are described below or may be downloaded at orgsyn.org. Submitters are also encouraged to consult this volume of *Organic Syntheses* for models with regard to style, format, and the level of experimental detail expected in *Organic Syntheses* procedures. Manuscripts should be submitted to the Associate Editor. Electronic submissions are encouraged; procedures will be accepted as e-mail attachments in the form of Microsoft Word files with all schemes and graphics also sent separately as ChemDraw files.

Procedures that do not conform to the Instructions to Authors with regard to experimental style and detail will be returned to authors for correction. Authors will be notified when their manuscript is approved for checking by the Editorial Board, and it is the goal of the Board to complete the checking of procedures within a period of no more than six months.

Additions, corrections, and improvements to the preparations previously published are welcomed; these should be directed to the Associate Editor. However, checking of such improvements will only be undertaken when new methodology is involved.

NOMENCLATURE

Both common and systematic names of compounds are used throughout this volume, depending on which the Volume Editor felt was more appropriate. The Chemical Abstracts indexing name for each title compound, if it differs from the title name, is given as a subtitle. Systematic

Chemical Abstracts nomenclature, used in the Collective Indexes for the title compound and a selection of other compounds mentioned in the procedure, is provided in an appendix at the end of each preparation. Chemical Abstracts Registry numbers, which are useful in computer searching and identification, are also provided in these appendices.

ACKNOWLEDGMENT

Organic Syntheses wishes to acknowledge the contributions of Amgen, Inc. and Merck & Co. to the success of this enterprise through their support, in the form of time and expenses, of members of the Board of Editors.

INSTRUCTIONS TO AUTHORS

All organic chemists have experienced frustration at one time or another when attempting to repeat reactions based on experimental procedures found in journal articles. To ensure reproducibility, *Organic Syntheses* requires experimental procedures written with considerably more detail as compared to the typical procedures found in other journals and in the "Supporting Information" sections of papers. In addition, each *Organic Syntheses* procedure is carefully "checked" for reproducibility in the laboratory of a member of the Board of Editors.

Even with these more detailed procedures, the experience of *Organic Syntheses* editors is that difficulties often arise in obtaining the results and yields reported by the submitters of procedures. To expedite the checking process and ensure success, we have prepared the following "Instructions for Authors" as well as a *Checklist for Authors* and *Characterization Checklist* to assist you in confirming that your procedure conforms to these requirements. These checklists, which are available at *www.orgsyn.org,* should be completed and submitted together with your procedure. Procedures submitted to *Organic Syntheses* will be carefully reviewed upon receipt. Procedures lacking any of the required information will be returned to the submitters for revision.

Scale and Optimization

The appropriate scale for procedures will vary widely depending on the nature of the chemistry and the compounds synthesized in the procedure. However, some general guidelines are possible. For procedures in which the principal goal is to illustrate a synthetic method or strategy, it is expected, in general, that the procedure should result in at least 5 g and no more than 50 g of the final product. In cases where the point of the procedure is to provide an efficient method for the preparation of a useful reagent or synthetic building block, the appropriate scale may be larger, but in general should not exceed 100 g of final product. Exceptions to these guidelines may be granted in special circumstances. For example, procedures describing the preparation of reagents employed as catalysts will often be acceptable on a scale of less than 5 g.

In considering the scale for an *Organic Syntheses* procedure, authors should also take into account the cost of reagents and starting materials. In general, the Editors will not accept procedures for checking in which the cost of any one of the reactants exceeds $500 for a single full-scale run. Authors are requested to identify the most expensive reagent or starting material on the procedure submission checklist and to estimate its cost per run of the procedure.

It is expected that all aspects of the procedure will have been optimized by the authors prior to submission, and that each reaction will have been carried out at least twice on exactly the scale described in the procedure. It is appropriate to report the weight, yield, and purity of the product of each step in the procedure as a range. In any case where a reagent is employed in significant excess, a Note should be included explaining why an excess of that reagent is necessary. If possible, the Note should indicate the effect of using amounts of reagent less than that specified in the procedure.

Reaction Apparatus

Describe the size and type of flask (number of necks) and indicate how *every* neck is equipped.

"A 500-mL, three-necked, round-bottomed flask equipped with an overhead mechanical stirrer, 250-mL pressure-equalizing addition funnel fitted with an argon inlet, and a rubber septum is charged with...."

Indicate how the reaction apparatus is dried and whether the reaction is conducted under an inert atmosphere. This can be incorporated in the text of the procedure or included in a Note.

"The apparatus is flame-dried and maintained under an atmosphere of argon during the course of the reaction."

In the case of procedures involving unusual glassware or especially complicated reaction setups, authors are encouraged to include a photograph or drawing of the apparatus in the text or in a Note (for examples, see *Org. Syn.*, Vol. 82, 99 and Coll. Vol. X, pp 2, 3, 136, 201, 208, and 669).

Reagents and Starting Materials

All chemicals employed in the procedure must be commercially available or described in an earlier *Organic Syntheses* or *Inorganic Syntheses* procedure. For other compounds, a procedure should be included either as one or more steps in the text or, in the case of relatively straightforward preparations of reagents, as a Note. In the latter case, all requirements with regard to characterization, style, and detail also apply.

In one or more Notes, indicate the purity or grade of each reagent, solvent, etc. It is desirable to also indicate the source (company the chemical was purchased from), particularly in the case of chemicals where it is suspected that the composition (trace impurities, etc.) may vary from one supplier to another. In cases where reagents are purified, dried, "activated" (e.g., Zn dust), etc., a detailed description of the procedure used should be included in a Note. In other cases, indicate that the chemical was "used as received".

"Diisopropylamine (99.5%) was obtained from Aldrich Chemical Co., Inc. and distilled under argon from calcium hydride before use. THF (99+%) was obtained from Mallinckrodt, Inc. and distilled from sodium benzophenone ketyl. Diethyl ether (99.9%) was purchased from Aldrich Chemical Co., Inc. and purified by pressure filtration under argon through activated alumina. Methyl iodide (99%) was obtained from Aldrich Chemical Co., Inc. and used as received."

The amount of each reactant should be provided in parentheses in the order mL, g, mmol, and equivalents with careful consideration to the correct number of significant figures.

The concentration of solutions should be expressed in terms of molarity or normality, and not percent (e.g., 1 N HCl, 6 M NaOH, not "10% HCl").

Reaction Procedure

Describe every aspect of the procedure clearly and explicitly. Indicate the order of addition and time for addition of all reagents and how each is added (via syringe, addition funnel, etc.).

Indicate the temperature of the reaction mixture (preferably internal temperature). Describe the type of cooling (e.g., "dry ice-acetone bath") and heating (e.g., oil bath, heating mantle) methods employed. Be careful to describe clearly all cooling and warming cycles, including initial and final temperatures and the time interval involved.

Describe the appearance of the reaction mixture (color, homogeneous or not, etc.) and describe all significant changes in appearance during the course of the reaction (color changes, gas evolution, appearance of solids, exotherms, etc.).

Indicate how the reaction can be monitored to determine the extent of conversion of reactants to products. In the case of reactions monitored by TLC, provide details in a Note, including eluent, R_f values, and method of visualization. For reactions followed by GC, HPLC, or NMR analysis, provide details on analysis conditions and relevant diagnostic peaks.

"The progress of the reaction was followed by TLC analysis on silica gel with 20% EtOAc-hexane as eluent and visualization with *p*-anisaldehyde. The ketone starting material has $R_f = 0.40$ (green) and the alcohol product has $R_f = 0.25$ (blue)."

Reaction Workup

Details should be provided for reactions in which a "quenching" process is involved. Describe the composition and volume of quenching agent, and time and temperature for addition. In cases where reaction mixtures are added to a quenching solution, be sure to also describe the setup employed.

"The resulting mixture was stirred at room temperature for 15 h, and then carefully poured over 10 min into a rapidly stirred, ice-cold aqueous solution of 1 N HCl in a 500-mL Erlenmeyer flask equipped with a magnetic stirbar."

For extractions, the number of washes and the volume of each should be indicated.

For concentration of solutions after workup, indicate the method and pressure and temperature used.

"The reaction mixture is diluted with 200 mL of water and transferred to a 500-mL separatory funnel, and the aqueous phase is separated and extracted with three 100-mL portions of ether. The combined organic layers are washed with 75 mL of water and 75 mL of saturated NaCl solution, dried over MgSO₄, filtered, and concentrated by rotary evaporation (25°C, 20 mmHg) to afford 3.25 g of a yellow oil."

"The solution is transferred to a 250-mL, round-bottomed flask equipped with a magnetic stirbar and a 15-cm Vigreux column fitted with a short path distillation head, and then concentrated by careful distillation at 50 mmHg (bath temperature gradually increased from 25 to 75°C)."

In cases where solid products are filtered, describe the type of filter funnel used and the amount and composition of solvents used for washes.

"... and the resulting pale yellow solid is collected by filtration on a Büchner funnel and washed with 100 mL of cold (0°C) hexane."

When solid or liquid compounds are dried under vacuum, indicate the pressure employed (rather than stating "reduced pressure" or "dried *in vacuo*").

"... and concentrated at room temperature by rotary evaporation (20 mmHg) and then at 0.01 mmHg to provide. ..."

"The resulting colorless crystals are transferred to a 50-mL, round-bottomed flask and dried overnight in a 100°C oil bath at 0.01 mmHg."

Purification: Distillation

Describe distillation apparatus including the size and type of distillation column. Indicate temperature (and pressure) at which all significant fractions are collected.

"... and transferred to a 100-mL, round-bottomed flask equipped with a magnetic stirbar. The product is distilled under vacuum through a 12-cm, vacuum-jacketed column of glass helices (Note 16) topped with a Perkin triangle. A forerun (ca. 2 mL) is collected and discarded, and the desired product is then obtained, distilling at 50–55°C (0.04–0.07 mmHg). ..."

Purification: Column Chromatography

Provide information on TLC analysis in a Note, including eluent, R_f values, and method of visualization.

Provide dimensions of column and amount of silica gel used; in a Note indicate source and type of silica gel.

Provide details on eluents used, and number and size of fractions.

"The product is charged on a column (5 × 10 cm) of 200 g of silica gel (Note 15) and eluted with 250 mL of hexane. At that point, fraction collection (25-mL fractions) is begun, and elution is continued with 300 mL of 2% EtOAc-hexane (49:1 hexanes:EtOAc) and then 500 mL of 5% EtOAc-hexane (19:1 hexanes:EtOAc). The desired product is obtained in fractions 24–30, which are concentrated by rotary evaporation (25°C, 15 mmHg). ..."

Purification: Recrystallization

Describe procedure in detail. Indicate solvents used (and ratio of mixed solvent systems), amount of recrystallization solvents, and temperature protocol. Describe how crystals are isolated and what they are washed with.

"The solid is dissolved in 100 mL of hot diethyl ether (30°C) and filtered through a Büchner funnel. The filtrate is allowed to cool to room temperature, and 20 mL of hexane is added. The solution is cooled at −20°C overnight and the resulting crystals are collected by suction filtration on a Büchner funnel, washed with 50 mL of ice-cold hexane, and then transferred to a 50-mL, round-bottomed flask and dried overnight at 0.01 mmHg to provide. . . ."

Characterization

Physical properties of the product such as color, appearance, crystal forms, melting point, etc. should be included in the text of the procedure. Comments on the stability of the product to storage, etc. should be provided in a Note.

In a Note, provide data establishing the identity of the product. This will generally include IR, MS, ^{1}H-NMR, and ^{13}C-NMR data, and in some cases UV data. Copies of the proton and carbon NMR spectra for the products of each step in the procedure should be submitted showing integration for all resonances. Submission of copies of NMR spectra for other nuclei are encouraged as appropriate.

In the same Note, provide quantitative analytical data establishing the purity of the product. Elemental analysis for carbon and hydrogen (and nitrogen if present) agreeing with calculated values within 0.4% is preferred. However, GC data (for distilled or vacuum-transferred samples) and/or HPLC data (for material isolated by column chromatography) may be acceptable in some cases. Provide details on equipment and conditions for GC and HPLC analyses.

In procedures involving non-racemic, enantiomerically enriched products, optical rotations should generally be provided, but enantiomeric purity must be determined by another method such as chiral HPLC or GC analysis.

In cases where the product of one step is used without purification in the next step, a Note should be included describing how a sample of the product can be purified and providing characterization data for the pure material. Copies of the proton NMR spectra of both the product both *before* and *after* purification should be submitted.

Hazard Warnings

Any significant hazards should be indicated in a statement at the beginning of the procedure in italicized type. Efforts should be made to avoid the use of toxic and hazardous solvents and reagents when less hazardous alternatives are available.

Discussion Section

The style and content of the discussion section will depend on the nature of the procedure.

For procedures that provide an improved method for the preparation of an important reagent or synthetic building block, the discussion should focus on the advantages of the new approach and should describe and reference all of the earlier methods used to prepare the title compound.

In the case of procedures that illustrate an important synthetic method or strategy, the discussion section should provide a mini-review on the new methodology. The scope and limitations of the method should be discussed, and it is generally desirable to include a table of examples. Competing methods for accomplishing the same overall transformation should be described and referenced. A brief discussion of mechanism may be included if this is useful for understanding the scope and limitations of the method.

Style and Format

Articles should follow the style guidelines used for organic chemistry articles published in the ACS journals such as *J. Am. Chem. Soc., J. Org. Chem., Org. Lett.* etc. as described in the ACS Style Guide (3rd Ed.). The text of the procedure should be constructed using a standard word processing program, like MS Word, with 14-point Times New Roman font. Chemical structures and schemes should be drawn using the standard ACS drawing parameters (in ChemDraw, the parameters are found in the "ACS Document 1996" option) with a maximum width of 6 inches. The graphics files should be inserted into the document at the correct location and the graphics files should also be submitted separately. All Tables that include structures should be entirely prepared in the graphics (ChemDraw) program and inserted into the word processing file at the appropriate location. Tables that include multiple, separate graphics files prepared in the word processing program will require modification.

Biographies and Photographs of Authors

Photographs and 100-word biographies of all authors should be submitted as separate files at the time of the submission of the procedure. The format of the biographies should be similar to those in the Volume 84 procedures found at the orgsyn.org website. Photographs can be accepted in a number of electronic formats, including tiff and jpeg formats.

HANDLING HAZARDOUS CHEMICALS

A Brief Introduction

General Reference: *Prudent Practices in the Laboratory*; National Academy Press; Washington, DC, 1995.

Physical Hazards

Fire. Avoid open flames by use of electric heaters. Limit the quantity of flammable liquids stored in the laboratory. Motors should be of the nonsparking induction type.

Explosion. Use shielding when working with explosive classes such as acetylides, azides, ozonides, and peroxides. Peroxidizable substances such as ethers and alkenes, when stored for a long time, should be tested for peroxides before use. Only sparkless "flammable storage" refrigerators should be used in laboratories.

Electric Shock. Use 3-prong grounded electrical equipment if possible.

Chemical Hazards

Because all chemicals are toxic under some conditions, and relatively few have been thoroughly tested, it is good strategy to minimize exposure to all chemicals. In practice this means having a good, properly installed hood; checking its performance periodically; using it properly; carrying out all operations in the hood; protecting the eyes; and, since many chemicals can penetrate the skin, avoiding skin contact by use of gloves and other protective clothing at all times.

a. Acute Effects. These effects occur soon after exposure. The effects include burn, inflammation, allergic responses, damage to the eyes, lungs, or nervous system (e.g., dizziness), and unconsciousness or death (as from overexposure to HCN). The effect and its cause are usually obvious and so are the methods to prevent it. They generally arise from inhalation or skin contact, so should not be a problem if one follows the admonition "work in a hood and keep chemicals off your hands". Ingestion is a rare route, being generally the result of eating in the laboratory or not washing hands before eating.

b. Chronic Effects. These effects occur after a long period of exposure or after a long latency period and may show up in any of numerous organs. Of the chronic effects of chemicals, cancer has received the most attention lately. Several dozen chemicals have been demonstrated to be carcinogenic in man and hundreds to be carcinogenic to animals. Although there is no simple correlation between carcinogenicity in animals and in man, there is little doubt that a significant proportion of the chemicals used in laboratories have some potential for carcinogenicity in man. For this and other reasons, chemists should employ good practices at all times.

The key to safe handling of chemicals is a good, properly installed hood, and the referenced book devotes many pages to hoods and ventilation. It recommends that in a laboratory where people spend much of their time working with chemicals there should be a hood for each two people, and each should have at least 2.5 linear feet (0.75 meter) of working space at it. Hoods are more than just devices to keep undesirable vapors from the laboratory atmosphere. When closed they provide a protective barrier between chemists and chemical operations, and they are a good containment device for spills. Portable shields can be a useful supplement to hoods, or can be an alternative for hazards of limited severity, e.g., for small-scale operations with oxidizing or explosive chemicals.

Specialized equipment can minimize exposure to the hazards of laboratory operations. Impact resistant safety glasses are basic equipment and should be worn at all times. They may be supplemented by face shields or goggles for particular operations, such as pouring corrosive liquids. Because skin contact with chemicals can lead to skin irritation or sensitization or, through absorption, to effects on internal organs, protective gloves should be worn at all times.

Laboratories should have fire extinguishers and safety showers. Respirators should be available for emergencies. Emergency equipment should be kept in a central location and must be inspected periodically.

MSDS (Materials Safety Data Sheets) sheets are available from the suppliers of commercially available reagents, solvents, and other chemical materials; anyone performing an experiment should check these data sheets before initiating an experiment to learn of any specific hazards associated with the chemicals being used in that experiment.

DISPOSAL OF CHEMICAL WASTE

General Reference: *Prudent Practices in the Laboratory* National
Academy Press, Washington, D.C. 1996.

Effluents from synthetic organic chemistry fall into the following
categories:

1. **Gases**

    1a. Gaseous materials either used or generated in an organic
    reaction.
    1b. Solvent vapors generated in reactions swept with an inert gas
    and during solvent stripping operations.
    1c. Vapors from volatile reagents, intermediates and products.

2. **Liquids**

    2a. Waste solvents and solvent solutions of organic solids (see
    item 3b).
    2b. Aqueous layers from reaction work-up containing volatile
    organic solvents.
    2c. Aqueous waste containing non-volatile organic materials.
    2d. Aqueous waste containing inorganic materials.

3. **Solids**

    3a. Metal salts and other inorganic materials.
    3b. Organic residues (tars) and other unwanted organic materials.
    3c. Used silica gel, charcoal, filter aids, spent catalysts and the like.

The operation of industrial scale synthetic organic chemistry in an
environmentally acceptable manner* requires that all these effluent categories be dealt with properly. In small scale operations in a research or

*An environmentally acceptable manner may be defined as being both in
compliance with all relevant state and federal environmental regulations *and*
in accord with the common sense and good judgment of an environmentally
aware professional.

academic setting, provision should be made for dealing with the more environmentally offensive categories.

1a. Gaseous materials that are toxic or noxious, e.g., halogens, hydrogen halides, hydrogen sulfide, ammonia, hydrogen cyanide, phosphine, nitrogen oxides, metal carbonyls, and the like.

1c. Vapors from noxious volatile organic compounds, e.g., mercaptans, sulfides, volatile amines, acrolein, acrylates, and the like.

2a. All waste solvents and solvent solutions of organic waste.

2c. Aqueous waste containing dissolved organic material known to be toxic.

2d. Aqueous waste containing dissolved inorganic material known to be toxic, particularly compounds of metals such as arsenic, beryllium, chromium, lead, manganese, mercury, nickel, and selenium.

3. All types of solid chemical waste.

Statutory procedures for waste and effluent management take precedence over any other methods. However, for operations in which compliance with statutory regulations is exempt or inapplicable because of scale or other circumstances, the following suggestions may be helpful.

Gases

Noxious gases and vapors from volatile compounds are best dealt with at the point of generation by "scrubbing" the effluent gas. The gas being swept from a reaction set-up is led through tubing to a large trap to prevent suck-back and into a sintered glass gas dispersion tube immersed in the scrubbing fluid. A bleach container can be conveniently used as a vessel for the scrubbing fluid. The nature of the effluent determines which of four common fluids should be used: dilute sulfuric acid, dilute alkali or sodium carbonate solution, laundry bleach when an oxidizing scrubber is needed, and sodium thiosulfate solution or diluted alkaline sodium borohydride when a reducing scrubber is needed. Ice should be added if an exotherm is anticipated.

Larger scale operations may require the use of a pH meter or starch/iodide test paper to ensure that the scrubbing capacity is not being exceeded.

When the operation is complete, the contents of the scrubber can be poured down the laboratory sink with a large excess (10–100 volumes) of water. If the solution is a large volume of dilute acid or base, it should be neutralized before being poured down the sink.

Liquids

Every laboratory should be equipped with a waste solvent container in which *all* waste organic solvents and solutions are collected. The contents of these containers should be periodically transferred to properly labeled waste solvent drums and arrangements made for contracted disposal in a regulated and licensed incineration facility.**

Aqueous waste containing dissolved toxic organic material should be decomposed *in situ*, when feasible, by adding acid, base, oxidant, or reductant. Otherwise, the material should be concentrated to a minimum volume and added to the contents of a waste solvent drum.

Aqueous waste containing dissolved toxic inorganic material should be evaporated to dryness and the residue handled as a solid chemical waste.

Solids

Soluble organic solid waste can usually be transferred into a waste solvent drum, provided near-term incineration of the contents is assured.

Inorganic solid wastes, particularly those containing toxic metals and toxic metal compounds, used Raney nickel, manganese dioxide, etc. should be placed in glass bottles or lined fiber drums, sealed, properly labeled, and arrangements made for disposal in a secure landfill.** Used mercury is particularly pernicious and small amounts should first be amalgamated with zinc or combined with excess sulfur to solidify the material.

Other types of solid laboratory waste including used silica gel and charcoal should also be packed, labeled, and sent for disposal in a secure landfill.

Special Note

Since local ordinances may vary widely from one locale to another, one should always check with appropriate authorities. Also, professional disposal services differ in their requirements for segregating and packaging waste.

**If arrangements for incineration of waste solvent and disposal of solid chemical waste by licensed contract disposal services are not in place, a list of providers of such services should be available from a state or local office of environmental protection.

PREFACE

I have not failed, I've just found 10,000 ways that don't work.

THOMAS A. EDISON

Obviously, and sadly, Mr. Edison did not yet have access to *Organic Syntheses*. Otherwise, he could have consulted one of the thousands of experimental protocols that have been reproduced and thus validated independently in select laboratories after their submission to the Board of Editors of *Organic Syntheses*. For 89 years, this mechanism has created a "gold standard" of experimentation in synthetic organic chemistry, and thus it has greatly diminished the number of experimental trials and errors which frustrate the novice and the expert alike.

As noted by my predecessor Volume Editors, the impact of *Organic Syntheses* is as steadily increasing as the exponential growth of new synthetic transformations continues unabated. Nobody has the opportunity anymore to receive training in all experimental methods during their undergraduate, graduate, and postdoctoral careers, and there is less and less time in discovery and process research set aside to prepare the Product. Thus, we all have to rely on the accuracy of synthetic procedures published in the primary literature, while, nonetheless, the traditional peer review of experimental protocols has significantly diminished as these sections are relegated to the ubiquitous "Supporting Information". *Organic Syntheses* retains the focus on the quality and reproducibility of crucial experimental procedures. All transformations are described in extensive detail, thereby eliminating most of the barriers toward quick dissemination and straightforward adaptation in the field.

Volume 87 contains thirty-eight procedures that reflect the current breadth and ingenuity of synthetic organic chemistry. The order of presentation is chronological in the order that they were posted on the *Organic Syntheses* website (www.orgsyn.org). In contrast, my description of the contents of this volume attempts to highlight common features that some of these protocols share and that might distinguish this particular issue from past and future volumes.

Interestingly, and somewhat unexpectedly, procedures dealing with the introduction of halogen atoms form a majority of procedures in this volume of *Organic Syntheses*, edging out the usual array of enantioselective methods. Zhong and Bulger present a practical *N*-chlorination protocol (p. 8), and Li, Buzon, and Zhang report on a regioselective 4-bromination of oxazoles (p. 16). The preparation of the SALDIPAC catalyst and its use in a cobalt-catalyzed hydrochlorination of a terminal alkene is demonstrated by Gaspar, Waser, and Carreira (p. 88). An (*E*)-iodoalkene is prepared from a benzyl bromide by Bull, Mousseau and Charette (p. 170), and Chalker, Thompson, and Davis report a safe and scalable preparation of the mild brominating agent pioneered by Barluenga (p. 288). Fluorinations are industrially very significant, and Kong, Meng, Ting and Wong as well as Chang, Lee and Bae showcase electrophilic fluorinations with NFSI and DEOXO-FLUOR®, respectively (p. 137 and 245). A trifluoromethyl group is conveniently introduced by the copper(I)-mediated coupling of a vinyl bromide at C-2 of a steroidal enone by Fei, Tian, Ding, Wang and Chen (p. 126).

Among enantioselective methods for the preparation of chiral products, reductions, 1,2-additions, 1,4-additions, and S_N2-reactions, as well as diaminations, are prominently featured. Stepanenko, Huang, and Ortiz-Marciales present the preparation of a spiroborate ester and its application in the catalytic enantioselective reduction of oximes (p. 26 and 36). Turlington, DeBerardinis, and Pu apply a chiral binaphthol diamine for the addition of *in-situ* generated arylzinc organometallics to aldehydes (p. 59 and 68). For the preparation of a beta-amino carboxylate, Davies, Fletcher and Roberts utilize a homochiral ammonia equivalent in a conjugate addition reaction (p. 143). A chiral phosphoramidite is used by Zhao et al. in the Pd-catalyzed diamination of alkenes with a sterically hindered diaziridinone (p. 263). The preparation and application of a PYBOX ligand is featured by Lou and Fu in two Ni-catalyzed asymmetric S_N2−displacements (p. 310, 317, and 330). These researchers also contribute a Pd-catalyzed achiral alkyl-alkyl Suzuki cross-coupling (p. 299). Other Pd-catalyzed cross-couplings include a methylation of bromoarenes with air-stable DABAL-Me$_3$ by Vinogradov and Woodward (p. 104), and a one-pot diazotization and Heck reaction of anthranilate by Zaragoza (p. 226). A Cu-diamine complex is used by Coste, Couty, and Evano for the synthesis of ynamides from 1,1-dibromoalkenes (p. 231).

C,H-Bond activations of arenes also make a prominent appearance in volume 87. A Cu-catalyzed arylation of *m*-iodotoluene is described by Alvarado, Do, and Daugulis (p. 184), and a catalytic Ru-complex is used by Kitazawa, Kochi, and Kakiuchi for the arylation of the *o*-C,H-bond in a tetralone (p. 209).

A growing percentage of the current literature is dedicated to organo-catalytic methods, and several of these, as well as the synthesis of important types of organocatalysts, are featured in volume 87. Beno-houd, Erkkilä and Pihko use simple pyrrolidine to perform an efficient alpha-methylenation of aldehydes (p. 201). Another extremely useful compound class for many organocatalytic transformations are the tri-azolium carbene precursors prepared by Rovis and coworkers, as well as Struble and Bode (p. 350 and 362).

New methods for the preparation of fundamental functional groups, such as carboxamides, are introduced by Ekoue-Kovi and Wolf in the metal-free oxidative amination of aldehydes (p. 1), and Ju and Bode in the decarboxylative condensation of alpha-keto acids and hydroxylamines (p. 218). Another functional group interconversion of general significance is presented by Zhao et al. in the synthesis of enamides from ketones (p. 275). A stereoselective Cu-catalyzed *syn*-hydroarylation of alkynoates by Kirai and Yamamoto (p. 53) and the use of an iodonium ylide for the cyclopropanation of styrene by Goudreau, Marcoux, and Charette (p. 115) round off the collection of modern synthetic transformations.

In addition to showcasing efficient synthetic methods, *Organic Syntheses* also maintains a stable of preparations of useful and interesting compounds, and volume 78 provides several examples in this category. (Chloromethyl)diphenylsilane is reported by Murakami, Yorimitsu, and Oshima (p. 178), and Alaimo et al. describe a simple protocol for 1,3,5-triacetylbenzene (p. 192). Microwave acceleration is utilized by Hans and Delaude in the synthesis of 1,3-dimesitylimidazolium chloride (p. 77). Cyclohexene imine can be readily accessed from cyclohexene oxide, as shown by Watson, Afagh and Yudin (p. 161). A highly func-tionalized cyclobutene derivative is prepared by Shubinets, Schramm, and Kozmin by a [2+2] cycloaddition process (p. 253).

The need for new synthetic methods for heterocycle synthesis contin-ues unabated, and in this volume, indazoles and indazolobenzoxazines are featured. Shi and Larock use a [3+2] cycloaddition to prepare an indazole building block (p. 95). Finally, Solano et al. report two effi-cient protocols for the synthesis of tetracyclic indazolobenzoxazines (p. 339).

As it becomes amply evident by this summary, the wealth of synthetic know-how and detailed preparative guidance in past, current, and future volumes of *Organic Syntheses* provides the databank which is needed to avoid or minimize many frustrating empirical reaction optimization and troubleshooting operations in the experimental laboratory. Although the printed publications will soon be displaced by electronic files, in one way or another, *Organic Syntheses* will therefore continue to occupy a prominent place on the bookshelves or computer desktops of synthetic organic chemists. The publication of this work would not be possible without the dedicated efforts of the authors and their coworkers (the Submitters), the Board of Editors and their coworkers (the Checkers), as well as the guidance of Jerry Freeman and Rick Danheiser (past Secretary to the Board and current Editor in Chief, respectively, during my tenure), Chuck Zercher (Associate Editor), and Carl Johnson (Treasurer). I would like to thank them as well as my past and current colleagues on the Board of Editors and Board of Directors for pointing out the way and generating the enduring legacy of *Organic Syntheses*. It is a very special privilege to be part of this group and this tradition.

PETER WIPF
Pittsburgh, Pennsylvania

CONTENTS

$$n\text{-}C_5H_{11}\text{---}\!\!\equiv\!\!\text{---}CO_2Me \;+\; \text{(phenyl)}\text{---}B(OH)_2 \quad \xrightarrow[\text{MeOH, 28 °C}]{\text{cat. CuOAc}} \quad \text{(product)}\; n\text{-}C_5H_{11}\quad CO_2Me$$

$$\text{(octahydro-binaphthol)} \;+\; HO\!-\!CH_2\!-\!N(\text{morpholine}) \quad \xrightarrow{\text{Dioxane, 60 °C, 18 h}} \quad \text{(bis-morpholinomethyl product)}$$

a. ZnEt$_2$, Li(acac), NMP, 0 °C, 12 h

b. cat (*S*)-1, THF, 0 °C, 1 h

c. cyclohexanecarboxaldehyde, rt, 16 h

(*S*)-1

SALDIPAC

PALLADIUM-CATALYZED CROSS-COUPLING USING AN AIR-STABLE TRIMETHYLALUMINUM SOURCE. PREPARATION OF ETHYL 4-METHYLBENZOATE

Andrej Vinogradov and Simon Woodward

SYNTHESIS OF DIMETHYL 2-PHENYLCYCLOPROPANE-1,1-DICARBOXYLATE USING AN IODONIUM YLIDE DERIVED FROM DIMETHYL MALONATE

Sébastien R. Goudreau, David Marcoux and André B. Charette

NEW, CONVENIENT ROUTE FOR TRIFLUOROMETHYLATION OF STEROIDAL MOLECULES

Xiang-Shu Fei, Wei-Sheng Tian, Kai Ding, Yun Wang and Qing-Yun Chen

DIRECT FLUORINATION OF THE CARBONYL GROUP OF BENZOPHENONES USING DEOXO-FLUOR®: PREPARATION OF BIS(4-FLUOROPHENYL) DIFLUOROMETHANE

Ying Chang, Hyelee Lee, Chulsung Bae

PREPARATION AND [2+2] CYCLOADDITION OF 1-TRIISOPROPYLSILOXY-1-HEXYNE WITH METHYL CROTONATE: 3-BUTYL-4-METHYL-2-TRIISOPROPYLSILOXY-CYCLOBUT-2-ENECARBOXYLIC ACID METHYL ESTER

Valeriy Shubinets, Michael P. Schramm, and Sergey A. Kozmin

Pd(0)-CATALYZED ASYMMETRIC ALLYLIC AND HOMOALLYLIC DIAMINATION OF 4-PHENYL-1-BUTENE WITH DI-*TERT*-BUTYLDIAZIRIDINONE

Baoguo Zhao, Haifeng Du, Renzhong Fu, and Yian Shi

SYNTHESIS OF ENAMIDES FROM KETONES: PREPARATION OF *N*-(3,4-DIHYDRONAPHTHALENE-1-YL)ACETAMIDE

Hang Zhao, Charles P. Vandenbossche, Stefan G. Koenig, Surendra P. Singh, and Roger P. Bakale

SAFE AND SCALABLE PREPARATION OF BARLUENGA'S REAGENT

Justin M. Chalker, Amber L. Thompson, and Benjamin G. Davis

PALLADIUM-CATALYZED ALKYL-ALKYL SUZUKI CROSS-COUPLINGS OF PRIMARY ALKYL BROMIDES AT ROOM TEMPERATURE: (13-CHLOROTRIDECYLOXY)TRIETHYLSILANE

Sha Lou and Gregory C. Fu

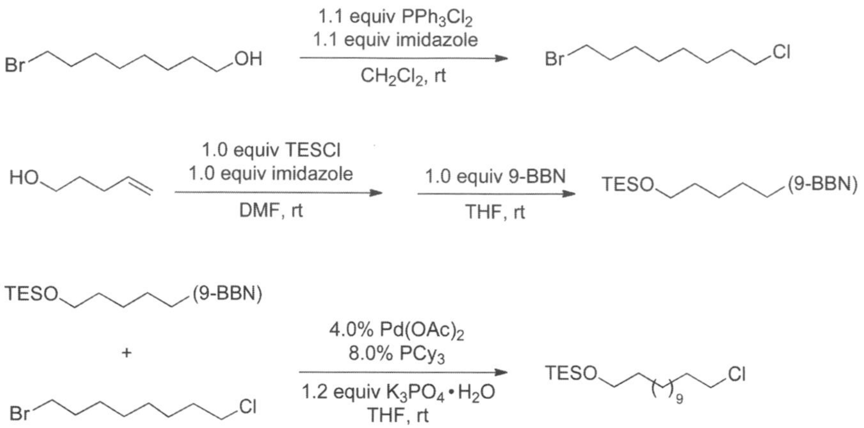

SYNTHESIS OF CHIRAL PYRIDINE BIS(OXAZOLINE) LIGANDS FOR NICKEL-CATALYZED ASYMMETRIC NEGISHI CROSS-COUPLINGS OF SECONDARY ALLYLIC CHLORIDES WITH ALKYLZINCS: 2,6-BIS[(4R)-4,5-DIHYDRO-4- (2-PHENYLETHYL)-2-OXAZOLYL]-PYRIDINE

Sha Lou and Gregory C. Fu

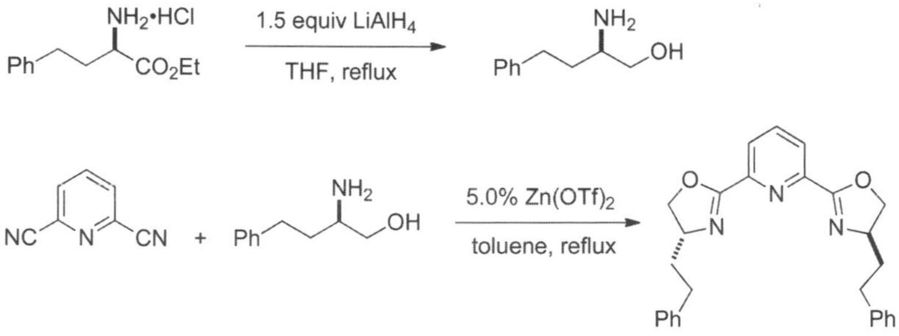

COUPLINGS OF RACEMIC SECONDARY ALLYLIC CHLORIDES WITH ALKYLZINCS: (*S,E*)-ETHYL 6-(1,3-DIOXOLAN-2-YL)-4-METHYLHEX-2-ENOATE
Sha Lou and Gregory C. Fu

CROSS-COUPLINGS OF RACEMIC SECONDARY α-BROMO AMIDES WITH ALKYLZINC REAGENTS: (*S*)-*N*-BENZYL-7-CYANO-2-ETHYL-*N*-PHENYLHEPTANAMIDE
Sha Lou and Gregory C. Fu

SYNTHESIS OF A *N*-MESITYL SUBSTITUTED AMINOINDANOL-DERIVED TRIAZOLIUM SALT

Justin R. Struble and Jeffrey W. Bode

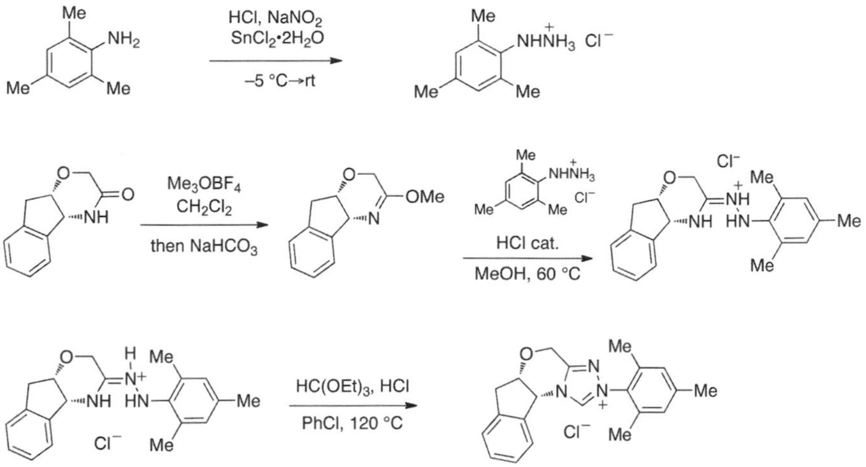

ERRATA

PREPARATION OF *O*-ALLYL-*N*-(9-ANTHRACENYL-METHYL)-CINCHONIDINIUM BROMIDE AS A PHASE TRANSFER CATALYST

E. J. Corey and Mark C. Noe

METAL-FREE ONE-POT OXIDATIVE AMINATION OF AROMATIC ALDEHYDES: CONVERSION OF BENZALDEHYDE TO *N*-BENZOYL PYRROLIDINE

Submitted by Kekeli Ekoue-Kovi and Christian Wolf.[1]
Checked by Vikram Bhat and Viresh Rawal.

1. Procedure[2]

A 250-mL two-necked, round-bottomed flask equipped with a stirring bar, a septum, and a reflux condenser fitted with a nitrogen adapter connected to a dual manifold is charged with nitrogen. The flask is immersed in an oil bath equipped with an external thermometer. An inert atmosphere is maintained during the course of the reaction. Anhydrous acetonitrile (50 mL) (Note 1) is added through the rubber septum via the repeated use of a 20-mL syringe. Benzaldehyde (5.71 mL, 6.0 g, 57 mmol, 1.0 equiv) (Note 2) is then added through the rubber septum using both a five-mL and a one-mL syringe. Pyrrolidine (5.66 mL, 4.82 g, 68 mmol, 1.0 equiv) (Note 3) is added next by a five-mL syringe followed by a solution of *tert*-butyl hydroperoxide in decane (12.4 mL, 62–74 mmol, 1.09–1.30 equiv) (Notes 4 and 5) using a 10-mL syringe. While maintaining a positive nitrogen pressure, the rubber septum is replaced with a glass stopper. After stirring for 5 min, the yellow reaction mixture is heated in the oil bath to 70 °C for 6 h, and the reaction progress is monitored by TLC (Notes 6 and 7). Upon complete consumption of benzaldehyde, the flask is removed from the oil bath and the reaction mixture cooled to room temperature. The reaction mixture is tested for the presence of peroxides using starch-KI paper, which indicated the absence of peroxides. The solution is then diluted with 100 mL of water and transferred to a 500-mL round-bottomed flask. The reaction flask is rinsed with dichloromethane (2 x 10 mL), and the rinses are transferred to the 500-mL flask. The resulting mixture is concentrated to approximately two-thirds of its initial volume by rotary evaporation (25 °C, 30 mmHg). The concentrated solution is then transferred to a 250-mL separatory funnel and extracted with dichloromethane (4 x 40 mL). The combined organic phases are washed with brine, dried over $MgSO_4$ (7 g), filtered and concentrated by rotary evaporation (25 °C, 30 mmHg) to afford an orange oil. The crude product is purified by flash chromatography (Note 8) using a glass column and flash-grade silica gel. The column is eluted with a total of 3.5 L of ethyl acetate:hexanes (1.2:1) and the fractions containing the desired product are combined and concentrated by rotary evaporation (25 °C, 30 mmHg) and then under high vacuum (25 °C, 1mmHg) for 2 h to give 8.74 g (49.9 mmol, 88%) of *N*-benzoyl pyrrolidine as a light yellow oil (Note 9).

2. Notes

1. Anhydrous acetonitrile (99.8%, Cat. No. 271004) was obtained from Sigma-Aldrich, Inc. and used as received.
2. Benzaldehyde (99.5%, Cat. No. 418099) was obtained from Sigma-Aldrich, Inc. and used as received.
3. Checkers obtained pyrrolidine (99%, Cat. No. P73803) from Sigma-Aldrich, Inc. and used it as received. Submitters purchased pyrrolidine (99%, Cat. No. 13208-2500) from Acros, Inc. and employed it without further purification.
4. *tert*-Butyl hydroperoxide (5–6 M in decane, Cat. No. 416665) was obtained from Sigma-Aldrich, Inc. and used as received.
5. The reaction was carried out in duplicate on a 6.0 g scale.
6. Checkers carried out TLC analysis using Whatman 60 Å glass plates with ethyl acetate:hexanes (1.2:1) as the eluent and UV lamp (254 nm) for visualization. The R*f* values of benzaldehyde and the product (*N*-benzoyl pyrrolidine) were found to be 0.9 and 0.3, respectively. Submitters conducted TLC on silica gel 60 plates (Fisher Scientific) using dichloromethane as mobile phase and a UV lamp (254 nm) for visualization. The R*f* of benzaldehyde and the product (*N*-benzoyl pyrrolidine) were determined as 0.9 and 0.5, respectively.
7. Submitters also followed the reaction by GC/MS: Varian GC Mass Saturn 2100T spectrometer using a Varian FactorFour capillary column VF-5ms (30 m x 0.25 mm) coated with 5% phenylpolysiloxane and 95% dimethylpolysiloxane. The oven temperature was held at 80 °C for 2 min and then increased to 260 °C over 14 min. The product was eluted after 10.7 min and detected by EI/MS.

2

8.   Checkers preformed flash chromatography on Kieselgel 60, particle size 0.032–0.063 mm using 3.5 L ethyl acetate:hexanes (1.2:1) as mobile phase. A glass column (8-cm diameter) was packed with a slurry of 300 g silica gel (18 cm height). Fractions of 25 mL were collected from the column and analyzed by TLC with UV detection as described above. Desired product was found in fractions 40–118. Submitters used dichloromethane:ethyl acetate:hexanes (17.5:4:1) as mobile phase. The product has an Rf of 0.3 with this mobile phase. The fractions were collected using 20 mL test tubes. Elution of the product started after 700 mL of the mobile phase was collected and the chromatographic process was continued with 2.5 L of the same mobile phase.

9.   N-Benzoyl pyrrolidine can be stored under air atmosphere at room temperature. Checkers found that during storage under ambient conditions, the color of the product deepens to reddish-amber with no change in the NMR or HRMS: UV-Vis (dichloromethane, 1.57 μg/mL) λ_{max} 230.1 nm; IR (neat) 2971, 2874, 1626, 1576, 1446, 1419 cm^{-1}; ^{1}H NMR (500 MHz, CDCl$_3$) δ: 1.86–1.90 (m, 2 H), 1.94–1.98 (m, 2 H), 3.42 (t, J = 6.6 Hz, 2 H), 3.65 (t, J = 7.0 Hz, 2 H), 7.38–7.41 (m, 3 H), 7.50–7.52 (m, 2 H); ^{13}C NMR (125 MHz, CDCl$_3$) δ: 24.6, 26.5, 46.3, 49.7, 127.2, 128.4, 129.9, 137.4, 169.9; HRMS (ESI) m/z calcd. for C$_{11}$H$_{13}$NO [M+H]$^+$ 176.107, found 176.107; MS (EI, 70 eV) 176.0 (66, [M+H]$^+$), 146 (11, [M+H-C$_2$H$_4$]$^+$), 105 (100, [M+H-C$_4$H$_8$N]$^+$), 77 (45, [Ph]$^+$); Anal. calcd. for C$_{11}$H$_{13}$NO: C, 75.40; H, 7.48; N, 7.99; Found: C, 75.08; H, 7.29; N, 7.62.

Hazard Warnings and Waste Disposal

Pyrrolidine is corrosive and highly flammable. Acetonitrile is toxic and flammable. TBHP is a strong oxidizer and corrosive and should be kept tightly closed and stored at 2–8 °C. All hazardous materials were disposed of in accordance with "Prudent Practices in the Laboratory"; National Academy Press; Washington, DC, 1995.

3. Discussion

The conversion of aldehydes to amides is typically accomplished by two separate steps, i.e. oxidation to a carboxylic acid intermediate and subsequent reaction with an amine in the presence of a coupling agent. Several groups have demonstrated that formation of the amide can be accomplished more efficiently through integration of both reaction steps (oxidation and *C-N* bond formation) into a single operation. One-pot oxidative aminations of aldehydes can be accomplished with readily available starting materials and this approach avoids the formation and isolation of free carboxylic acid intermediates that often interfere with sensitive functional groups present in the substrate.

Using ammonia and stoichiometric amounts of nickel peroxide, Nakagawa and coworkers were first to demonstrate that aldehydes can be directly transformed into amides.[3] Since then, one-pot oxidative aminations have received increasing attention and several other methods based on the Cannizzaro reaction,[4] the Beckmann rearrangement,[5] and catalytic reactions that require either precious transition metal complexes[6] or *N*-heterocyclic carbenes[7] have been reported.

Table 1. Oxidative amination of aromatic aldehydes with secondary amines.

Entry	Aldehyde	Amine	Product	Yield %
1	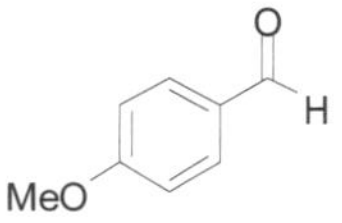	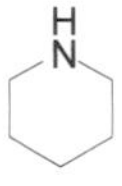	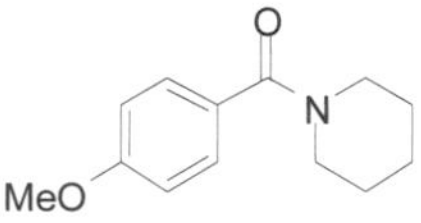	96
2	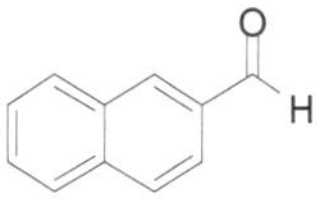		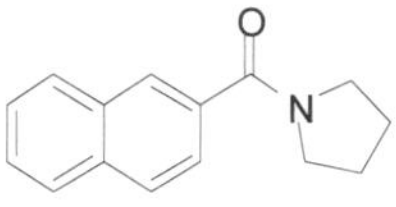	85
3	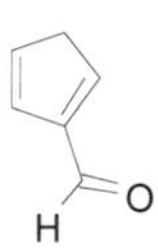	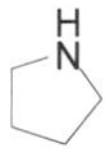	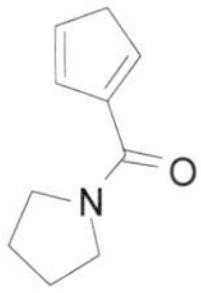	90
4	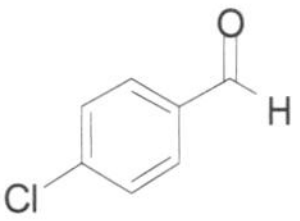	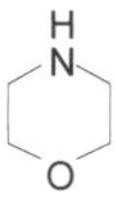	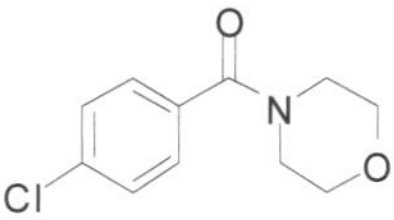	92
5	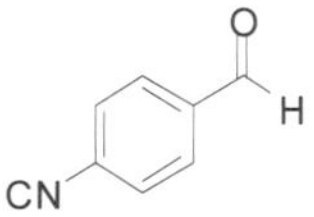	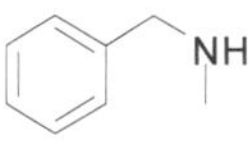	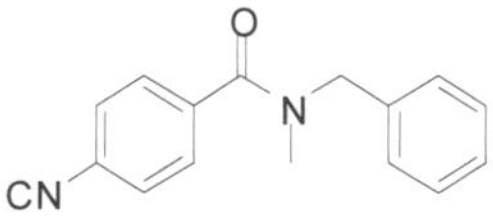	90

4

The procedure described herein is most applicable to aromatic aldehydes and secondary amines. It has been successfully used to prepare electron-rich and electron-deficient benzamides derived from pyrrolidine, morpholine or *N*-methyl benzylamine (see Table for selected examples). While excellent results are generally obtained with aromatic aldehydes, aliphatic substrates give only moderate yields due to competing aldol condensation.

In contrast to previously reported methods, this oxidative amination procedure does not utilize expensive catalysts or additives and proceeds within 6 hours in the presence of readily available *tert*-butyl hydroperoxide. The mechanism probably involves intermediate formation of a carbinolamine that is oxidized *in situ* to the corresponding amide while water and *tert*-butyl alcohol are formed as the side products (Scheme 1).

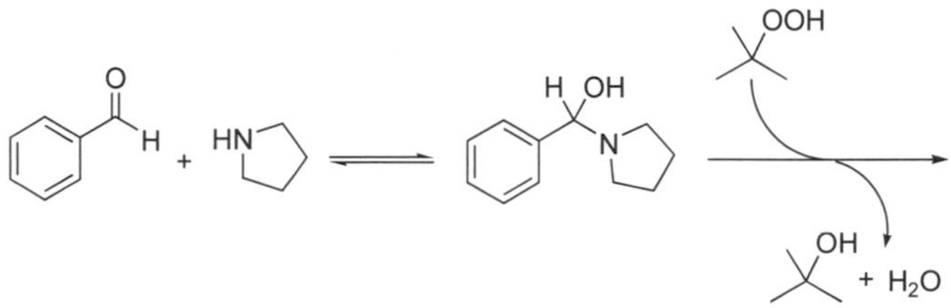

Scheme 1. Mechanism of the oxidative amination of benzaldehyde to *N*-benzoyl pyrrolidine using *tert*-butyl hydroperoxide as oxidant.

1. Georgetown University, Department of Chemistry, 37th and "O" Streets, Washington, DC 20057, USA. Email: cw27@georgetow.edu
2. This procedure was adapted from: Ekoue-Kovi, K.; Wolf, C. *Org. Lett.* **2007**, *9*, 3429–3432.
3. (a) Nakagawa K.; Inoue, H.; Minami, K. *Chem. Commun. (London)* **1966**, 17–18. (b) Nakagawa K.; Mineo, S.; Kawamura, S.; Horikawa, M.; Tokumoto, T.; Mori, O. *Syn. Commun.* **1979**, *9*, 529–534. (c) Ekoue-Kovi, K.; Wolf, C. *Chem. Eur. J.* **2008**, *14*, 6302–6315.
4. (a) Zhang, L.; Wang, S.; Zhou S.; Yang, G.; Sheng, E. *J. Org. Chem.* **2006**, *71*, 3149–3153. (b) Ishihara, K.; Yano T. *Org. Lett.* **2004**, *6*, 1983–1984. (c) Abaee, S. M.; Sharifi, R.; Mojtahedi, M. *Org. Lett.* **2005**, *7*, 5893–5895. (d) Seo, S; Marks, T. J. *Org. Lett.* **2008**, *10*, 317–319.

5. (a) Hosseini-Sarvari, M.; Sharghi, H. *Synthesis* **2002**, 1057–1060. (b) Sharghi, H.; Hosseini-Sarvari, M.; *Tetrahedron* **2002**, 58, 10323–10328.
6. (a) Naota, T.; Murahashi, S. I. *Synlett* **1991**, 693–695. (b) Tamaru, Y.; Yamada, Y.; Yoshida, Z. *Synthesis* **1983**, 474–476. (c) Tillack, A.; Rudloff, I.; Beller, M. *Eur. J. Org. Chem.* **2001**, 523–528. (d) Yoo, W. J.; Li, C. J. *J. Am. Chem. Soc.* **2006**, *128*, 13064–13065.
7. (a) Alcaide, B.; Almendros, P.; Redondo, M. C. *Org. Lett.* **2004**, *6*, 1765–1767. (b) He, M.; Bode, J. W. *Org. Lett.* **2005**, *7*, 3131–3134. (c) Li, G.-Q.; Li, Y.; Dai, L.-X.; You, S.-L. *Org. Lett.* **2007**, *9*, 3519–3521. (d) Vora, H. U.; Rovis, T. *J. Am. Chem. Soc.* **2007**, *129*, 13796–13797. (e) Bode, J. W.; Sohn, S. S. *J. Am. Chem. Soc.* **2007**, *129*, 13798–13799.

Appendix
Chemical Abstracts Nomenclature; (Registry Number)

Benzaldehyde; (100-52-7)

Pyrrolidine; (123-75-1)

N-Benzoyl pyrrolidine: Methanone, phenyl-1-pyrrolidinyl-; (3389-54-6)

tert-Butyl hydroperoxide: Hydroperoxide, 1,1-dimethylethyl; (75-91-2)

Christian Wolf was born in Hamburg, Germany in 1968. He received his Ph.D. under the direction of Professor Wilfried A. König from the University of Hamburg in 1995. After working as a postdoctoral Feodor-Lynen Fellow with Professor William H. Pirkle at the University of Illinois in Urbana he accepted an R&D position at SmithKline Beecham Pharmaceuticals in 1997. In 2000, he began his academic career as Assistant Professor in the Chemistry Department at Georgetown University in Washington, DC where he was promoted to Associate Professor with tenure in 2006. His research interests comprise stereodynamics of chiral compounds, asymmetric synthesis, stereoselective sensing, chiral recognition, transition metal-catalyzed cross-coupling reactions, development of antimalarial drugs, and chiral chromatography.

Kekeli Ekoue-Kovi Adjoa Sika was born in Lome, Togo. She attended Berea College in Kentucky where she received a B.A. in Chemistry in 2003. In the summer of 2001 and 2002, she gained her first lab experience in David Atwood's laboratory at the University of Kentucky. She is now pursuing a PhD degree in Chemistry under the supervision of Prof. Christian Wolf at Georgetown University in Washington, DC. Her research interests include palladium-phosphinous acid-catalyzed cross-coupling reactions and the development of heme-targeted antimalarial drugs.

Vikram Bhat was born in Ajmer, India. He received his undergraduate education at the Indian Institute of Technology, Bombay, Mumbai. In 2005 he entered the graduate program at the University of Chicago where he joined the research group of Professor Viresh H. Rawal. Currently, his graduate research focuses on the total synthesis of welwitindolinone alkaloids.

A PRACTICAL AND SCALABLE SYNTHESIS OF *N*-HALO COMPOUNDS: 2-CHLORO-6,7-DIMETHOXY-1,2,3,4-TETRAHYDROISOQUINOLINE

Submitted by Yong-Li Zhong[1] and Paul G. Bulger.
Checked by Stephen G. Newman and Mark Lautens.

1. Procedure

2-Chloro-6,7-dimethoxy-1,2,3,4-tetrahydroisoquinoline (**2**). A 250-mL, three-necked, round-bottomed flask is equipped with an overhead stirrer, an addition funnel that is fitted with a tap adaptor connected to a nitrogen line, and a rubber septum through which a digital thermometer probe is inserted. The septum is temporarily removed and the apparatus is purged with nitrogen for 5 min and then the flask is charged with 6,7-dimethoxy-1,2,3,4-tetrahydroisoquinoline hydrochloride **1** (5.16 g, 21.8 mmol, 1.0 equiv) (Notes 1 and 2), *tert*-butanol (2.11 mL, 21.8 mmol, 1.0 equiv) (Note 3), water (15 mL) (Note 4), and toluene (40 mL) (Note 5) at room temperature. The septum is replaced and the mixture is stirred at 120 rpm, and the resulting biphasic solution is cooled to −5 °C (internal temperature) in an ice/salt bath over which time the aqueous layer freezes to became a white slurry in the organic layer. The addition funnel is charged with 5% aqueous sodium hypochlorite (30.9 mL, 25.1 mmol, 1.15 equiv, Note 6), which is then added dropwise over 30 min to the reaction mixture, maintaining the internal temperature between −5 and 0 °C. Once the addition is complete, stirring is continued at this temperature for a further 15 min (Note 7). Once the mixture reaches >15 °C, it is transferred to a 125 mL separatory funnel (washing with 2 × 10 mL of toluene). The layers are separated and the lower, aqueous layer is discarded. The colorless organic layer is washed with water (1 × 15 mL), and then with brine (1 × 15 mL) (Note 8). The organic layer is concentrated by rotary evaporation (20 mmHg, keeping the bath temperature below 30 °C) to a volume of approximately 12 mL (Note 9). A stir bar is added, and to the stirred toluene

8

solution is then added *n*-heptane (40 mL) (Note 10) dropwise *via* an addition funnel over 20 min at room temperature, and the resulting white slurry is then cooled to 0 °C in an ice bath and stirred at that temperature for a further 3 h. The crystalline product is then collected by filtration through a 60-mL fritted-glass funnel of medium porosity. The wet solid is washed with *n*-heptane (2 × 10 mL) and then dried under vacuum with a nitrogen sweep (Note 11) for 1 h to give 2-chloro-6,7-dimethoxy-1,2,3,4-tetrahydroisoquinoline **2** (4.70–4.83 g, 95–97%) (Notes 12, 13, and 14) as a white powder (Note 15).

2. Notes

1.   6,7-Dimethoxy-1,2,3,4-tetrahydroisoquinoline hydrochloride **1** (97%) was purchased from Sigma-Aldrich, Inc. and used as received.
2.   If the starting material is an amide or a neutral amine, then the addition of one equivalent of acetic acid to the reaction is also required.
3.   *tert*-Butanol (anhydrous, ≥ 99.5%) was purchased from Sigma-Aldrich, Inc. and used as received (submitters used 99+% grade from Sigma-Aldrich, Inc.).
4.   Distilled water was used (submitters used deionized water).
5.   Toluene was purchased from ACP Chemicals, Canada and used as received (submitters used 99% grade from Sigma-Aldrich, Inc.).
6.   5% Aqueous sodium hypochlorite (bleach) was purchased from Acros Organics Ltd. and used as received.
7.   The progress of the reaction was followed by TLC analysis on silica gel with visualization under UV (254 nm) and with a ceric ammonium molybdate stain. Using an eluent comprising 10% methanol in dichloromethane, product **2** has R_f = 0.87, while the starting amine **1** has R_f = 0.22. Alternatively, the reaction was monitored by reverse-phase HPLC employing the following conditions: Zorbax Eclipse Plus C18 Rapid Resolution HT column (4.6 × 50 mm, 1.8 μm particle size, Agilent part number: 253583-0191); column temperature 25 °C; flow rate: 1.5 mL/min; linear gradient of 10:90 to 95:5 acetonitrile:0.1% v/v aqueous H_3PO_4 in 5 min, then hold at 95:5 for 1 min, then back to 10:90 in 0.1 min, then hold at 10:90 for 1.9 min; UV detection at 210 nm. Retention times for compounds **1** and **2** are 0.83 and 4.53 min, respectively.
8.   At this stage the organic layer may be dried using sodium sulfate (2 g) for 0.5 h if necessary.

9.   The product may begin to crystallize prior to the addition of heptane, depending upon the exact volume of the toluene solution and ambient temperature.

10.   *n*-Heptane was purchased from ACP Chemicals, Canada and used as received (submitters used 99% grade from Sigma-Aldrich, Inc.).

11.   The product is dried for 1 h under house vacuum (20 mmHg) with a nitrogen sweep, using the apparatus illustrated below:

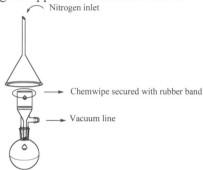

12.   Physical properties and spectroscopic data for **2** are as follows: white powder; mp >76 °C (dec.); IR (film): 2935, 2830, 1610, 1520, 1465, 1360, 1260, 1220, 1120 cm^{-1}; ^{1}H NMR (CDCl$_3$, 400 MHz) δ: 2.96 (t, J = 6.0 Hz, 2 H), 3.43 (t, J = 6.0 Hz, 2 H), 3.84 (s, 3 H), 3.85 (s, 3 H), 4.24 (s, 2 H), 6.47 (s, 1 H), 6.59 (s, 1 H); ^{13}C NMR (CDCl$_3$, 100 MHz) δ: 28.1, 55.9 (2 peaks), 59.2, 63.4, 109.1, 111.1, 124.1, 125.1, 147.6, 148.1; HRMS (ESI) *m/z* calcd. for C$_{11}$H$_{15}$ClNO$_2$ [M+H]$^+$ 228.0785; found 228.0780; Anal. calcd. for C$_{11}$H$_{14}$ClNO$_2$: C, 58.05; H, 6.20; N, 6.15; found: C, 58.05; H, 6.29; N, 6.20.

13.   During melting-point analysis, the solid product **2** decomposed rapidly at 76 °C to give a brown oil. Differential scanning calorimetry (DSC) studies suggested that the product was stable at temperatures below 61 °C. Handling and manipulation of this, and other, *N*-halo compounds at or below 30 °C is therefore recommended.

14.   The crystalline product **2** gave negative results in drop-weight tests, indicating that the material is not shock-sensitive.

15.   The white product slowly colorized to pale yellow upon standing at room temperature. However, the product was stable at room temperature for several weeks without noticeable degradation.

10

Safety and Waste Disposal Information

All hazardous materials should be handled and disposed of in accordance with "Prudent Practices in the Laboratory"; National Academy Press; Washington, DC, 1995.

3. Discussion

N-Halo compounds are versatile reagents in synthesis. For many organic chemists, experience with *N*-chloro or *N*-bromo derivatives is limited to the use of the halogenating agents *N*-chlorosuccinimide (NCS) and *N*-bromosuccinimide (NBS). However, this class of compounds is employed in a variety of synthetically useful transformations, including radical cyclizations,[2] the Hofmann–Löffler–Freytag reaction,[3] and elimination to give imines,[4] amongst other processes.[5]

The synthesis of *N*-halo compounds is most readily achieved by the reaction of amines or amides with electrophilic halogen sources. The most commonly employed reagents for this oxidation are sodium hypohalites (bleach and NaOBr),[6] the *N*-halo succinimides NCS and NBS,[7] and *tert*-butyl hypochlorite;[8] each has its own advantages and limitations. Sodium hypochlorite is a cheap, safe, commodity chemical, though oxidation of secondary amines (R_2NH to R_2NCl) can be plagued by modest yields and prolonged reaction times. NCS and NBS are widely utilized, but removal of the succinimide by-product is often problematic.[9] Yields with *tert*-butyl hypochlorite are generally good to excellent, but the reagent is expensive, hazardous, and suitable only for small-scale research applications.[5a,10]

We have demonstrated a practical and scalable preparation of *N*-halo compounds based on the *in situ* generation of *tert*-butyl hypohalite (*t*-BuOCl or *t*-BuOBr) under biphasic conditions.[11] Organic solutions of amines or amides are treated with aqueous sodium hypohalite (0.5–0.75 M, 1–1.5 equiv) in the presence of *tert*-butanol (0.25–1 equiv) and acetic acid (1–1.5 equiv). The *tert*-butyl hypohalite is generated slowly by the dropwise addition of the oxidant, and then reacts rapidly in the organic phase with the substrate to give the desired N–Cl or N–Br product. Slow addition of the sodium hypohalite maintains a low concentration of *tert*-butyl hypohalite present at any given time, a feature which, combined with elimination of the need for handling and manipulation of the latter reagents, makes this protocol attractive for larger-scale processing.

Table 1. Synthesis of *N*-halo compounds

Entry	Substrate	Product	Conditions[a-c]	Yield (%)[d]
1		**3**	1/1/1/toluene/2	100
2 3	BnO₂C— (ring) N-H	BnO₂C— (ring) N-X **4**: X = Cl **5**: X = Br	1.5/0.5/1.5/IPAc/1 1.5/0.5/1.5/MTBE/1	100 98
4		**6**	1/0.5/1/MTBE/0.5	100
5 6	(ring) N-H · HCl CO₂Bn	(ring) N-X CO₂Bn **7**: X = Cl **8**: X = Br	1/1/0/MTBE/0.25 1/1/0/MTBE/0.5	100 90
7 8	H-N (ring) CO₂Et	X-N (ring) CO₂Et **9**: X = Cl **10**: X = Br	1/0.25/1/MTBE/0.5 1.2/1/1.2/MTBE/0.5	100 94
9 10	H-N (ring) CONH₂	X-N (ring) CONH₂ **11**: X = Cl **12**: X = Br	1/1/1/EtOAc/0.5 1/1/1/EtOAc/1	90 90
11	Ph NH₂ · HCl CO₂Et	Ph Cl-NH CO₂Et **13**	1/0.5/0/MTBE/1	96
12	(ring) N-H S(=O)₂	(ring) N-Cl S(=O)₂ **14**	1.5/1/1.5/IPAc/0.5	92
13	(complex structure) Cl⁻	(complex structure) **15**: Ar = *p*-F-C₆H₄	1.1/1/1.1/IPAc/1	100

[a] equiv NaOX/equiv *t*-BuOH/equiv AcOH/solvent/time (h); [b] 0.75 M aqueous NaOCl or 0.5 M aqueous NaOBr solutions were used; [c] IPAc = isopropyl acetate; MTBE = *tert*-butyl methyl ether; EtOAc = ethyl acetate; [d] Isolated yields; obtained after workup and solvent evaporation to give the desired products, which were determined to be > 95% pure by ¹H NMR

The conditions are mild, high yielding, and general for a variety of substrates (Table 1). Reactions are typically complete within 15–30 minutes for amines and 1–2 hours for amides. This rate differential allows for chemoselective oxidation (Entries 9 and 10). A primary amine is selectively monochlorinated (Entry 11). Amine hydrochloride salts are sufficiently acidic to promote the reaction (Entries 5, 6 and 11); no acetic acid is required for these substrates. This protocol has been demonstrated on multi-kilogram scale (Entry 13).[5a]

Following a simple workup procedure, the *N*-halo products are isolated in high purity by concentration and, if applicable, crystallization. For the *N*-halo amines, the crystalline solids (e.g. **2** and Table 1, **11** and **12**) are generally quite stable at room temperature, but the stability of the neat oils (**6–10, 13**) is variable, with some (e.g. **6** and **13**) decomposing within thirty minutes. Manipulation and storage of the more sensitive non-crystalline *N*-halo amine products as solutions, in which degradation is much slower, is recommended. In contrast, all the *N*-halo amide (**3–5**) or sulfonamide (**14**) products were found to be very stable either in crystalline form or as neat oils.

1. Department of Process Research, Merck & Co., Inc., Rahway, NJ 07065, USA. Email: yongli_zhong@merck.com.
2. Liu, J. F.; Heathcock, C. H. *J. Org. Chem.* **1999**, *64*, 8263–8266.
3. Stocking, E. M.; Sanz-Cervera, J. F.; Unkefer, C. J.; Williams, R. M. *Tetrahedron* **2001**, *57*, 5303–5320.
4. Maughan, M. A. T.; Davies, I. G.; Claridge, T. D. W.; Courtney, S.; Hay, P.; Davis, B. G. *Angew. Chem. Int. Ed.* **2003**, *42*, 3788–3792.
5. For further selected examples, see: (a) Zhong, Y.-L., Krska, S. W.; Zhou, H.; Reamer, R. A.; Lee, J.; Sun, Y.; Askin, D. *Org. Lett.* **2009**, *11*, 369–372. (b) Drouin, A.; Lessard, J. *Tetrahedron Lett.* **2006**, *47*, 4285–4288. (c) Cossy, J.; Tresnard, L.; Pardo, D. G. *Tetrahedron Lett.* **1999**, *40*, 1125–1128. (d) Stella, L. *Angew. Chem. Int. Ed. Engl.* **1983**, *22*, 337–350. (e) Kovacic, P.; Lowery, M. K.; Field, K. W. *Chem. Rev.* **1970**, *70*, 639–655.
6. (a) Smith, J. R. L.; McKeer, L. C.; Taylor, J. M. *Org. Synth. Coll. Vol. 8* **1993**, 167–173. (c) Gassman, P. G.; Dygos, D. K.; Trent, J. E. *J. Am. Chem. Soc.* **1970**, *92*, 2084–2090.

7. (a) Zhao, M. M.; McNamara, J. M.; Ho, G.-J.; Emerson, K. M.; Song, Z. J.; Tschaen, D. M.; Brands, K. M. J.; Dolling, U. H.; Grabowski, E. J. J.; Reider, P. J. *J. Org. Chem.* **2002**, *67*, 6743–6747. (b) Snider, B. B.; Liu, T. *J. Org. Chem.* **1997**, *62*, 5630–5633. (c) Guillemin, J. C.; Denis, J. M. *Synthesis* **1985**, 1131–1133.

8. (a) Durham, T. B.; Miller, M. J. *J. Org. Chem.* **2003**, *68*, 27–34. (b) Herranz, E.; Sharpless, K. B. *Org. Synth. Coll. Vol. 7* **1990**, 223–226. (c) Gassman, P. G.; Campbell, G. A.; Frederick. R. C. *J. Am. Chem. Soc.* **1972**, *94*, 3884–3891.

9. Fieser, M. In *Reagents for Organic Synthesis*, John Wiley and Sons, New York, **1982**, Vol. 10, p 67.

10. (a) Mintz, M. J.; Walling, C. *Org. Synth. Coll. Vol. 5*, **1973**, 184–187. (b) Paquette, L. A. In *Encyclopedia of Reagents for Organic Synthesis*, John Wiley and Sons, New York, **1995**, Vol. 2, p 890.

11. Zhong, Y.-L.; Zhou, H.; Gauthier, D. R., Jr.; Lee, J.; Askin, D.; Dolling, U. H.; Volante, R. P. *Tetrahedron Lett.*, **2005**, *46*, 1099–1101.

Appendix
Chemical Abstracts Nomenclature; (Registry Number)

2-Chloro-6,7-dimethoxy-1,2,3,4-tetrahydroisoquinoline
6,7-Dimethoxy-1,2,3,4-tetrahydroisoquinoline hydrochloride: Isoquinoline, 1,2,3,4-tetrahydro-6,7-dimethoxy-, hydrochloride (1:1); (2328-12-3)
Sodium hypochlorite; (7681-52-9)
tert-Butanol: 2-Methyl-2-propanol; (75-65-0)

Yong-Li Zhong was born in Guangzhou, China. He received his B.S. and M.S. degrees in Chemistry from Zhongshan University under the direction of Professors Jingyu Su and Longmei Zeng. He obtained his Ph.D. in organic synthesis from the Chinese University of Hong Kong in 1998 under the supervision of Professor Tony K. M. Shing. After three years of postdoctoral studies with Professor K. C. Nicolaou at The Scripps Research Institute, he joined the Process Research Department of Merck & Co., Inc. in 2001. His research interests include the development of practical and efficient synthesis of heterocycles, and new synthetic methodologies.

Paul G. Bulger was born in London, England, in 1978. He received his undergraduate M.Chem degree in 2000 from the University of Oxford, completing his Part II project under the supervision of Dr. Mark G. Moloney. He remained at Oxford for his graduate studies, obtaining his D. Phil. in chemistry in 2003 for research conducted under the supervision of Professor Sir Jack E. Baldwin. After an enjoyable three-year stint as a postdoctoral researcher in Professor K. C. Nicolaou's group at The Scripps Research Institute, he joined the Process Research Department of Merck & Co., Inc. in the fall of 2006.

Stephen Newman was born in 1985 in Grand Falls-Windsor, Newfoundland, Canada. He studied chemistry at Dalhousie University, where he obtained his B. Sc. in 2008. During this time, he worked in the laboratory of Professor D. Jean Burnell on iron and copper catalyzed oxidative cleavage of cyclic ketones. He is currently pursuing his Ph.D. in the laboratory of Professor Mark Lautens at the University of Toronto where his research is focused on the development of methodologies for the synthesis of novel heterocycles.

REGIOSELECTIVE C-4 BROMINATION OF OXAZOLES:
4-BROMO-5-(THIOPHEN-2-YL)OXAZOLE

A.

B.

Submitted by Bryan Li,[1] Richard A. Buzon, and Zhijun Zhang.
Checked by Takahiro Koshiba and Tohru Fukuyama.

1. Procedure

A. 5-(Thiophen-2-yl)oxazole (**1**).[2] A 1-L three-necked, round-bottomed flask equipped with a reflux condenser fitted with an argon inlet, a thermocouple probe (range -200 to +1370 °C) and a mechanical stirrer is charged with 2-thiophenecarboxaldehyde (11.21 g, 100.0 mmol, 1.00 equiv), K_2CO_3 (24.19 g, 175.0 mmol, 41.75 equiv), tosylmethylisocyanide (21.12 g, 100.0 mol, 1.00 equiv) and methanol (448 mL) (Note 1). The reaction mixture is heated at reflux (66 °C) for 4 h (Note 2). After cooling to room temperature, water (224 mL) is added and the mixture is stirred for 10 min at the same temperature. The resulting solution is transferred to a single-necked 2-L flask, and methanol is removed by rotary evaporation (45 °C, 75 mmHg). The residual liquid is transferred to a 1-L separatory funnel, and then extracted with MTBE (3 × 120 mL). The combined organic layers are washed with water (1 × 50 mL) and a saturated NaCl solution (1 × 50 mL), dried over anhydrous $MgSO_4$, filtered, and concentrated by rotary evaporation (35 °C, 50 mmHg) (Note 3) to give a brown oil (15.31 g) as a crude product, which is vacuum distilled (60–62 °C, 0.4 mmHg) to afford the oxaozle **1** as a pale yellow oil (12.70 g, 84%) (Note 4).

B. 4-Bromo-5-(thiophen-2-yl)oxazole (**2**). A dried 300-mL, three-necked, round-bottomed flask equipped with a 75-mL pressure equalizing addition funnel fitted with an argon inlet, a temperature probe and a mechanical stirrer (stirring rate >250 rpm) is charged with 5-(thiophen-2-

yl)oxazole (**1**) (9.14 g, 60.5 mmol, 1.00 equiv) and anhydrous DMF (36.6 mL) (Note 5). The resulting solution is cooled to a –15 °C to –10 °C range (Note 6), and 1 N lithium bis(trimethylsilyl)amide in THF (1.0 M solution, 63.5 mL, 1.05 equiv) (Note 7) is then added slowly via a syringe while keeping the reaction temperature at –15 °C to –10 °C. The reaction mixture is stirred at –15 °C for 30 min and then cooled to –78 °C (dry ice-acetone bath). A solution of *N*-bromosuccinimide (NBS) (10.76 g, 60.45 mmol, 1.00 equiv) in anhydrous DMF (36.6 mL) (Note 8) is added while maintaining the temperature below –65 °C (Note 9). The mixture is stirred for 30 min at the same temperature, and quenched by addition of 2 N aqueous NaOH solution (160 mL) while allowing the solution to warm to above 0 °C during the quench (Note 10). After diluting with water (160 mL), the reaction is extracted with MTBE (3 × 120 mL). The combined organic layers are washed successively with a 0.5 N NaOH solution (3 × 50 mL) and a saturated NaCl solution (1 × 50 mL), dried over anhydrous MgSO$_4$, filtered, and concentrated by rotary evaporation (35 °C, 115 mmHg) (Note 11) to afford a brown solid as the crude product. The solid is then dissolved in isopropyl alcohol (30 mL) with gentle warming (60 °C water bath). The resulting solution is cooled to room temperature over 20 min and further cooled in an ice-bath (without agitation). After stirring for 30 min with ice bath cooling, the resulting crystals are collected by suction filtration on a Kiriyama funnel, rinsed with ice-cold isopropyl alcohol (5 mL), and transferred to a 250-mL Erlenmeyer flask. The filtrate is transferred to a 50-mL, single-necked, round-bottomed flask, and concentrated to ~10 mL by rotary evaporation (40 °C, 50 mmHg). The concentrated filtrate is cooled to room temperature over 20 min and further cooled in a (water)ice-acetone bath for 30 min (without agitation). The resulting crystals are collected by suction filtration on a Kiriyama funnel, rinsed with (water)ice-acetone cold isopropanol (3 mL), and transferred to the same 250-mL Erlenmeyer flask. Heptanes (40 mL) are charged to the Erlenmeyer flask, the resulting mixture is stirred for 10 min, and filtered with a Kiriyama funnel. The collected product is dried in a desiccator (room temperature/ < 1.0 mmHg) (Note 12) for 18 h to give 4-bromo-5-(thiophen-2-yl)oxazole (**2**) (11.01 g, 79%) (Note 13) as an off-white solid.

2. Notes

1. Unless otherwise noted, all reagents were purchased from Aldrich Chemical Company and used without further purification. The checkers purchased anhydrous methanol and TosMIC from Wako chemicals and Tokyo chemical industry company, respectively, and used them as received.

2. The reaction turned brown after 4 h at reflux. According to the submitters, HPLC analysis of a reaction aliquot showed complete consumption of 2-thiophenecarbaldehyde. HPLC method: Agilent® Eclipse XDB–C8 columns (4.6 mm × 150 mm). Mobile phase: acetonitrile (A) / 0.2% perchloric acid in water (B), 0–2 min, isocratic 70% B; 2–10 min, gradient from 70% B to 10% B. Flow rate: 2 mL/min. Detection wavelength: 210 nm. The HPLC retention times for 5-(thiophen-2-yl)oxazole (**1**) and 4-bromo-5-(thiophen-2-yl)oxazole (**2**) were 6.37 min and 7.84 min, respectively. The checkers could detect the consumption of the starting material by TLC.

3. The submitters reported the pressure at 20 mmHg.

4. The physical properties of 5-(thiophen-2-yl)oxazole (**1**) are as follows: bp 60–62 °C (0.40 mmHg) [submitter: bp 66–68 °C (0.60 mmHg)]; IR (neat): 3111, 1614, 1593, 1517, 1486, 1425, 1364, 1305, 1258, 1241, 1210, 1197, 1101, 991, 899, 848, 818, 700, 637 cm^{-1}; ^{1}H NMR (400 MHz, CDCl$_3$) δ: 7.09 (dd, J = 4.8, 3.6 Hz, 1 H), 7.22 (s, 1 H), 7.33 (dd, J = 3.6, 1.1 Hz, 1 H), 7.35 (dd, J = 4.8, 1.1 Hz, 1 H), 7.86 (s, 1 H); ^{13}C NMR (100 MHz, CDCl$_3$) δ: 121.0, 124.4, 125.7, 127.7, 129.3, 146.8, 149.7; HRMS (DART) m/z calcd. for C$_7$H$_6$NOS [M+H]$^+$ 152.0170; found 152.0165; Anal. calcd. for C$_7$H$_5$NOS: C, 55.61; H, 3.33; N, 9.26; found: C, 55.44; H, 3.57; N, 9.22. The submitters determined the purity of **2** to be 98.8% by HPLC analysis (210 nm). The HPLC method is the same as in Note 2.

5. Anhydrous DMF was purchased from Wako Chemical.

6. Dry ice-acetone bath with limited amount of dry ice to keep the temperature in range.

7. Lithium bis(trimethylsilyl)amide (1.0 M in THF) supplied in SureSeal bottles was purchased from Aldrich Chemical Company. The submitters recommend that charging the reagent by weight is preferred over charging by volume for higher accuracy.

8. To minimize the transfer of the NBS in DMF solution, solid NBS is directly added to the 100-mL pressure equalizing addition funnel with the bottom of the funnel attached to a 10-mL single-necked round bottom flask

18

and the top capped with a septum fitted with an argon inlet, DMF is charged to the funnel via syringe. The mixture is swirled into a complete solution, the 10-mL single-necked round bottom flask is removed, and the addition funnel is attached to the 300-mL, three-necked reaction flask.

9. Although the freezing point of DMF is –61 °C, the reaction mixture does not freeze at –78 °C. Best regioselectivity is obtained when the NBS in DMF solution is allowed to trickle down the sidewall of the reaction flask with good agitation of the reaction mixture (stirring rate >250 rpm). The reaction is exothermic, and addition is completed in approx. 30 min.

10. The quench was complete in 3 min. The dry ice-acetone bath was removed immediately after the quench.

11. The submitters reported that the resulting solution was concentrated by rotary evaporation (35 °C, 20 mmHg).

12. The submitters used a vacuum oven (40 to 50 °C/ ca. 75 mmHg) to dry the collected product.

13. The physical properties of 4-bromo-5-(thiophen-2-yl)oxazole (2) are as follows: mp 85–87 °C; IR (neat): 3120, 1603, 1501, 1486, 1427, 1319, 1169, 1116, 1062, 1027, 932, 915, 850, 831, 706, 642 cm^{-1};^{1}H NMR (400 MHz, CDCl$_3$) δ: 7.14 (dd, J = 5.0, 3.7 Hz, 1 H), 7.43 (dd, J = 5.0, 1.4, 1 H), 7.61 (dd, J = 3.7, 1.4 Hz, 1 H), 7.80 (s, 1 H); ^{13}C NMR (100 MHz, CDCl$_3$) δ: 110.0, 125.9, 126.8, 127.7, 127.8, 144.0, 149.0; HRMS (ESI?) m/z calcd. for C$_7$H$_5$BrNOS [M+H]$^+$ 229.9275; found 229.9271; Anal. calcd. for C$_7$H$_4$BrNOS: C, 36.54; H, 1.75; N, 6.09; found: C, 36.61; H, 1.91; N, 6.10. The submitters determined the purity of 2 to be 98.9% by HPLC analysis (210 nm). The HPLC method is the same as in Note 2.

Waste Disposal Information

All toxic materials were disposed of in accordance with "Prudent Practices for Disposal of Chemicals from Laboratories"; National Academy Press; Washington, DC, 1995.

3. Discussion

Oxazoles with substitutions at both C-4 and C-5 have recently attracted increased attention in medicinal chemistry research for their therapeutic potential in treating inflammation, cancer, and asthma.[3] We recently engaged in the development of a scaleable preparation of 3, a potent and selective inhibitor of the stress-activated kinase p38α.[4] Our strategy was to employ a Suzuki-Miyaura or Negishi coupling as shown in **Scheme 1**. In this approach, a robust and efficient synthesis of 4-halo-5-substituted oxazoles is a critical component.

Scheme 1

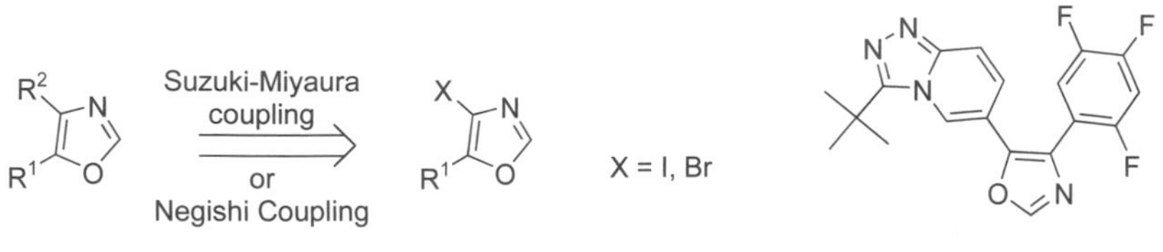

3

C-4 iodination of 5-substituted oxazoles was first reported by Vedejs et al.[5] However, the regioselectivity (C-4/C-2) of the reaction was modest, both 2-iodooxazoles and 2,4-diiodooxazoles were formed as by-products at significant levels. It is known in the literature that the regioselectivity of this reaction is determined by the equilibrium between the 2-lithiooxazole and its acyclic tautomer (**Scheme 2**).[6,7] In polar aprotic solvents, the acyclic form is expected to predominate because of improved solvation. We envisioned that DMF, as a strongly polar aprotic solvent, would be an ideal solvent in this reaction. It is cheap, commercially available in anhydrous form and most importantly, it has a low freezing point of –61 °C. Other common polar aprotic solvents were ruled out because of their high freezing points (DMAC, NMP, DMSO, DMI, acetonitrile), ability to react with anions (acetone, acetonitrile, DMSO), cost (DMI, DMPU), and worker safety and environmental concerns (HMPT, HMPA).[8]

Scheme 2

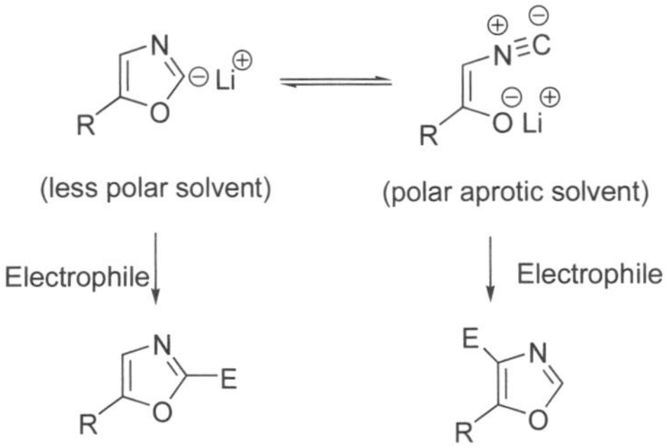

(less polar solvent)   (polar aprotic solvent)

Electrophile ↓   ↓ Electrophile

Our initial attempts employing DMF as solvent at –70 °C gave poor to modest C-4/C-2 regioselectivity. We reasoned that the initially formed 2-lithiooxazole did not have ample opportunity to fully equilibrate to the acyclic form at low temperatures within the time scale of the reaction. We therefore decided to perform the lithiation at –15 °C, equilibrate the anion for 30 min before cooling to –70 °C and quenching with NBS. Employing these conditions, a regioselectivity of 98:2 favoring the desired C-4 bromide was obtained. The product (**Table 1**, entry 1) was isolated by crystallization (MTBE/hexanes) in 87% yield and contained less than 0.2% of 2-bromooxazole as an impurity. As expected, the lithium anions behave similarly to stabilized enolates in that we did not detect any formylated products (from nucleophilic addition to DMF) in the reaction mixtures. The regioselective C-4 bromination is general (**Table 1**); excellent regioselectivity (> 97:3) was observed using other C-5 substituted aryl and heteroaryl oxazoles as substrates under the same reaction conditions.

Table 1. C-4 Bromination of 5-Substituted Oxazoles via 2-Lithiooxazoles.

Entry	5-Substituted Oxazole	4-Bromo-5-Substituted Oxazole	Isolated Yield
1	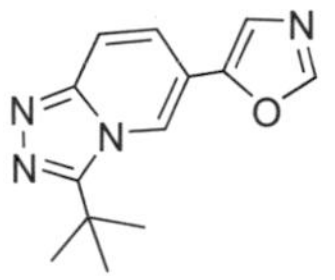	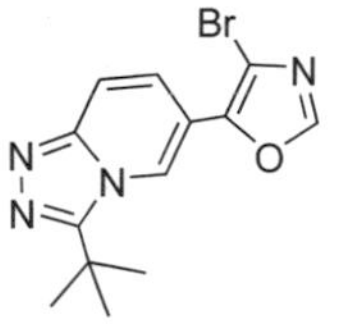	87%
2	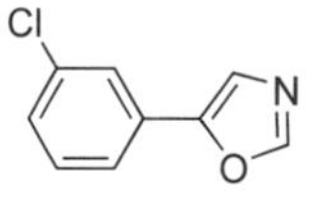	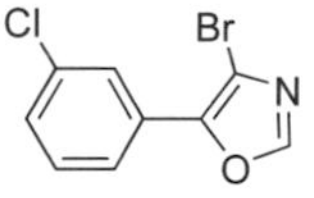	82%
3	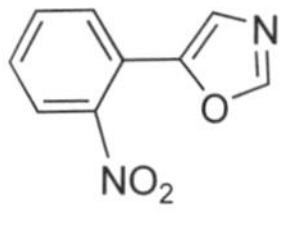	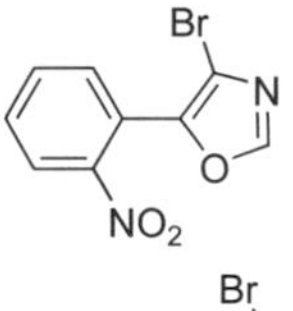	78%
4	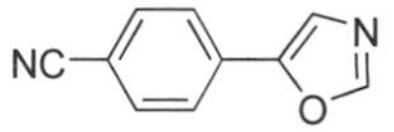	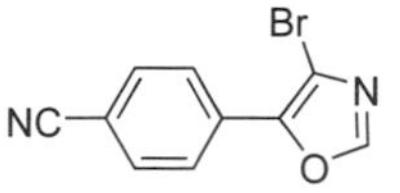	78%

Table 2. Suzuki-Miyaura Coupling Reactions of 4-Bromooxazoles.

Entry	4-Bromooxazole	Suzuki-Miyaura Coupling Product	Isolated Yield
1			81%
2			61%
3			54%

4-Bromooxazoles are known to be good substrates for Suzuki-Miyaura coupling reactions. Coupling with aryl boronic acids in the presence of Pd(dppf)Cl$_2$ proceeded uneventfully (**Table 2**).

In conclusion, a highly regioselective bromination at C-4 of 5-substituted oxazoles is described. The use of DMF as solvent and aging of the lithiated oxazole are critical to drive the equilibrium in favor of the acyclic isonitrile enolate, resulting in significantly improved C-4/C-2 regioselectivity. These 4-bromooxazoles were shown to be good Suzuki-Miyaura coupling partners with arylboronic acids. The bromination protocol was demonstrated on multi-kilogram scale supporting the development of **3**.

1. Research API – Pharmaceutical Science, Pfizer Global Research and Development, Groton Laboratories, Connecticut 06340. email: bryan.li@pfizer.com.
2. Saikachi, H.; Kitagawa, T.; Sasaki, H.; van Leusen, A. M. *Chem. Pharm. Bull.* **1979**, *27*, 793-796.
3. (a) Révész, L.; Schlapbach, A. PCT Int. Appl. 2000, WO 0063204. *Chem. Abstr. 133*:321897. (b) Lipton, S. A. PCT Int. Appl. 2001, WO 0101986. (c) Andersson, M.; Hansen, P.; Lönn, H.; Nikitidis, A.; Sjölin, P. PCT Int. Appl. 2005, WO 2005026123 *Chem. Abstr. 142*:316705. (d)

Org. Synth. **2010**, *87*, 16-25

Revesz, L.; Blum, E.; Di Padova, F. E.; Buhl, T.; Feifel, R.; Gram, H.; Hiestand, P.; Manning, U.; Rucklin, G. *Bioorg. Med. Chem. Lett.* **2004**, *14*, 3595-3599. (e) Blumberg, L. C.; Munchhof, M. J. PCT Int. Appl. 2004, WO 2004026863 *Chem. Abstr. 140*:287371. (f) Svenstrup, N.; Kuhl, A.; Flubacher, D.; Brands, M.; Ehlert, K.; Ladel, C.; Otteneder, M.; Keldenich, J. PCT Int. Appl. 2003, WO 03072574 *Chem. Abstr. 139*:230786. (g) Alcaraz, L.; Furber, M.; Purdie, M.; Springthorpe, B. PCT Int. Appl. 2003, WO 03068743. *Chem. Abstr.* 139:197375. (h) Iwanowicz, E. J.; Watterson, S. H.; Guo, J.; Pitts, W. J.; Murali Dhar, T. G.; Shen, Z.; Chen, P.; Gu, H. H.; Fleener, C. A.; Rouleau, K. A.; Cheney, D. L.; Townsend, R. M.; Hollenbaugh, D. L. *Bioorg. Med. Chem. Lett.* **2003**, *13*, 2059-2063.

4. (a) McClure, K. F.; Abramov, Y. A.; Laird, E. R.; Barberia, J. T.; Cai, W.; Carty, T. J.; Cortina, S. R.; Danley, D. E.; Dipesa, A. J.; Donahue, K. M.; Dombroski, M. A.; Elliott, N. C.; Gabel, C. A.; Han, S.; Hynes, T. R.; LeMotte, P. K.; Mansour, M. N.; Marr, E. S.; Letavic, M. A.; Pandit, J.; Ripin, D. B.; Sweeney, F. J.; Tan, D.; Tao, Y. *J. Med. Chem.* **2005**, *48*, 5728-5737. (b) McClure, K. F.; Letavic, M. A.; Kalgutkar, A. S.; Gabel, C. A.; Audoly, L.; Barberia, J. T.; Braganza, J. F.; Carter, D.; Carty, T. J.; Cortina, S. R.; Dombroski, M. A.; Donahue, K. M.; Elliott, N. C.; Gibbons, C. P.; Jordan, C. K.; Kuperman, A. V.; Labasi, J. M.; LaLiberte, R. E.; McCoy, J. M.; Naiman, B. M.; Nelson, K. L.; Nguyen, H. T.; Peese, K. M.; Sweeney, F. J.; Taylor, T. J.; Trebino, C. E.; Abramov, Y. A.; Laird, E. R.; Volberg, W. A.; Zhou, J.; Bach, J.; Lombardo, F. *Bioorg. Med. Chem. Lett.* **2006**, *16*, 4339–4344.

5. Vedejs, E.; Luchetta, L. M. *J. Org. Chem.* **1999**, *64*, 1011-1014.

6. Schröder, R.; Schöllkopf, U.; Blume, E.; Hoppe, I. *Justus Liebigs Ann. Chem.* **1975**, 533-546.

7. Dondoni, A.; Fantin, G.; Fogagnolo, M.; Medici, A.; Pedrini, P. *J. Org. Chem.* **1987**, *52*, 3413-3420.

8. (a) Sarrif, A. M.; Krahn, D. F.; Donovan, S. M.; O'Neil, R. M. *Mutat. Res.* **1997**, *380*, 167-177. (b) Ashby, J.; Styles, J. A.; Paton, D. *Br. J. Cancer* **1978**, *38*, 418-427.

Appendix
Chemical Abstracts Nomenclature; (Registry Number)

5-(Thiophen-2-yl)oxazole: Oxazole, 5-(2-thienyl)-; (70380-70-0)

2-Thiophenecarboxaldehyde; (98-03-3)

Tosylmethylisocyanide: Benzene, [(isocyanomethyl)sulfonyl]-; (36635-63-9)

4-Bromo-5-(thiophen-2-yl)oxazole: Oxazole, 4-bromo-5-(2-thienyl)-;
    (959977-82-3)

Lithium bis(trimethylsilyl)amide; (4039-32-1)

N-Bromosuccinimide: 2,5-Pyrrolidinedione, 1-bromo-; (128-08-5)

Bryan Li was born in 1966. He received a B.S. from the Department of Fine Chemical Technology, East China University of Chemical Technology in 1986. After three years of working in Amoy Pharmaceutical Inc. at Xiamen, China. He pursued his graduate education in chemistry at the University of Rhode Island under the guidance of Prof. Elie Abushanab, and received a Ph.D. in 1993. In the following two years, he worked in Prof. Gary Posner's group at the Johns Hopkins University at Baltimore, Maryland as a postdoctoral fellow. In 1996, he moved overseas and worked in the National University of Singapore, and returned to the U.S. in 1998. He has since worked in Pfizer Inc (Groton, CT) as a process chemist, and serves as a program lead responsible for exploratory development of drug candidates.

Rich Buzon started at Pfizer in 1979 in the Specialty Chemical Group as a chemical operator. After seven years in various positions in Specialty Chemicals, he began taking Organic Chemistry classes at the University of Connecticut and also at Connecticut College. He transferred to the Process Research and Development department in 1992 and began enabling process development for project chemistry to 22L glassware and 500 gal Pilot Plant reactor size. He is currently in the Research API group at Pfizer Global Research and Development.

Zhijun Zhang (born 1971) received his B.S. from Nanjing University, China in 1993 and M.A. from Bryn Mawr College, Pennsylvania in 2000. He has since worked with Pfizer Inc. in a few roles. He started as a medicinal chemist with Pharmacia Corp, Kalamazoo, Michigan. After the acquisition, he transferred to Pfizer Groton and later joined the Research API group and worked on process chemistry. He is now in the Research Analytical group, working on API/drug release and PAT (Process Analytical Technology) projects.

Takahiro Koshiba was born in 1981 in Saitama, Japan. He received B.S. in 2004 from university of the Meiji Pharmaceutical University and M.S. in 2006 from the University of Tokyo. Presently, he is pursuing Ph.D. degree at the Graduate School of Pharmaceutical Sciences, the University of Tokyo, under the guidance of Professor Tohru Fukuyama. His research interests are in the area of the total synthesis of natural product.

SYNTHESIS OF SPIROBORATE ESTERS FROM 1,2-AMINOALCOHOLS, ETHYLENE GLYCOL AND TRIISOPROPYL BORATE: PREPARATION OF (S)-1-(1,3,2-DIOXABOROLAN-2-YLOXY)-3-METHYL-1,1-DIPHENYLBUTAN-2-AMINE.

Submitted by Viatcheslav Stepanenko, Kun Huang and Margarita Ortiz-Marciales.[1]

Checked by David Hughes.[2]

CAUTION: Benzene is generated during the quench of Step A. All subsequent handling should be carried out in a well-ventilated hood.

1. Procedure

A.   *(2S)-2-Amino-3-methyl-1,1-diphenyl-butan-1-ol (1).*[3]   An oven-dried, 1-L, three-necked round-bottomed flask is fitted with rubber septa on two of the three necks and a hose adapter connected to a nitrogen line and bubbler on the third neck. A thermocouple temperature probe is inserted through one of the septa.   Nitrogen is flowed through the flask while cooling to room temperature. To the flask is added via cannula, 2.0 M phenylmagnesium chloride in THF (300 g, 288 mL, 0.576 mol, 5.0 equiv) (Notes 1, 2, and 3).   The mixture is stirred using a 2.5-cm oval Teflon-coated magnetic stir bar and cooled to –2 °C using an ice/acetone bath. *L*-Valine methyl ester hydrochloride (19.2 g, 114 mmol) is added portion-wise as a solid over 50 min while keeping the temperature below 5 °C. After addition, the cooling bath is removed. The solution is allowed to warm to room temperature over an hour and is stirred at room temperature for 2 h (Note 4). The mixture is cooled to 5 °C with an ice-bath, then carefully

Org. Synth. **2010**, *87*, 26-35
Published on the Web 12/1/2009

hydrolyzed with half-saturated aqueous NH$_4$Cl (250 mL) (Note 5). Ethyl acetate (150 mL) and saturated NaCl (50 mL) are added to the flask and the mixture is stirred vigorously for 5 min, then the solids are allowed to settle. The aqueous and organic layers are decanted into a 1-L separatory funnel and the layers separated. The aqueous layer is returned to the reaction flask and extracted in a similar fashion with two 150-mL portions of EtOAc (Note 6). The organic layers are combined, washed with saturated NaCl (100 mL), and filtered through a bed of Na$_2$SO$_4$ (50 g) followed by an EtOAc rinse (100 mL) of the bed.   The resulting clear solution is concentrated by rotary evaporation (20 mmHg, 40 °C bath) to afford a yellow solid (26.5 g), which is purified by column chromatography on silica gel (Note 7). The product is obtained as an off-white solid (18.1 g, 62% yield), which is used without further purification in the next step (Notes 8 and 9).

B.   (-)-(S)-1-(1,3,2-Dioxaborolan-2-yloxy)-3-methyl-1,1-diphenylbutan-2-amine (2). An oven-dried, 3-necked 250-mL round-bottomed flask is fitted with a short reflux condenser (10 cm) on one outer neck, a stoppered 50-mL addition funnel on the middle neck, and a rubber septum on the other outer neck. A thermocouple temperature probe is inserted through the septum. A 2.5-cm oval Teflon-coated magnetic stirrer is added to the flask. A 24/40 hose adapter is attached to the top of the reflux condenser and connected to a nitrogen line and gas bubbler. The flask is charged with toluene (60 mL) followed by triisopropyl borate (3.88 g, 4.7 mL, 20.6 mmol) and ethylene glycol (1.26 g, 20.3 mmol), each added via syringe (Note 10). The mixture is stirred and heated with a heating mantle over 20 min to 80 °C, at which time the mixture becomes homogeneous. The solution is cooled to 60 °C, then a solution of (2S)-2-amino-3-methyl-1,1-diphenyl-butanol (5.20 g, 20.4 mmol) (1) in toluene (25 mL) is added via the addition funnel over 3 min, during which time the temperature decreases to 55 °C and white crystalline product is formed. The addition funnel is rinsed with toluene (5 mL). The resulting mixture is cooled to ambient temperature over 30 min. The mixture is transferred to a 500-mL round-bottomed flask and the mixture is concentrated to dryness by rotary evaporation (20 mmHg, 50 °C bath), and then dried for 15 h at 80 °C in a vacuum oven (20 mmHg) to yield 6.56 g (99%) of spiroborate ester 2 as a white solid (Notes 11 and 12).

2. Notes

1.   Reagents and solvents used in this preparation were sourced from Sigma-Aldrich and used without further purification, including 2.0 M phenylmagnesium chloride in THF, L-valine methyl ester hydrochloride (99%), ethyl acetate (ACS reagent grade, >99.5%), hexanes (ACS reagent grade, >98.5%), n-heptane (anhydrous, 99%), ethylene glycol (99.8%, anhydrous), toluene (ACS reagent grade, >99.5%, dried over pelleted 4 Å molecular sieves), triisopropyl borate (>98%, Sure/Seal™), and silica gel (200-400 mesh, 60 Å).

2.   The 1-L flask is marked prior to oven drying at a 280-mL fill to help estimate the amount added to the flask during cannulation. The amount of Grignard reagent added is determined by weighing the bottle before and after addition.

3.   This reaction can also be carried out with 4 equiv PhMgCl. A slightly lower yield is obtained (57% vs 62%).

4.   The reaction is followed by ^{1}H NMR as follows: a sample (0.2 mL) is quenched into saturated aqueous NH_4Cl (1 mL), extracted with $CDCl_3$, and the bottom organic layer is filtered through a cotton plug (to remove water) into an NMR tube. The set of 2 doublets of valine methyl ester at 0.9 ppm is diagnostic to assess complete reaction. All starting material is reacted (<1 %) after 1 h at room temperature.

5.   The initial addition of NH_4Cl is highly exothermic. The first 25 mL is added dropwise with the temperature controlled below 40 °C. Total addition time is 20 min.

6.   The solids (magnesium salts) cause poor layer separation, therefore it is important to decant the solution from the solids for improved separation.

7.   The chromatography is carried out using silica gel (150 g) and a 5-cm diameter flash column. The column is packed using 10:1 hexanes:EtOAc and topped with 0.5 cm of sea sand. The solid product from concentration is dissolved with sonication in CH_2Cl_2 (20 mL) and loaded onto the column. The product is eluted with hexane/ethyl acetate (8:1 v/v), collecting 75 mL fractions. TLC is used to monitor fraction collection, using 1:3 EtOAc:hexanes, UV and iodine visualization, with the product having an R_f of 0.3. The product tails off the column and is collected in about 1.5 L of eluent. The column is then eluted with 1:1 EtOAc:hexanes (1 L) which affords a mixture of by-products (4.3 g), which are discarded. Fractions 6-

25 containing pure product by TLC are combined and concentrated by rotary evaporation at 100 mmHg and 50 °C. Due to the formation of solids during concentration, bumping occurs if a higher vacuum is used.

8. An analytically pure sample is prepared by recrystallization from *n*-heptane, as follows. The product (5.0 g) from the chromatography is added to *n*-heptane (50 mL) and warmed to 80 °C to dissolve the solids. The mixture is slowly cooled to ambient temperature over 1 h and held at ambient temperature for 2 h. The resulting solids are filtered and washed with *n*-heptane (10 mL) to afford 4.45 g (89%) of product.

9. *(-)-(2S)-2-Amino-3-methyl-1,1-diphenyl-butan-1-ol (1)*: mp 98–99 °C, Lit[3a] 94–95 °C; $[\alpha]^{20}_D$ –126 (*c* 2, CHCl$_3$), Lit[3a] $[\alpha]^{25}_D$ –127.5 (*c* 0.639, CHCl$_3$); ^{1}H NMR (400 MHz, CDCl$_3$) δ: 0.90 (d, 3 H, *J* = 6.8 Hz, CH$_3$), 0.95 (d, 3 H, *J* = 7.0 Hz, CH$_3$), 1.10 (br s, 2 H, NH$_2$), 1.77 (septet of doublets, 1H, *J* = 2.2, 6.9 Hz, CH) 3.85 (d, 1 H, *J* = 2.2 Hz, NCH), 4.40 (br s, 1 H, OH), 7.15–7.21 (m, 2 H, Ph), 7.27–7.34 (m, 4 H, Ph), 7.50–7.53 (m, 2 H, Ph), 7.61–7.65 (m, 2 H, Ph); ^{13}C NMR (100 MHz, CDCl$_3$) δ: 16.3, 23.2, 28.0, 60.4, 79.9, 125.7, 126.1, 126.5, 126.8, 128.2, 128.6, 145.1, 148.2; Anal. Cald. for C$_{17}$H$_{21}$NO: C, 79.96; H, 8.29; N, 5.50. Found: C, 79.69; H, 8.62, N, 5.50.

10. The amount added is determined by weighing the syringe before and after addition.

11. The spiro-borate ester product can be isolated by filtration instead of concentration to dryness. A 65% isolated yield of analytically pure material is obtained when the reaction mixture is vacuum-filtered through a 60-mL sintered glass funnel at ambient temperature. Catalyst isolated in this manner is purer by NMR than material concentrated to dryness, but both perform comparably in the oxime ether asymmetric reduction (see accompanying procedure).

12. *(-)-(S)-1-(1,3,2-Dioxaborolan-2-yloxy)-3-methyl-1,1-diphenylbutan-2-amine (2)*: mp 209–211 °C; $[\alpha]^{20}_D$ –112 (*c* 2.0, DMSO); ^{1}H NMR (400 MHz, d$_6$-DMSO) δ: 0.38 (d, 3 H, *J* = 6.8 Hz, CH$_3$), 0.98 (d, 3 H, *J* = 6.8 Hz, CH$_3$), 1.86 (septet of doublets, 1 H, *J* = 3.0, 6.9 Hz, CH), 3.67 (m, 4 H, 2 × OCH$_2$), 3.80 (dt, 1 H, *J* = 3.0, 7.0, NCH), 5.0 (br s, 1H, NH), 6.0 (br s, 1H, NH), 7.12–7.29 (m, 6 H, Ph), 7.41–7.46 (m, 4 H, Ph); ^{13}C NMR (100 MHz, d$_6$-DMSO) δ: 16.2, 21.8, 27.5, 63.2, 64.0, 83.1, 126.2, 126.3, 126.8, 127.2 (2 degenerate C), 127.5, 145.3, 148.1; ^{11}B NMR (d$_6$-DMSO) δ: 12.9 (s); the catalyst is not stable in CDCl$_3$, therefore, dry d$_6$-DMSO is selected as the solvent for NMR; HRMS *m/z* calc for

$[C_{19}H_{25}O_3NB]^+$ 326.1928 ($[M+H]^+$), found 326.1924; Anal. Cald. for $C_{19}H_{24}BNO_3$: C, 70.17; H, 7.44, N, 4.31. Found: C, 70.25; H, 7.79, N, 4.26. The compound showed no change by NMR or decrease in reaction performance when kept in a sealed bottle for a month at ambient temperature.

Safety and Waste Disposal Information

All hazardous materials should be handled and disposed of in accordance with "Prudent Practices in the Laboratory"; National Academy Press; Washington, DC, 1995.

3. Discussion

Chiral organoborane reagents, in particular [1,3,2]-oxazaborolidines, have been extensively studied and applied as efficient Lewis acid catalysts to a wide range of asymmetric transformations.[4] The B-H oxazaborolidine-borane complexes are frequently reported as convenient catalysts for enantioselective borane reduction of prochiral ketones, imines and oximes, since they are readily prepared from the corresponding aminoalcohols and either borane-THF or borane-DMS complexes.[4a,c-d,5] However, the extreme sensitivity of these reagents to atmospheric moisture makes them difficult to isolate and purify. Consequently, they are prepared *in situ* prior to use in asymmetric reductions. Moreover, B-H oxazaborolidines can form dimers and other species which can affect the nature of catalyst.[6] Impurities present in the catalyst may lead to irreproducible results.[7] On the other hand, B-substituted oxazaborolidines show excellent synthetic utility due to their highly reproducible enantioselectivity, but require careful purification procedures to eliminate traces of boronic acid and their esters. Moreover, the commercially available reagents are expensive, moisture sensitive and unstable during extended storage. Accordingly, the design of new stable, easily available and efficient catalysts is always a challenging task in synthetic organic chemistry.

Recently, we prepared new oxazaborolidine-like spiroborate esters that were successfully applied for asymmetric borane reduction of prochiral ketones[8] and oxime ethers.[9] The spiroborate esters **2-10** were obtained by a simple procedure from commercially available chiral 1,2-aminoalcohols, ethylene glycol and triisopropyl borate with good purity and in essentially quantitative yields (Scheme 1). White crystalline spiroborate complexes **2** and **3** were found to be particularly stable, since no changes were observed after exposure to moist air for 24 h at 25 °C, as evidenced by their ^{11}B, ^{1}H and ^{13}C NMR spectra.

Scheme 1

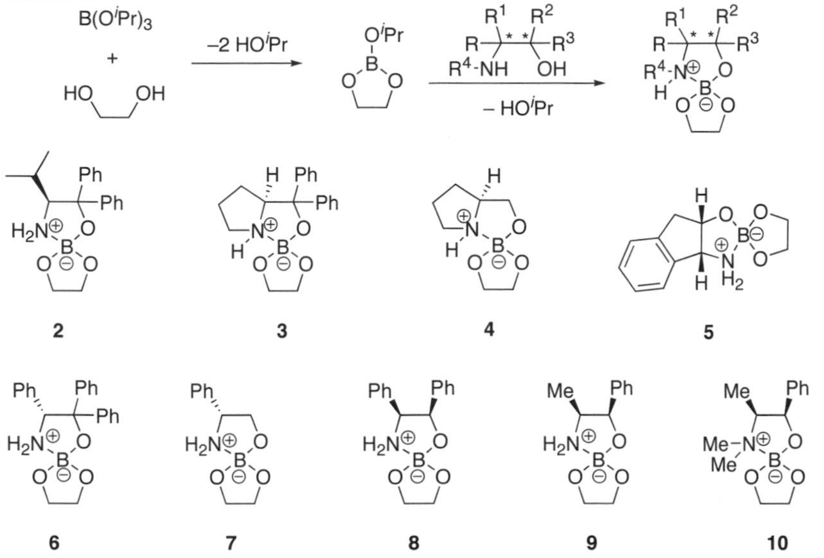

Scheme 2

Catalysts **2** and **3** proved to be most valuable due to their convenience of handling and outstanding enantioselectivity in the asymmetric borane reduction of *O*-benzyl oximes ethers and ketones, respectively (Scheme 2). The application of spiroborate **2** for the synthesis of non-racemic secondary amines is presented in the accompanying procedure.

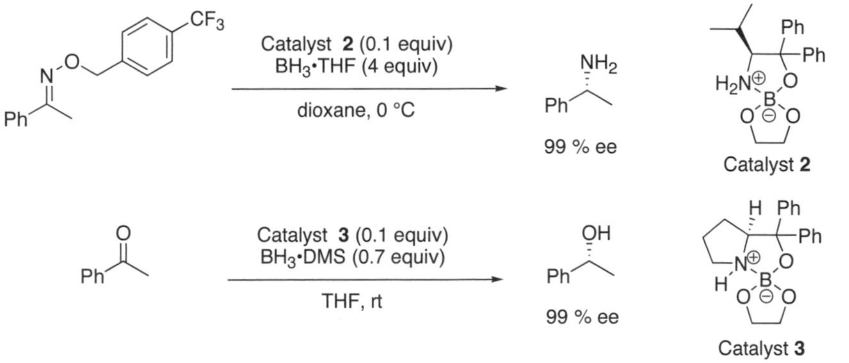

Table 1. Preparation of Chiral Alcohols via Borane Reduction Catalyzed by Spiroborate Ester **3**[a]

Reaction scheme: R–C(=O)–R' → (Catalyst **3** (0.01 -0.1 equiv), BH₃·DMS (0.7 equiv), THF, rt) → R–CH(OH)–R' (chiral, *)

Entry	Cat. 3 (mol %)	Product	Yield[b] (%)	ee[c] (%)
1	10	(1-(naphthalen-2-yl)ethanol)	84	99
2	10	(1,2,3,4-tetrahydronaphthalen-1-ol)	99	>99
3	10	(1-(adamantan-1-yl)ethanol)	98	99
4	10	(1-(4-chlorophenyl)ethanol)	98	99
5	1	(1-(2,4-difluorophenyl)ethanol)	88	98
6	1	(1-(benzofuran-2-yl)ethanol)	95	96
7	1	(6-chlorothiochroman-4-ol)	92	99
8[d]	10	(1-(10H-phenothiazin-2-yl)ethanol)	97	>99
9[d]	1	(1-(pyridin-3-yl)ethanol)	96	98
10[d]	1	(1-(6-methoxypyridin-3-yl)ethanol)	77	98
11[d]	1	(1-(pyridin-4-yl)ethanol)	92	99

[a] The reactions were carried out using 1 equiv of ketone (10 mmol), 0.1 or 0.01 equiv of catalyst **3** and 0.7 equiv of borane-DMS complex in THF at rt. [b] Isolated yield after chromatography or distillation. [c] Determined by GC of O-acetates on a chiral column (CP-Chirasil-Dex-CB). [d] 1.7 equiv of borane-DMS complex was used.

In addition, catalyst **3** demonstrated an excellent alternative for asymmetric reduction of a variety of ketones (Table 1), similar in enantioselectivity to those reported for the B-methyl CBS reagent. Moreover, the amount of catalyst load can be decreased to 1 mol % without significantly affecting the enantioselectivity.

1. Department of Chemistry, University of Puerto Rico-Humacao, CUH Station, Humacao, Puerto Rico 00791, USA. E-mail: ortiz@quimica.uprh.edu.
2. The checker would like to thank Dr. Peter Dormer for assistance with the NMR spectra.
3. (a) Itsuno, S.; Ito, K. *J. Org. Chem.* **1984**, *49*, 555–557. (b) Kelsen, V.; Pierrat, P.; Cros, P. *Tetrahedron* **2007**, *63*, 10693-10697.
4. For reviews see: (a) Glushkov, V. A.; Tolstikov, A.G. *Russ. Chem. Rev.* **2004**, *73*, 581-608; (b) Corey, E. J. *Angew. Chem., Int. Ed.* **2002**, *41*, 1650-1667; (c) Fache, F.; Schulz, E.; Tomamasino, M.; Lemaire, M. *Chem. Rev.* **2000**, *100*, 2159-2231; (d) Corey, E. J.; Helal, C. J. *Angew. Chem., Int. Ed.* **1998**, *37*, 1986-2012; (e) Deloux, L.; Srebnik, M. *Chem. Rev.* **1993**, *93*, 763-784.
5. (a) Itsuno, S.; Nakano, M.; Miyazaki, K.; Masuda, H.; Ito, K. *J. Chem. Soc., Perkin Trans. 1* **1985**, 2039–2044. (b) Corey, E. J.; Bakshi, R. K.; Shibata, S. *J. Am. Chem. Soc.* **1987**, *109*, 5551-5553. (c) Itsuno, S.; Sakurai, Y.; Shimizu, K.; Ito, K. *J. Chem. Soc., Perkin Trans. 1* **1990**, 1859–1863. (d) Cho, B. T.; Ryu, M. H. *Bull. Korean Chem. Soc.* **1994**, *15*, 191. (e) Quallich, G. J.; Keavey, K. N.; Woodall, T. M. *Tetrahedron Lett.* **1995**, *36*, 4729-4732. (f) Yadav, J. S.; Reddy, P. T.; Hashim, S. R. *Synlett* **2000**, *7*, 1049-1051. (g) Fontaine, E.; Namane, C.; Meneyrol, J.; Geslin, M.; Serva, L.; Royssey, E.; Tissandie, S.; Maftouh, M.; Roger, P. *Tetrahedron: Asymmetry* **2001**, *12*, 2185-2189. (h) Krzeminski, M. P.; Zaildewicz, M. *Tetrahedron: Asymmetry* **2003**, *14*, 1463-1466.
6. (a) Lang, A.; Nöth, H.; Schmidt, M. *Chem. Ber.* **1997**, *130*, 241-246. (b) Ortiz-Marciales, M.; De Jesús, M.; González, E.; Raptis, R. G.; Baran, P. *Acta Crystallogr.* **2004**, *C60*, 173-175. (c) Stepanenko, V.; Ortiz-Marciales, M.; Barnes, C. E.; Garcia, C. *Tetrahedron Lett.* **2006**, *47*, 7603-7606.

7. (a) Berenguer, R.; Garcia, J.; Vilarrasa, J. *Tetrahedron: Asymmetry* **1994**, *5*, 165-168. (b) Jones, S.; Atherton, J. C. C. *Tetrahedron: Asymmetry* **2000**, *11*, 4543-4548.

8. (a) Stepanenko, V.; Ortiz-Marciales, M.; Correa, W.; De Jesús, M.; Espinosa, S.; Ortiz, L. *Tetrahedron: Asymmetry* **2006**, *17*, 112-115. (b) Ortiz-Marciales, M.; Stepanenko, V.; Correa, W.; De Jesús, M.; Espinosa, S. U.S. Patent Application 11/512,599, Aug. 30, 2006. (c) Stepanenko, V.; Ortiz-Marciales, M.; De Jesús, M.; Correa, W.; Vázquez, C.; Ortiz, L.; Guzmán, I.; De la Cruz, W. *Tetrahedron: Asymmetry* **2007**, *18*, 2738-2745.

9. (a) Huang, X.; Ortiz-Marciales, M.; Huang, K.; Stepanenko, V.; Merced, F. G.; Ayala, A. M.; De Jesús, M. *Org. Lett.* **2007**, *9*, 1793-1795. (b) Ortiz-Marciales, M.; Huang, X.; Huang, K.; Stepanenko, V.; De Jesús, M.; Merced, F. G. PCT/US07/76195, Aug 17, 2007. (c) Huang, K.; Ortiz-Marciales, M.; Merced, F. G.; Meléndez, H. J.; Correa, W.; De Jesús, M. *J. Org. Chem.* **2008**, *73*, 4017-4026. (d) Huang, K.; Ortiz-Marciales, M.; Stepanenko, V.; De Jesús, M.; Correa, W. *J. Org. Chem.* **2008**, *73*, 6928-693.

Appendix
Chemical Abstracts Nomenclature; (Registry Number)

(2*S*)-2-Amino-3-methyl-1,1-diphenyl-butanol; (78603-95-9)
Bromobenzene; (108-86-1)
L-Valine methyl ester hydrochloride; (6306-52-1)
Phenylmagnesium chloride, 2 M in THF; (100-59-4)
Ethylene glycol; (107-21-1)
Triisopropyl borate; (5419-55-6)
(–)-(*S*)-1-(1,3,2-Dioxaborolan-2-yloxy)-3-methyl-1,1-diphenylbutan-2-amine; (879981-94-9)

Margarita Ortiz-Marciales was born in 1943 in Bogotá, Colombia, and obtained her B.S. from Universidad Nacional de Colombia in 1968. She studied at Freiburg and Mainz universities, Germany, for two years with a DAAD fellowship. She received her M.S. from the University of Alabama in Huntsville under Prof. S. McManus supervision in 1973, and her Ph.D. in Organic Chemistry at the University of Alabama-Tuscaloosa in 1979 under the direction of Prof. McManus and Prof. R. Abramovitch. In 1980, she did postdoctoral studies in Prof. G. Larson's group at the University of Puerto Rico-Río Piedras. She joined the University of Puerto Rico-Humacao in 1981, where she is currently a professor. Her interests are in the development of new synthetic methodologies using boron and silicon compounds for the preparation of important organic intermediates and biologically-active amines.

Viatcheslav Stepanenko received his M.S. degree from Belarusian State University, Belarus in 1994. He obtained his Ph. D. in 2000 at Institute of Organic Chemistry, Polish Academy of Sciences, Poland, under Prof. J. Wicha supervision, working on the application of tandem Mukaiyama-Michael reaction for synthesis of Vitamin D. Then, he worked as a postdoctoral fellow at the University of Leeds, UK under the direction of Prof. P. J. Kocienski studying a copper-mediated rearrangement. In 2003 he moved to University of Puerto Rico where he was a postdoctoral research associate with Prof. Margarita Ortiz-Marciales until 2009, studying applications of new chiral organoborane reagents for asymmetric synthesis.

Kun Huang received his B.S. degree in 1996 from Sichuan University China. He obtained his Ph.D. in 2006 at Nanjing University, China, working on the asymmetric epoxidation and cyclopropanation of chiral sulfonium ylides. He worked as an advanced synthetic researcher in Wuxi Pharma Tech Co., Ltd, Shanghai, China until he moved to Puerto Rico University in Humacao for his postdoctoral studies with Professor Margarita Ortiz-Marciales on the asymmetric reduction of O-benzyl oximes and ketones catalyzed by spiroborate esters. Later, he had a postdoctoral research position at the chemistry department at Oregon State University where his research was focused on the total synthesis of natural products. Currently, he is a postdoctoral researcher at Peking University in China.

CATALYTIC ENANTIOSELECTIVE BORANE REDUCTION OF BENZYL OXIMES: PREPARATION OF (S)-1-PYRIDIN-3-YL-ETHYLAMINE BIS HYDROCHLORIDE

Submitted by Kun Huang and Margarita Ortiz-Marciales.[1]
Checked by David Hughes.[2]

1. Procedure

A. (E)-1-Pyridin-3-yl-ethanone oxime (1). A 250-mL, three-necked, round-bottomed flask is equipped with a 2-cm Teflon-coated magnetic stir bar, a reflux condenser, a pressure-equalizing dropping funnel and a rubber septum through which a thermocouple thermometer probe is inserted. The flask is charged with 3-acetyl pyridine (12.1 g, 100 mmol) (Note 1), EtOH (100 mL) and $NH_2OH \cdot Cl$ (11.85 g, 170 mmol), and the mixture is heated with a heating mantle to 55 °C. A pre-made solution of Na_2CO_3 (7.46 g, 70 mmol) in water (20 mL) is added dropwise over 10 min via the dropping funnel. At the end of the addition, the temperature has risen to 62 °C. The heterogeneous mixture is stirred at 60 °C for 2.5 h (Notes 2 and 3), then

Org. Synth. **2010**, *87*, 36-52
Published on the Web 12/1/2009

filtered through a 60-mL medium-porosity sintered-glass funnel to remove the inorganic salts. The solid is washed with EtOH (2 x 20 mL) and the combined filtrate is transferred to a 500-mL, round-bottomed flask and concentrated by rotary evaporation (50 °C, 20 mmHg) to afford 19 g of solid residue. Then, water (250 mL) is added along with a 2.5-cm Teflon-coated magnetic stir bar and the mixture is warmed to 70 °C (Note 4) using a heating mantle. The heat is turned off, allowing the mixture to slowly cool to ambient temperature over 3 h. The product is isolated by filtration on a 150-mL medium-porosity sintered glass funnel, washed with water (2 x 15 mL), and dried for 15 h under vacuum (60 °C, 20 mmHg) to afford (E)-1-pyridin-3-yl-ethanone oxime (**1**) (11.3 g, 83%) as a white crystalline solid (Notes 5, 6, 7, and 8). A second crop is obtained as follows. The filtrate from the crystallization is extracted with EtOAc (3 x 100 mL). The combined organic extracts are washed with brine (50 mL), filtered through a bed of sodium sulfate, and concentrated by rotary evaporation (40 °C, 20 mm Hg) to 1.9 g. This solid material is transferred to a single-necked 100-mL round-bottomed flask along with water (35 mL) and a 1.5-cm Teflon-coated magnetic stir bar. The mixture is warmed using a heating mantle to 80 °C with stirring to dissolve all the solids (Note 9). The heating mantle is turned off, allowing the solution to slowly cool to ambient temperature over 1 h. The mixture is held at room temperature an additional hour, then filtered through a 60-mL medium porosity sintered glass funnel, washed with water (2 x 5 mL), and dried for 15 h under vacuum (60 °C, 20 mmHg) to afford (E)-1-pyridin-3-yl-ethanone oxime (**1**) (1.40 g, 10%) containing 0.6% of the undesired Z-isomer. The first and second crops have similar purity by NMR and can be combined (12.7 g, 93%).

B. *(E)-1-Pyridin-3-yl-ethanone O-benzyl-oxime (2).* An oven-dried 500-mL, three-necked, round-bottomed flask is equipped with a 2.5-cm oval Teflon-coated magnetic stirrer, a pressure-equalizing dropping funnel on the middle neck, a rubber septum through which is inserted a thermocouple temperature probe, and a nitrogen inlet adapter connected to a nitrogen line and gas bubbler. Nitrogen is flowed through the flask while cooling. The flask is charged with anhydrous DMF (175 mL, dried with 3 Å 1.6 mm pelleted sieves) and cooled to –15 °C (internal temperature) using a dry-ice acetone bath at –25 °C. Sodium hydride (4.59 g, 60%, 115 mmol, 1.4 equiv) is added to the cold DMF solution. A solution of (E)-1-pyridin-3-yl-ethanone oxime (**1**) (10.95 g, 80.5 mmol) in anhydrous DMF (80 mL, dried with 3 Å 1.6 mm pelleted sieves is added dropwise via the dropping funnel over 15 min, followed by a rinse of DMF (5 mL). The temperature of the yellow heterogeneous mixture rises to –10 °C during the addition of the oxime. The resulting mixture is stirred for 30 min at –12 to –10 °C, then benzyl bromide (14.45 g, 85 mmol, 1.05 equiv) is added dropwise via syringe over 10 min (Note 10). After complete addition, the mixture is stirred for 40 min at –12 to –10 °C and checked by TLC for reaction completion (Note 11). A saturated aqueous NH$_4$Cl solution (100 mL) is added to quench the reaction. The first 20 mL is added slowly over 10 min with hydrogen evolution, with an exotherm of +5 °C being observed during those 10 min. The remaining 80 mL is added over the next 5 min with an exotherm leading to a temperature of 23 °C. The mixture is transferred to a 2-L separatory funnel. The flask is rinsed with water (100 mL) and EtOAc (200 mL), which are also transferred to the separatory funnel. The layers are separated. The aqueous layer is extracted with EtOAc (2 x 100 mL). The organic layers are combined, washed with water (2 x 100 mL), then with brine (50 mL). The organic layer is filtered through a bed of sodium sulfate (40 g), rinsing the bed with EtOAc (100 mL). The resulting organic filtrate is concentrated by rotary evaporation (40 °C, 20 mmHg) to give a yellow oil (22 g) (Note 12). The material is purified by chromatography on SiO$_2$ (Note 13). The resulting oil is further vacuum dried at room temperature (20 mmHg) for 15 h to afford (E)-1-pyridin-3-yl-ethanone O-benzyl-oxime (2) (18.02 g, 99%) as a pale yellow oil (Note 14).

C. (S)-1-Pyridin-3-yl-ethylamine (4). An oven-dried, 1-L, 3-necked, round-bottomed flask, marked at the 465 mL fill prior to drying, is equipped with a 3-cm oval Teflon-coated magnetic stir bar, two rubber septa on the outer necks, and a 100-mL gas equilibrating addition funnel that is connected by a gas adapter to a nitrogen inlet and gas bubbler. A thermocouple thermometer probe is inserted through one of the septa. Nitrogen is flowed through the system as the flask is cooling. Spiroborate ester 3[3] (5.80 g, 18 mmol) and anhydrous dioxane (230 mL, dried with 3 Å 1.6 mm pelleted sieves) are added to the flask and the heterogeneous mixture is stirred at ambient temperature. Borane-tetrahydrofuran complex (1.0 M, 235 mL, 235 mmol, 3.9 equiv) (Note 15) is added via cannula to the reaction flask over 5 min, resulting in gentle hydrogen evolution, which cools the reaction mixture by 2 °C. The resulting mixture is stirred for 0.5 h at room temperature, whereby most of the solids dissolve to give a hazy solution. This solution is cooled to 3 °C using an ice-bath, then a solution of (E)-1-pyridin-3-yl-ethanone O-benzyl-oxime (2) (13.54 g, 60 mmol) in anhydrous

38

dioxane (40 mL) is added over 1 h via the addition funnel. The addition funnel is rinsed with dioxane (5 mL). The mixture is stirred for 30 h at 0–5 °C in an ice bath (Note 16). The cold reaction is carefully quenched by the dropwise addition of methanol (100 mL) over 15 min. The flask is equipped with a reflux condenser and heated under reflux (67–69 °C) for 15 h. Then the solution is concentrated by rotary evaporation (50 °C bath, 20 mmHg) to afford 30 g of crude solid (Note 17). The residue is purified by chromatography on SiO$_2$ (Note 18) to provide (S)-1-pyridin-3-yl-ethylamine (4) (8.03 g) as a hazy oil that is approximately 83 wt % pure (91% yield corrected for purity) (Note 19). The enantiomeric excess is 98% by chiral HPLC of the acetyl derivative of (S)-1-pyridin-3-yl-ethylamine (4) (Notes 20 and 21).

D. (S)-1-Pyridin-3-yl-ethylamine hydrochloride (5). An oven-dried 250-mL, 3-necked, round-bottomed flask is equipped with 2-cm Teflon-coated magnetic stirrer, sealed with two rubber septa on each outer neck, through one of which is inserted a thermocouple thermometer probe, and a 100-mL gas-equilibrating addition funnel on the center neck. (S)-1-Pyridin-3-yl-ethylamine 4 (7.49 g, 83% pure, 50.9 mmol) is dissolved in MeOH (25 mL) and the hazy solution is filtered by syringe through a 0.45 micron Teflon syringe filter into the addition funnel. Hydrochloric acid in ether (2.0 M, 60 mL, 120 mmol, 2.4 equiv) is added to the flask via syringe and the solution is stirred vigorously (500 rpm) at ambient temperature. The amine in methanol solution is added dropwise to the HCl solution over 20 min, during which time crystallization occurs and the temperature rises to 30 °C (Note 22). The dropping funnel is rinsed with MeOH (3 mL). The mixture is stirred at ambient temperature for 2 h, then filtered through a pressure filter (Note 23), washed with diethyl ether (2 x 10 mL), and dried for 15 h (60 °C, 20 mmHg,) to afford (S)-1-pyridin-3-yl-ethylamine bis-hydrochloride (5) (8.90 g, 89% yield) as an analytically pure white solid (Note 24). Enantiomeric excess determined by chiral HPLC analysis of the acetyl derivative is 99% (Notes 25 and 26).

2. Notes

1. The following reagents and solvents used in this preparation were sourced from Sigma-Aldrich and used without further purification, including 3-acetyl pyridine (98 %), ethyl acetate (ACS reagent, >99.5%), NH$_2$OH·HCl (ReagentPlus 99%), NaH (60% dispersion in mineral oil),

dimethylformamide (ACS spectrophotometric grade, 99.8%), benzyl bromide (98%), hexanes (ACS reagent, >98.5%), 1,4-dioxane (ACS reagent, >99%), 1.0 M BH$_3$·THF stabilized with NaBH$_4$, methanol (ACS reagent, >99.8%), dichloromethane (ACS reagent, >99.5%), 2.0M HCl in diethyl ether, silica gel (200-400 mesh, 60 Å). Absolute ethanol was obtained from Pharmaco. Ammonium chloride, sodium carbonate, and sodium sulfate were sourced from Fisher. De-ionized tap water is used throughout.

2. The oxime and ketone are not separable by TLC. The reaction is followed by ^{1}H NMR as follows: A sample (0.1 mL) is added to CDCl$_3$ (1 mL) and filtered through glass wool into an NMR tube. The methyl group of 3-acetylpyridine is masked by the large OH peak of EtOH, but the aromatic protons are clearly distinguishable as markers of unreacted ketone. The reaction mixture sampled after 2 h at 60 °C contained no starting material.

3. The oxime formation at 60 °C generates a 97:3 ratio of E/Z isomers. The reaction conducted at ambient temperature results in a ratio of 88:12.

4. The mixture remains heterogeneous at 70 °C. Further product crystallization occurs upon cooling.

5. The amount of the Z-isomer of the isolated material ranged from 0.2 to 0.7%. The submitters analyzed the E/Z ratio by GC-MS using the conditions described in Note 8. The checker analyzed by 400 MHz ^{1}H NMR by integration of the Z-isomer (upfield from the E-isomer by 0.06 ppm) and comparing to the integration of both ^{13}C-^{1}H satellites (0.55% each) of the E-isomer. Z-Isomer levels could be detected and accurately integrated at the 0.2% level. An enriched sample of the Z-isomer (8%) was obtained by concentration of the filtrate from the crystallization. For further details on the use of ^{13}C satellites for quantitative analysis of low-level components, see Claridge, T. D. W.; Davies, S. G.; Polywka, M. E. C.; Roberts, P. M.; Russell, A. J.; Savory, E. D.; Smith, A. D. *Org. Lett.* **2008**, *10*, 5433-5436.

6. The submitters carried out a recrystallization from EtOAc as follows. The oxime (12.9 g) is added to ethyl acetate (70 mL) in a 250-mL single-necked flask and warmed to 75 °C with stirring to dissolve all solids. After cooling to ambient temperature, the flask is placed in a freezer at −18 °C for 5 h. The product is collected by filtration on a Büchner funnel and dried in a round-bottomed flask for 2 h at 80 °C under high vacuum (0.1 mmHg) to yield 11.2 g.

7. An analytically pure sample was prepared by recrystallization from water as follows. Oxime (2.0 g) is added to water (25 mL) in a 100-mL

round-bottomed flask containing a 1.5-cm oval Teflon-coated magnetic stir bar. The mixture is warmed to 80 °C with stirring using a heating mantle and rapidly hot filtered through a 60-mL medium-porosity sintered glass funnel that has been pre-heated to 110 °C in an oven. The resulting filtrate is re-heated to 80 °C to re-dissolve all solids, then allowed to slowly cool to ambient temperature over 1 h. After an additional 30 min at ambient temperature, the solids are collected by filtration on a 15-mL medium-porosity sintered glass funnel, washed with water (2 x 5 mL) and dried for 15 h (60 °C, 20 mmHg) to afford analytically pure product (1.5 g, 75%).

8. Physical data for (E)-1-pyridin-3-yl-ethanone oxime (1): mp 118–119 °C; ^{1}H NMR (400 MHz, CDCl$_3$) δ: 2.32 (s, 3 H, CH$_3$), 7.32 (dd, 1 H, J = 4.8, 8.0 Hz, H(5)-Py), 7.98 (ddd, 1 H, J = 2.0, 1.9, 8.0 Hz, H(4)-Py), 8.61 (dd, 1 H, J = 1.5, 4.8 Hz, H(6)-Py), 8.97 (d, 1 H, J = 2.0 Hz, H(2)-Py), 10.50 (br s, 1 H, NOH); ^{13}C NMR (100 MHz, CDCl$_3$) δ: 11.9, 123.6, 133.1, 133.8, 147.4, 149.6, 153.0; GC-MS m/z 136.1 ([M]$^+$); Anal. Cald. for C$_7$H$_8$N$_2$O: C, 61.75; H, 5.92; N, 20.57. Found: C, 61.55; H, 5.96; N, 20.41. GC-MS analysis (1 μL sample) was carried out on a Thermo Finnigan PolarisQ, GC/MS (EI), Trace GC 2000 using a Restek RTX-5MS (5% phenylsilicon) column (30 m, 0.25 mm diameter, 0.25 μm): gas carrier He, flow 50 mL/min, split set to 0.7 mL/min; oven gradient conditions 70 °C initial Temp, 1 min, ramp of 11 °C/min until final Temp 250 °C, and hold for 10 min. Post run Temp: 300 °C for 5 min. Max Temp 350 °C, prep run timeout 10 min, equilibration time 0.5 min. MS method: source Temp 200 °C, 3 micro scans, max ion time 25 (E-isomer t$_R$ 10.56 min).

9. Heating to 80 °C results in equilibration of the E/Z ratio from the original 92:8 to 97:3. The second crop material is comparable in quality to the first crop (0.5 – 0.7% Z-isomer).

10. Benzyl bromide is a lachrymator and should only be handled in a well vented hood.

11. A sample (0.2 mL) of the mixture is removed by syringe, quenched with water (0.5 mL) and extracted with ethyl acetate (0.5 mL). The ethyl acetate layer is analyzed by TLC, eluting with hexane/ethyl acetate (1:1 v/v) and visualized by UV: R$_f$ 0.6 (benzyl oxime), R$_f$ 0.3 (oxime). In both runs by the checker, the reaction was complete using the original charge of NaH. The submitters note that, if unreacted oxime is present, NaH (2.24 g, 56 mmol, 60% suspension in mineral oil) can be added to the mixture to drive the reaction to completion.

12. The crude material contains about 10 mol% EtOAc and 10 mol% DMF by ^{1}H NMR analysis.

13. The crude material is purified by chromatography on SiO$_2$ (150 g) in a 6-cm diameter column, wet-packed using hexanes. The column is topped with sea sand (0.5 cm). The product oil is loaded onto the column and is eluted with hexanes (400 mL) followed by 1:1 EtOAc:hexanes (1.3 L), collecting 75 mL fractions. The fractions are analyzed by TLC as described in Note 11. Fractions 7-18 are combined and concentrated by rotary evaporation (40 °C, 20 mmHg).

14. Physical data for *(E)*-1-pyridin-3-yl-ethanone *O*-benzyl-oxime (**2**): ^{1}H NMR (400 MHz, CDCl$_3$) δ: 2.28 (s, 3 H, CH$_3$), 5.27 (s, 2 H, OCH$_2$), 7.26 (ddd, 1 H, J = 0.8, 4.8, 8.0 Hz, H(5)-Py), 7.30–7.45 (m, 5 H, Ar), 7.94 (ddd, 1 H, J = 1.8, 1.8, 8.0 Hz, H(4)-Py), 8.59 (dd, 1 H, J = 1.7, 4.7 Hz, H(6)-Py), 8.88 (dd, 1 H, J = 0.6, 2.0 Hz, H(2)-Py); ^{13}C NMR (100 MHz, CDCl$_3$) δ: 12.7, 76.7, 123.4, 128.1, 128.5, 128.6, 132.4, 133.4, 137.9, 147.7, 150.2, 152.7; GC-MS according to the method described in Note 8: m/z 226.3 ([M]$^+$) (t_R 16.44 min). Only the *E*-isomer was observed by GC/MS and ^{1}H NMR. A sample enriched in the *Z*-isomer (4:1 *E*:*Z* ratio) was prepared by reaction of 3-acetylpyridine with *O*-benzylhydroxylamine.

15. BH$_3$•THF (1.0 M) is added to the pre-marked line representing a 465-mL fill (dioxane (230 mL) and BH$_3$•THF (235 mL)). The actual amount of reagent added is determined by weighing the reagent bottle before and after addition (density 0.867 g/mL). The exact amount of borane added is not critical as comparable results are obtained with 3 to 5 equiv.

16. Completion of reaction is determined by ^{1}H NMR as follows. An aliquot (0.1 mL) is quenched into CD$_3$OD (0.6 mL) and 37% DCl in D$_2$O (0.1 mL) is added. The uncapped NMR tube is held for 1 h at ambient temperature until the hydrogen evolution ceases. The *O*-benzyl oxime resonances at 5.3 and 2.3 ppm are diagnostic of unreacted oxime. When 4 to 5 equiv of borane are used, the reaction is complete (<3% oxime) within 24 h. With 3 equiv borane, 8% oxime remained unreacted after 24 h and did not react further upon stirring an additional day.

17. The amine product can co-distill with dioxane if the temperature and vacuum are too high during concentration.

18. The amine is purified by chromatography on SiO$_2$ (260 g), wet packed with 10% MeOH/CH$_2$Cl$_2$ and topped with sea sand (0.5 cm). The product is dissolved with sonication in CH$_2$Cl$_2$ (40 mL), loaded, and eluted using 10% MeOH/CH$_2$Cl$_2$ (500 mL), 25% MeOH/CH$_2$Cl$_2$ (500 mL), 50%

Org. Synth. **2010**, *87*, 36-52

MeOH/CH$_2$Cl$_2$ (500 mL), and 4% Et$_3$N/MeOH (1.5 L), collecting 250-mL fractions. The product fractions 3-8 are concentrated by rotary evaporation (40 °C, 20 mmHg) to afford 8.03 g of the product. ^{1}H NMR analysis indicated the presence of 13 wt% ethylene glycol and 4 wt% MeOH, indicating the product was approximately 83 wt % pure (91% yield corrected for purity)

19. *(S)-1-Pyridin-3-yl-ethylamine (4)*: ^{1}H NMR (400 MHz, CDCl$_3$) δ: 1.40 (d, 3 H, *J* = 6.6 Hz, CH$_3$), 2.0 (br s, 2 H, NH$_2$), 4.17 (q, 1 H, *J* = 6.6 Hz, NCH), 7.26 (dd, 1 H, *J* = 4.9, 8.0 H(5)-Py), 7.71 (ddd, 1 H, *J* = 1.6, 2.3, 7.8 Hz, H(4)-Py), 8.47 (dd, 1 H, *J* = 1.6, 4.8 Hz, H(6)-Py), 8.57 (d, 1 H, *J* = 2.2 Hz, H(2)-Py) ^{13}C NMR (100 MHz, CDCl$_3$) δ: 25.7, 49.3, 123.7, 133.6, 142.7, 148.2, 148.5; GC-MS conditions from Note 8) *m/z* 223.2 ([M]$^+$) (t$_R$ 6.56 min).

20. ***Procedure for the preparation of racemic amine and acetamide:*** An oven-dried, 100-mL round-bottom flask equipped with a 1-cm Teflon-coated magnetic stirrer and a reflux condenser connected via a nitrogen adapter and a gas bubbler, is charged with benzyl oxime **2** (1.15 g, 5.1 mmol) and 1.0 M BH$_3$•THF (20 mL, 20 mmol). The solution is heated at reflux for 3 h, then cooled to ambient temperature and quenched by the dropwise addition of methanol (5 mL). The resulting mixture is heated at reflux for 14 h, then concentrated by rotary evaporation (40 °C bath, 20 mm Hg). The crude amine product is purified by column chromatography using 15 g silica gel wet packed with 10% MeOH/CH$_2$Cl$_2$, eluting with 50 ml MeOH/CH$_2$Cl$_2$, 50 mL MeOH, and 100 mL 5% Et$_3$N/MeOH, taking 10-15 mL fractions. Fractions 3-10 are combined and concentrated by rotary evaporation (40 °C bath, 20 mm Hg) to afford an oil (0.56 g, 90% yield). The acetamide is prepared by dissolution of the racemic amine (0.22 g, 1.8 mmol) in CH$_2$Cl$_2$ followed by addition of Et$_3$N (0.3 g, 3.0 mmol), Ac$_2$O (0.20 g, 2.0 mmol) and 4-dimethylyaminopyridine (DMAP) (15 mg). The mixture is stirred at ambient temperature for 30 min. The solvent is removed by rotary evaporation (40 °C bath, 20 mm Hg) and the residue is purified by chromatography on SiO$_2$ (15 g), eluting with 100 mL CH$_2$Cl$_2$:MeOH (97:3 v/v), collecting 10 mL fractions. Fraction 3 is concentrated by rotary evaporation (40 °C bath, 20 mm Hg) to afford the acetamide product (0.19 g, 64%). Fraction 4 is likewise concentrated to afford additional product (100 mg, approx 80% pure by NMR, 27% yield; overall yield from both fractions, 91%) TLC conditions: 10% MeOH/CH$_2$Cl$_2$, R$_f$ 0.5.

21. The submitters developed a chiral GC assay to analyze the ee of the derivatized amine as follows: Crompack Chirasil-Dex-CB column (30 m × 0.25 mm × 0.25μm). Conditions: 90 °C, 2 °C/min to 120 °C, hold 20 min; 2 °C/min to 130 °C, hold 20 min; 2 °C/min to 140 °C, hold 20 min, gives one enantiomer t_R 70.70 min, other enantiomer t_R 73.47 min. The checkers developed a chiral HPLC assay for the derivatized amine as follows: Chiralpak AD-H column (150 x 4.6 mm, 5 micron), A: Heptane, B: 1:1 MeOH:EtOH, 5% B for 4 min, then to 40% B over 18 min, hold 3 min, then to 5% B over 3 min, 20 min post time, 1.0 mL/min, ambient temperature, 210 nm. The undesired (R)-enantiomer elutes at 9.5 min, the desired (S)-enantiomer at 12.5 min.

22. Crystallization of the bis-HCl salt begins immediately and a high stirring rate is maintained to prevent clumping.

23. The bis-hydrochloride salt is hygroscopic, especially as a solvent-wet solid, and requires isolation under a nitrogen atmosphere. The checker used a pressure filter (cf, Sigma-Aldrich Z147656 or Z422886) under nitrogen to isolate the bis-HCl salt.

24. (S)-1-Pyridin-3-yl-ethylamine bis-hydrochloride (5): mp 191–193°C; $[\alpha]^{20}_D$ +5 (c 1.5, CH$_3$OH); ^{1}H NMR (400 MHz, D$_2$O) δ: 1.66 (d, 3 H, J = 7.0 Hz, CH$_3$), 4.79 (q, 1 H, J = 7.0 Hz, NCH), 8.09 (dd, 1 H, J = 5.9, 8.2 Hz, Py), 8.66 (d, 1 H, J = 8.3 Hz, Py), 8.77 (d, 1 H, J = 5.8 Hz, Py), 8.87 (d, 1 H, J = 1.4 Hz, Py); ^{13}C NMR (100 MHz, D$_2$O) δ: 18.5, 48.0, 128.0, 137.7, 140.5, 142.1, 145.5. Anal. Cald. for C$_7$H$_{12}$Cl$_2$N$_2$: C, 43.09; H, 6.20; Cl, 36.35; N, 14.36. Found: C, 43.31; H, 6.45; Cl, 36.35; N, 14.26.

25. The ee of the bis-HCl salt was determined by the derivatization method and HPLC analysis described in Notes 20 and 21. The derivatization method using the bis-HCl salt required an additional 2 equiv of triethylamine. Formation of the hydrochloride salt does not substantially enrich the ee of the product.

26. The table below summarizes three asymmetric reduction experiments carried out by the checker at the 60-mmol scale. The data suggest slightly improved enantioselectivity using 3-4 equiv vs 5 equiv of borane. The reaction with 3 equiv of borane did proceed to completion. Crystallization as the bis-HCl salt affords little to no ee upgrade.

Table 1. Summary of Checkers Results for the Enantioselective Reduction of Benzyl Oxime **2**, 60 mmol scale

	Equiv BH₃·THF	Unreacted oxime	Yield of amine **4**	ee of amine **4**	Yield of HCl salt **5**	ee of HCl salt **5**
Run 1	5	<3%	88%	94%	85%	94%
Run 2	4	3%	91%	98%	89%	99%
Run 3	3	8 %	86%	99%	89%	99%

Safety and Waste Disposal Information

All hazardous materials should be handled and disposed of in accordance with "Prudent Practices in the Laboratory"; National Academy Press; Washington, DC, 1995.

3. Discussion

The catalytic enantioselective borane reduction of C=N bonds is of great interest due to the formation of nonracemic primary amines which are widely used as key intermediaries in the synthesis of a large variety of pharmaceuticals, chiral auxiliaries, catalysts and resolving agents.[4] Although the asymmetric reduction of oxime ethers with borane-based catalysts offers a facile and direct approach to obtain enantioenriched primary amines, more than a stoichiometric amount of *in situ* prepared oxazaborolidines was previously employed to obtain a high degree of enantioselectivity.[5] Fontaine, et al.[6] used 2.5 equiv of the diphenylvalinol-derived B-H oxazaborolidine to achieve complete reduction with high selectivity. Itsuno and co-workers[7] reported the first catalytic borane-based reduction of acetophenone *O*-benzyl oxime using 10 mol % of the B-H oxazaborolidine, generated *in situ* from (*S*)-diphenylvalinol, obtaining (*S*)-1-phenylethanamine in 52% ee. Thus, the development of highly enantioselective reagents for the catalytic reduction of benzyl oximes is highly desirable.

Recently, we reported the reduction of benzyl oxime ethers affording the desired primary amines in good to high yield and excellent enantioselectivities using catalytic amounts (10-30%) of the stable spiroborate ester **3**,[3] which has been previously discovered in our group as a new class of catalysts.[8,9] Optically active pyridine-derived amines have attracted a strong interest, primarily, due to their existence in naturally occurring compounds, such as tobacco alkaloids,[10] or as potential drug candidates.[11] The procedure described here presents the first catalytic asymmetric reduction of (E)-1-pyridin-3-yl-ethanone O-benzyl-oxime (2) to (S)-1-pyridin-3-yl-ethylamine (4), used as a representative method for the rapid access of primary amines with a high degree of enantiopurity and good yield using a simple and convenient approach. Since the enantiofacial selectivity in the reduction of C=N bonds depends not only on the chirality of the transfer agent but also on the E/Z isomeric purity,[12] the present procedure affords a high ratio of E/Z isomer (>95%) in the crude product, and the E- isomer is readily obtained by a simple recrystallization from either water or EtOAc with >99% purity as analyzed by GC/MS and ^{1}H NMR. Pure (E)-benzyl oxime ether is obtained in high yield (>95%) from the (E)-oxime by the reaction with NaH and benzyl bromide in DMF at –10 °C. In contrast, use of O-benzylhydroxylamine to directly access oxime ether 2 in one step from 3-acetylpyridine results in a 4:1 ratio of E/Z isomers, which are not readily separable by flash chromatography and are not crystalline.

To achieve excellent enantioselectivity and high yield in the spiroborate borane-mediated reduction of benzyl oxime ether 2, moisture has to be rigorously excluded from the reaction medium and BH$_3$•THF of high purity is also required. The enantioselectivity of the primary amine is slightly affected by the reaction solvent: in dioxane a 97–99% ee was achieved, while in THF a 95% ee was observed. However, similar chemical yields were afforded after column chromatography (>80%) in both solvents. The amine from the chromatography contained up to 15% ethylene glycol as indicated by NMR analysis. Therefore, the bis-hydrochloride salt 5 was readily prepared with high purity from diethyl ether/methanol in 85–90% yield. The amine bis-hydrochloride salt is very hygroscopic as a solvent-wet solid and has to be handled under nitrogen during isolation, but the dry solid is less hygroscopic and picks up water slowly over several hours when exposed to ambient air. The bis-hydrochloride salt is more stable to decomposition by oxidation and more convenient to handle than the free amine.

An analogous procedure can be applied to other heteroaryl, heterocyclic and pyridyl alkyl O-benzyl oxime ethers, and the results are summarized in Tables 2 and 3.

Table 2. Asymmetric Reduction of (*E*)-Heteroaryl and Heterocyclic *O*-Benzyloximes

1. cat **3** (0.1 equiv), BH$_3$·THF (4 equiv) dioxane, 0 °C, 36 – 48 h

2. Ac$_2$O, DMAP CH$_2$Cl$_2$, rt

Entry	Benzyl oxime	Acetamide[a]	Yield (%)[b]	ee (%)[c]
1			85	98
2			92	96
3			76	94
4			73	95
5			71	99
6			70	94

[a] The reactions were carried out using 4 equiv of borane stabilized with NaBH$_4$.
[b] Isolated yield of amides purified by column chromatography. [c] Determined by GC of acetyl derivatives on chiral column (CP-Chirasil-DexCB).

Table 3. Preparation of other Chiral Pyridyl Amines via Borane Reduction Catalyzed by Spiroborate Ester **3**

Entry	Benzyl oxime	Amine[a]	Yield (%)[b]	ee (%)[c]
1			89	99
2			83	96[d]
3			88	98
4			84	99
5			85	96
6			91	98
7			82	95
8			95	95[e]

[a] The reactions were carried out using 1 equiv of oxime ether (1 mmol), 0.3 equiv of catalyst **3** and 5 equiv of borane stabilized with NaBH$_4$ in dioxane at 10 °C. [b] Isolated yield based on the acetyl derivative of amines. [c] Determined by GC on a chiral column (CP-Chirasil-Dex-CB). [d] Determined by chiral HPLC (Chiralcel OD-H column). [e] Determined by chiral HPLC (Chiralcel IB column).

1. Department of Chemistry, University of Puerto Rico-Humacao, CUH Station, Humacao, Puerto Rico 00791, USA. E-mail: ortiz@quimica.uprh.edu.
2. The checker would like to thank Tanja Brkovic for development of the chiral HPLC assay of the acetamide derivative of amine **4** described in Note 21.
3. (a) Ortiz-Marciales, M.; Stepanenko, V.; Correa, W.; De Jesús, M.; Espinosa, S. U.S. Patent Application 11/512,599, Aug. 30, 2006. (b) Ortiz-Marciales, M.; Huang, X.; Huang, K.; Stepanenko, V.; De Jesús, M.; Merced, F. G. PCT/US07/76195, Aug 17, 2007.
4. (a) Lawrence, S. A. *Amines: Synthesis, Properties and Applications*; Cambridge University Press: Cambridge, U.K., 2004. (b) Breuer, M.; Ditrich, K.; Habicher, T.; Hauer, B.; Keâeler, M.; Sturmer, R.; Zelinski, T. *Angew. Chem., Int. Ed.* **2004**, *43*, 788–824. (c) Tanuwidjaja, J.; Peltier, H. M.; Ellman, J. A. *J. Org. Chem.* **2007**, *72*, 626–629. (d) Bloch, R. *Chem. Rev.* **1998**, *98*, 1407–1438. (e) Fache, F. S. E.; Tommasino, M. L.; Lemaire, M. *Chem. Rev.* **2000**, *100*, 2159–2232. (f) Kozma, D. *CRC Handbook of Optical Resolutions via Diastereomeric Salt Formation* 1st ed.; CRC Press LLC: Boca Raton, FL, 2001.
5. (a) Itsuno, S.; Nakano, M.; Miyazaki, K.; Masuda, H.; Ito, K. *J. Chem. Soc., Perkin Trans. 1* **1985**, 2039–2044. (b) Itsuno, S.; Sakurai, Y.; Shimizu, K.; Ito, K. *J. Chem. Soc., Perkin Trans. 1* **1990**, 1859–1863. (c) Bolm, C.; Felder, M. *Synlett* **1994**, 655–666. (d) Lantos, I.; Flisak, J.; Liu, L.; Matsunoka, R.; Mendelson, W.; Stevenson, D.; Tubman, K.; Tucker, L.; Zhang, W.-Y.; Adams, J.; Sorenson, M.; Garigipati, R.; Erhardt, K.; Ross, S. *J. Org. Chem.* **1997**, *62*, 5385–5391. (e) Inoue, T.; Sato, D.; Komura, K.; Itsuno, S.; *Tetrahedron Lett.* **1999**, *40*, 5379–5382. (f) Itsuno, S.; Matsumoto, T.; Sato, D.; Inoue, T. *J. Org. Chem.* **2000**, *65*, 5879–5881. (g) Sailes, H. E.; Watts, J. P.; Whiting, A. *J. Chem. Soc., Perkin Trans. 1* **2000**, 3362–3374. (h) Krzeminski, M. P.; Zaidlewicz, M. *Tetrahedron: Asymmetry* **2003**, *14*, 1463–1466. (i) Sakito, Y.; Yoneyoshi, Y.; Suzukamo, G. *Tetrahedron Lett.* **1988**, *29*, 223–224. (j) Shimizu, M.; Tsukamoto, K.; Matsutani, T.; Fujisawa, T. *Tetrahedron* **1998**, *54*, 10265–10274. (k) Masui, M.; Shioiri, T. *Tetrahedron Lett.* **1998**, *39*, 5195–5198. (l) Chu, Y.-B.; Shan, Z.-X.; Liu, D.-J.; Sun, N.-N. *J. Org. Chem.* **2006**, *71*, 3998–4001.

6. Fontaine, E.; Namane, C.; Meneyrol, J.; Geslin, M.; Serva, L.; Roussey, E.; Tissandie´, S.; Maftouh, M.; Roger, P. *Tetrahedron: Asymmetry* **2001**, *12*, 2185–2189.

7. Itsuno, S.; Sakurai, Y.; Ito, K.; Hirao, A.; Nakahama, S. *Bull. Chem. Soc. Jpn.* **1987**, *60*, 395–396.

8. (a) Huang, X.; Ortiz-Marciales, M.; Huang, K.; Stepanenko, V.; Merced, F. G.; Ayala, A. M.; De Jesús, M. *Org. Lett.* **2007**, *9*, 1793–1795. (b) Huang, K.; Ortiz-Marciales, M.; Merced, F. G.; Melendez, H. J.; Correa, W.; De Jesús, M. *J. Org. Chem.* **2008**, *73*, 4017–4026. (c) Huang, K.; Ortiz-Marciales, M.; Stepanenko, V.; De Jesús, M.; Correa, W. *J. Org. Chem.* **2008**, *73*, 6928–693.

9. (a) Stepanenko, V.; Ortiz-Marciales, M.; Correa, W.; De Jesús, M.; Espinosa, S.; Ortiz, L. *Tetrahedron: Asymmetry* **2006**, *17*, 112–115. (b) Stepanenko, V.; Ortiz-Marciales, M.; De Jesús, M.; Correa, W.; Vázquez, C.; Ortiz, L.; Guzmán, I.; De la Cruz, W. *Tetrahedron: Asymmetry* **2007**, *18*, 2738–2745.

10. Breining, S. R. *Curr. Top. Med. Chem.* **2004**, *4*, 609–621.

11. (a) Holladay, M. W.; Dart, M. J.; Lynch, J. K. *J. Med. Chem.* **1997**, *40*, 4169–4194. (b) Latli, B.; D'Amour, K.; Casida, J. E. *J. Med. Chem.* **1999**, *42*, 2227–2234. (c) Nielsen, S. F.; Nielsen, E. O.; Olsen, G. M.; Liljefors, T.; Peters, D. *J. Med. Chem.* **2000**, *43*, 2217–2226. (d) Mullen, G.; Napier, J.; Balestra, M.; DeCory, T.; Hale, G.; Macor, J.; Mack, R.; Loch, J.; Wu, E.; Kover, A.; Verhoest, P.; Sampognaro, A.; Phillips, E.; Zhu, Y.; Murray, R.; Griffith, R.; Blosser, J.; Gurley, D.; Machulskis, A.; Zongrone, J.; Rosen, A.; Gordon, J. *J. Med. Chem.* **2000**, *43*, 4045–4050. (e) Chelucci, G. *Tetrahedron: Asymmetry* **2005**, *16*, 2353–2383.

12. (a) Bolm, C.; Felder, M. *Synlett* **1994**, 655–656. (b) Lantos, I.; Flisak, J.; Liu, L.; Matsunoka, R.; Mendelson, W.; Stevenson, D.; Tubman, K.; Tucker, L.; Zhang, W.-Y.; Adams, J.; Sorenson, M.; Garigipati, R.; Erhardt, K.; Ross, S. *J. Org. Chem.* **1997**, *62*, 5385–5391. (c) Tillyer, R. D.; Boudreau, C.; Tschaen, D.; Dolling, U.-H.; Reider, P. J. *Tetrahedron Lett.* **1995**, *36*, 4337–4340. (d) Shimizu, M.; Kamei, M.; Fujisawa, T. *Tetrahedron Lett.* **1995**, *36*, 8607–8610. (e) Shimizu, M.; Tsukamoto, K.; Matsutani, T.; Fujisawa, T. *Tetrahedron* **1998**, *54*, 10265–10274. (f) Masui, M.; Shioiri, T. *Tetrahedron Lett.* **1998**, *39*, 5195-5198. (g) Cho, B. T.; Ryu, M. H. *Bull. Korean Chem. Soc.* **1994**, *15*, 191-192. (h) Demir, A. S. *Pure Appl. Chem.* **1997**, *69*, 105-108. (i)

Sakito, Y.; Yoneyoshi, Y.; Suzukamo, G. *Tetrahedron Lett.* **1988**, *29*, 223–224.

Appendix
Chemical Abstracts Nomenclature; (Registry Number)

(–)-(*S*)-1-(1,3,2-Dioxaborolan-2-yloxy)-3-methyl-1,1-diphenylbutan-2-amine; (879981-94-9)

3-Acetyl pyridine; (350-03-8)

Hydroxylamine hydrochloride; (5470-11-1)

Sodium hydride; (7646-69-7)

Benzyl bromide; (100-39-0)

Borane tetrahydrofuran complex solution, 1.0 M in tetrahydrofuran; (14044-65-6)

(*E*)-1-Pyridin-3-yl-ethanone oxime; (106881-77-0)

(*E*)-1-Pyridin-3-yl-ethanone *O*-benzyl-oxime: Ethanone, 1-(3-pyridinyl)-, *O*-(phenylmethyl)oxime, (1*E*)-; (1010079-98-7)

(*S*)-1-Pyridin-3-yl-ethylamine: 3-Pyridinemethanamine, α-methyl-, (α*S*)-; (27854-9)

(*S*)-1-Pyridin-3-yl-ethylamine hydrochloride: 3-Pyridinemethanamine, α-methyl-, dihydrochloride, (*S*)-; (40154-84-5)

Margarita Ortiz-Marciales was born in 1943 in Bogotá, Colombia, and obtained her B. S. from "Universidad Nacional de Colombia" in 1968. She studied at Freigburg and Mainz Universities, Germany, for two years with a DAAD fellowship. She received her M. S. from the University of Alabama in Hunstsville under Prof. S. McManus supervision in 1973, and her Ph. D. in Organic Chemistry at the University of Alabama-Tuscaloosa in 1979 under the direction of Prof. Macmanus and Prof. R. Abramovitch. In 1980, she did postdoctoral studies in Prof. G. Larson's group at the University of Puerto Rico-Río Piedras. She joined the University of Puerto Rico- Humacao in 1981, where she is currently a professor. Her interests are in the development of new synthetic methodologies using boron and silicon compounds for the preparation of important organic intermediaries and biological active amino compounds.

Kun Huang received his B.S. degree in 1996 from Sichuan University China. He obtained his Ph. D. in 2006 at Nanjing University, China, working on the asymmetric epoxidation and propanation of chiral sulfonium ylides. Then, he worked as an advanced synthetic researcher in Wuxi Pharma Tech Co., Ltd, Shanghai, China until he moved to Puerto Rico University in Humacao for his postdoctoral studies with Professor Margarita Ortiz-Marciales on the asymmetric reduction of O-benzyl oximes and ketones catalyzed by spiroborate esters. Later, he held a postdoctoral research position at the chemistry department at Oregon State University where his research was focused on the total synthesis of natural products. Currently, he is a postdoctoral researcher at Peking University in China.

STEREOSELECTIVE SYNTHESIS OF 3-ARYLACRYLATES BY COPPER-CATALYZED SYN HYDROARYLATION
[(E)-Methyl 3-phenyloct-2-enoate]

Submitted by Naohiro Kirai and Yoshihiko Yamamoto.[1]
Checked by Peter Wipf and Courtney L. Vowell.[2]

1. Procedure

(E)-Methyl 3-phenyloct-2-enoate (1). A 200-mL two-necked round-bottomed flask equipped with a magnetic stirring bar (2.5 cm × 0.8 cm) is charged with methyl 2-octynoate (5.00 mL, 4.60 g, 29.5 mmol) (Note 1), MeOH (60 mL) (Note 2) and phenylboronic acid (5.52 g, 44.3 mmol, 1.5 equiv) (Notes 3 and 4). One neck of the flask is equipped with rubber septum and nitrogen inlet, the other with a glass stopper, and the reaction mixture is degassed once at −78 °C (Note 5). The glass stopper is removed and CuOAc (84.0 mg, 0.606 mmol) (Note 6) is added. The reaction mixture is degassed three times at −78 °C (Note 5). After vigorous stirring for 24 h at 28 °C under an N_2 atmosphere (Note 7), the reaction mixture is filtered through a pad of Celite® (Note 8) and washed with a mixture of hexanes/EtOAc (150:1; 3x10 mL) to remove insoluble materials. The filtrate is concentrated by rotary evaporation (5.5 mmHg) at 30 °C to afford a crude oil. The crude product is purified by chromatography on SiO_2 (Note 9) to yield **1** (6.48 g, 94%) as a colorless oil (Notes 10 and 11).

2. Notes

1. Methyl 2-octynoate (99%) was purchased from Sigma-Aldrich (Checkers) or TCI (Submitters) and used as received.
2. Methanol was purchased from J.T. Baker (Checkers) or Wako (Submitters) and used as received.
3. Phenylboronic acid (98%) was purchased from Lancaster

(Checkers) or Sigma-Aldrich (Submitters) and used as received.

4. A slight excess of phenylboronic acid was used to ensure the complete consumption of the alkynoate.

5. The flask is placed in a -78 °C bath (dry ice/acetone), evacuated (1.45 mmHg, 30 sec), and backfilled with nitrogen gas.

6. CuOAc (90%) was purchased from Acros (Checkers) or Sigma-Aldrich (Submitters) and used as received.

7. The initial green reaction mixture turned yellow in several hours. Reaction progress was monitored by TLC analysis on EM Science precoated silica gel 60 F254 plates, visualized by a 254-nm UV lamp and stained with a $KMnO_4$ solution (5 wt.% in 1 M NaOH). TLC analysis showed the formation of acrylate **1** (yellow) (hexanes/ethyl acetate, 10:1, $R_f = 0.59$), while the R_f of methyl 2-octynoate (yellow) was 0.49.

8. Fisher Celite® 545 fine.

9. Column chromatography was carried out on a 4-cm diameter column packed with 93 g SiliaFlash P60 silica gel (230-400 mesh) using hexanes/ethyl acetate (24:1) as the eluent. Approximately 1 L of the solvent mixture was used. Fractions 20-48 (fraction size of 5 mL) were collected.

10. Spectroscopic and analytical data of **1** are as follows: ^1H NMR (300 MHz, $CDCl_3$) δ: 0.88 (t, $J = 7.2$ Hz, 3 H), 1.30–1.47 (m, 6 H), 3.14 (t, $J = 7.2$ Hz, 2 H), 3.76 (s, 3 H), 6.06 (s, 1 H), 7.36–7.46 (m, 5 H); ^{13}C NMR (75 MHz, $CDCl_3$) δ: 14.0, 22.5, 28.8, 31.0, 31.9, 51.0, 116.7, 126.7, 128.5, 128.9, 141.4, 161.4, 166.8; IR (oil) 2955, 2931, 2360, 2341, 1713, 1624, 1191, 1170 cm^{-1}; MS (EI) m/z (rel. intensity, %): 232 (76, M^+), 201 (66), 189 (75), 176 (100); HRMS (EI+) m/z Calcd for $C_{15}H_{20}O_2$: 232.146330, found: 232.146275

11. GC-analysis was carried out using an Agilent Technologies 6890 N Network GC System equipped with a capillary column (Agilent Technologies, HP-5) (30.0 m × 0.32 mm × 0.25 mm). Oven program for GC-analysis: starting temperature, 50 °C for 1 min; heating to 280 °C at a rate of 30 °C/min. Retention time of acrylate **1**: $R_t = 8.65$ min. GC-analysis indicated that **1** has >99% purity.

Safety and Waste Disposal Information

All hazardous materials should be handled and disposed of in accordance with "Prudent Practices in the Laboratory"; National Academy Press; Washington, DC, 1995.

3. Discussion

The conjugate additions of aryl metal reagents to alkynoates produce synthetically useful 3-arylacrylates.[3] For this purpose, organocopper reagents have been typically employed,[4,5] although their use has several limitations: (a) (semi)stoichiometric amounts of copper source are required, (b) the stereochemistry of the products depends on both the reaction conditions and nature of the organocopper reagents, and (c) they are incompatible with highly reactive functional groups. To address these issues, the transition-metal-catalyzed selective hydroarylations of alkynoates have recently been developed using arylboronic acids as bench-top stable, easy to handle arylating reagents.[5] These methods, however, require expensive precious metal catalysts together with phosphine ligands and/or relatively harsh reaction conditions. The present procedure presents a new hydroarylation method of alkynoates with arylboronic acids using inexpensive copper acetate as a catalyst. The reaction on a 30 mmol scale efficiently proceeds at 28 °C in methanol to afford 3-arylacrylates in good yields with excellent *syn*-selectivity. This method is also applicable to ethyl 3-phenylpropiolate as the substrate to stereoselectively deliver 3,3-diarylacrylates in good yields (Table 1).

Table 1. Hydroarylation of ethyl 3-phenylpropiolate.[a]

R^b	Cu(OAc) (mol%)	Time (h)	Yield (%)[c]
Me	4	24	77
OMe	2	24	78
Cl	6	12	92

[a]30 mmol scale. [b]1.5 equiv of arylboronic acids were used. [c]Isolate yield.

Table 2. Scope of Cu-catalyzed hydroarylation of alkynoates.[a]

Alkynoate Scale	Product	Cu(OAc) (mol %)	Time (h)	Yield (%)[b]
$n\text{-}C_5H_{11}$—≡—CO_2Me 0.5 mmol		5	3	91
$n\text{-}C_5H_{11}$—≡—CO_2Me 0.5 mmol		3	10	88
$MeO(CH_2)_2$—≡—CO_2Me 0.5 mmol		2	3	92
$Cl(CH_2)_3$—≡—CO_2Me 1.0 mmol		1	2	97
MeO—〈 〉—≡—CO_2Et 1.0 mmol		1	24	97
〈 〉—≡—CO_2Et OMe 0.5 mmol		3	1	88

[a]3 equiv. of arylboronic acids were used. [b]Isolated yield.

The generality of this method has been demonstrated on a smaller scale as shown in Table 2, although 3 equiv of arylboronic acids were employed in our previous study to ensure the complete consumption of the alkynoate substrates.[6] The newly optimized 30 mmol scale reaction can be carried out with 1.5 equiv of arylboronic acid. The Cu-catalyzed hydroarylation reaction is applicable to formyl- or iodo-substituted arylboronic acids and methoxy- or chloro-substituted alkynoates. When methoxy-substituted phenylpropiolates were used, 3,3-bis(methoxyphenyl)acrylates were obtained stereoselectively. The synthesis of 4-arylcoumarins has been accomplished by utilizing this stereoselective hydroarylation reaction.[7]

1. Department of Applied Chemistry, Graduate School of Science and Engineering, Tokyo Institute of Technology, Ookayama, Meguro-ku, Tokyo 152-8552, Japan.
2. Department of Chemistry, University of Pittsburgh, Pittsburgh, PA 15260, USA; pwipf@pitt.edu.
3. Nilsson, K.; Andersson, T.; Ullenius, C.; Gerold, A.; Krause, N. *Chem. Eur. J.* **1998**, *4*, 2051–2058.
4. (a) Corey, E. J.; Katzenellenbogen, J. A. *J. Am. Chem. Soc.* **1969**, *91*, 1851–1852. (b) Siddall, J. B.; Biskup, M.; Fried, J. H. *J. Am. Chem. Soc.* **1969**, *91*, 1853–1854. (c) Klein, J.; Turkel, R. M. *J. Am. Chem. Soc.* **1969**, *91*, 6186–6187.
5. (a) Hayashi, T.; Inoue, K.; Taniguchi, N.; Ogasawara, M. *J. Am. Chem. Soc.* **2001**, *123*, 9918–9919. (b) Oh, C. H.; Ryu, J. H. *Bull. Korean Chem. Soc.* **2003**, *24*, 1563–1564.
6. Yamamoto, Y.; Kirai, N.; Harada, Y. *Chem. Commun.* **2008**, 2010–2012.
7. Yamamoto, Y.; Kirai, N. *Org. Lett.* **2008**, *10*, 5513–5516.

Appendix
Chemical Abstracts Nomenclature; (Registry Number)

2-Octynoic acid, methyl ester; (111-12-6)
Boronic acid, phenyl-; (98-80-6)
Acetic acid, copper(1+) salt; (598-54-9)
2-Octenoic acid, 3-phenyl-, methyl ester, (2*E*)-; (189890-29-7)

Yoshihiko Yamamoto was born in Nagoya in 1968. He obtained his Ph.D. (1996) from Nagoya University, where he was appointed as an Assistant Professor in 1996 and promoted to an Associate Professor in 2003. In 2006, he tentatively moved to Tokyo Institute of Technology, and returned to Nagoya University in 2009. His research interests are focused on the development of organometallic reagents and catalysis.

Naohiro Kirai was born in Kagawa, Japan in 1984. He received his M.E. in 2009 from the Tokyo Institute of Technology under the supervision of Professor Yoshihiko Yamamoto. In the same year, he started his doctoral studies at the Department of Chemistry, Tokyo Institute of Technology, under the guidance of Professor Nobuharu Iwasawa. His current interest is catalytic CO_2-fixation reaction.

Courtney L. Vowell obtained a B.S. degree in Chemistry in 2007 from Millsaps College in Jackson, Mississippi, where she conducted chiral separations using the macrocyclic antibiotic vancomycin as a stationary phase under the supervision of Prof. Timothy J. Ward. She obtained her M.S. in Chemistry in 2009 from the University of Wyoming with Prof. Dean M. Roddick. Her research focused on the synthesis and oxidation properties of homoleptic perfluoroalkylphosphine platinum(0) complexes. In the summer of 2009, she joined the University of Pittsburgh Center for Chemical Methodologies and Library Development (UPCMLD) as a staff member.

PREPARATION OF (S)-3,3'-BIS-MORPHOLINOMETHYL-5,5',6,6',7,7',8,8'-OCTAHYDRO-1,1'-BI-2-NAPHTHOL

(S)-1

Submitted by Mark Turlington and Lin Pu.[1]
Checked by Kyle L. Kimmel and Jonathan A. Ellman.

1. Procedure

Paraformaldehyde (30 g, 1.0 mol) (Note 1) is added to a 250-mL round-bottomed flask, open to air, which is cooled with an ice bath. With vigorous stirring, morpholine (88 mL, 1.0 mol) (Note 2) is added dropwise over 1 h via addition funnel. The addition must be performed slowly with cooling because the reaction is exothermic. After the addition is complete, the ice bath is removed (Note 3), and with stirring, the solution is allowed to warm to room temperature over 2 h (Note 4). The round-bottomed flask is fitted with a reflux condenser and the reaction mixture is heated at 60 °C in an oil bath for 10 h (Note 5). A clear and viscous solution (107 mL) containing morpholinomethanol as well as other species is formed. The product is verified by ^{1}H and ^{13}C NMR analyses and the concentration of morpholinomethanol in the solution is determined to be 2.3 M (Notes 6-8).

To a 250-mL, 3-necked, round-bottomed flask is added (S)-5,5',6,6',7,7',8,8'-octahydro-1,1'-bi-2-naphthol [(S)-H$_8$BINOL, 5.0 g, 17.0 mmol] (Note 9). One neck of the flask is fitted with a reflux condenser which is fitted with a vacuum adaptor connected to a Schlenk line. The second neck of the flask is fitted with a glass stopper. The third neck is fitted with a thermometer to monitor the internal temperature. The flask is then evacuated (5 x 10^{-3} mmHg) and subsequently refilled with nitrogen (Note 10). This process is conducted three times. Maintaining a positive pressure of nitrogen, the glass stopper is then replaced with a rubber septum. To the flask is added 1,4-dioxane (30 mL) via syringe (Note 11).

The solution containing morpholinomethanol (Note 12) is bubbled with nitrogen for 30 min to remove any dissolved oxygen and then is transferred via cannula through the septum to the reaction flask containing (S)-H$_8$BINOL (Note 13). While maintaining a positive pressure of nitrogen, the septum is replaced with a glass stopper. The reaction flask is then heated in an oil bath for 18 h to maintain the internal temperature at 60 °C. Thin layer chromatography (TLC) shows complete consumption of the starting material, (S)-H$_8$BINOL (Note 14).

After allowing the flask to cool to room temperature, the reaction mixture is diluted with ethyl acetate (150 mL) in a 500-mL separatory funnel (Note 15). The organic layer is washed with a saturated NaHCO$_3$ solution (3 x 100 mL) and distilled water (2 x 100 mL). The organic layer is then dried with sodium sulfate (20 g) (Note 16), filtered, and transferred to a 1-L round-bottomed flask. The solvent is removed via rotary evaporation (30 °C, 10 mmHg) to yield a white solid.

The compound is purified via crystallization in ethanol. Hot ethanol (600 mL) is added to the 1-L round-bottomed flask to dissolve the product (Note 17). The solution is cooled to room temperature (Note 18), and then further cooled at –18 °C in a laboratory freezer for 44 h. Vacuum filtration (Note 19) with a Büchner funnel gives 7.48 g of (S)-**1** (90% yield) as white, fluffy, needle-like crystals (Notes 20–22). (S)-**1** is obtained with >99% ee and 99% purity as demonstrated by HPLC analysis (Note 23).

2. Notes

1.   Paraformaldehyde, powder, (95%) was used as purchased from Sigma-Aldrich.

2.   Morpholine, A.C.S. reagent, (≥99%) was used as purchased from Sigma-Aldrich.

3.   After removal of the ice bath, the solution was viscous, preventing stirring. A spatula was used to free the magnetic stir-bar, and stirring was resumed.

4.   During this time, the round-bottomed flask became warm to the touch as the paraformaldehyde fully dissolved to give a clear solution.

5.   The reaction can be allowed to proceed overnight without decomposition of the product.

Org. Synth. **2010**, *87*, 59-67

6.   The solution containing morpholinomethanol is also known to contain dimorpholinomethane, as well as other unidentified species.[2] The ^{1}H and ^{13}C NMR spectra clearly show both species present in solution. The analytical data corresponding to morpholinomethanol are as follows: ^{1}H NMR (300 MHz CDCl$_3$) δ: 2.02 (s, 1 H), 2.64–2.68 (m, 4 H), 3.67–3.70 (m, 4 H), 4.11 (s, 2 H); ^{13}C NMR (300 MHz CDCl$_3$) δ: 49.3, 66.5, 86.9. The peaks were assigned based on known chemical shifts for morpholinomethanol.[2] Dimorpholinomethane, which is also present in solution, gives the following analytical data: ^{1}H NMR (300 MHz, CDCl$_3$) δ: 2.47 (t, $J = 4.65$ Hz , 8 H), 2.88 (s, 2 H), 3.67–3.70 (m, 8 H); ^{13}C NMR (300 MHz, CDCl$_3$) δ: 51.6, 66.5, 81.2. These peaks were assigned based on known chemical shifts for dimorpholinomethane.[2] Other unknown species are present, as there are additional peaks in both the ^{1}H NMR and ^{13}C NMR spectra. These include: ^{1}H NMR (300 MHz, CDCl$_3$) δ: 4.23, 4.74, 4.80–4.93; ^{13}C NMR (300 MHz, CDCl$_3$) δ: 48.6, 85.8, 86.0, 86.2, 87.8, 89.6, 91.4. The unassigned peaks in the ^{13}C NMR spectra are much smaller than the assigned peaks. We propose that dimorpholinomethyl ether may be one of the unknown species.

7.   The concentrations of morpholinomethanol and dimorpholinomethane were determined to be approximately 2.3 M and 2.9 M, respectively. These concentrations were determined by using ^{1}H NMR spectroscopy with an added internal standard (*p*-chlorobenzaldehyde).

8.   Morpholinomethanol becomes unreactive after an undetermined amount of time, though it remains reactive for at least one to two months. It is advisable to prepare fresh morpholinomethanol. No difference in the ^{1}H NMR spectra is detectable for reactive verse unreactive morpholinomethanol.

9.   Both (*S*)-(-)-5,5',6,6',7,7',8,8'-octahydro-1,1'-bi-2-naphthol and (*R*)-(+)-5,5',6,6',7,7',8,8'-octahydro-1,1'-bi-2-naphthol can be purchased from TCI America. Alternatively, they can be prepared from BINOL.[3]

10.   BINOL and its derivatives are susceptible to oxidation.

11.   1,4-Dioxane (Certified A.C.S. grade) was used as purchased from Fisher Chemicals and passed through a column of activated alumina under nitrogen pressure immediately prior to use.

12.   A large excess of the morpholinomethanol solution was necessary to obtain a good yield. The yield was diminished when smaller amounts of the morpholinomethanol solution were used.

13. To bubble the morpholinomethanol solution with nitrogen gas, a needle connected to a nitrogen line was inserted through the rubber septum of the flask. A vent needle was also inserted through the septum. The nitrogen pressure was adjusted so that bubbles were clearly seen in the liquid.

14. Aluminum-backed silica gel 60 F_{254} TLC plates purchased from EMD Chemicals Inc. were used. Hexanes/ethyl acetate (2:1) was used as the eluent and the plate was visualized with ultraviolet light. The product moves only slightly off the baseline (R_f = 0.04), while H_8BINOL has an R_f = 0.28.

15. Ethyl acetate (Certified A.C.S. grade) was used as purchased from Fisher Chemicals.

16. Sodium sulfate, anhydrous, was purchased from Fisher Chemicals.

17. Ethanol, absolute, 200 proof, Acros Organics, was used as purchased from Fisher Chemicals. Because the entire solid does not readily dissolve even with increasing amounts of ethanol, the flask was fitted with a reflux condenser and placed under a nitrogen atmosphere. The solution was then stirred at 78 °C in an oil bath for approximately 1 h. During this time, the solid completely dissolved.

18. A small amount of precipitation occurs quickly as the flask cools to room temperature.

19. The product was rinsed with 25 mL of cold ethanol and allowed to air dry over the vacuum filter for 15 min.

20. On half scale, 2 crops of crystals were obtained, with a total yield of 3.48 g (83%). Both crops of crystals were obtained with >99% ee and 99% purity, as determined by HPLC analysis. The addition of a few seed crystals was required to initiate the second crystallization.

21. In order to remove trace ethanol from the catalyst, it was necessary to follow the drying procedure, wherein the compound (1.43 g) is dissolved in THF (20 mL) and evacuated with stirring under high vacuum (5 x 10^{-3} mmHg) at 40 °C for 2 h. A more detailed description of the drying procedure can be found in the following procedure.

22. The product, (S)-3,3'-bis-morpholinomethyl-5,5',6,6',7,7',8,8'-octahydro-1,1'-bi-2-naphthol, gives the following analytical data after following the drying procedure described in Note 21: $[\alpha]_D^{20}$ -15.4 (c 1.09, CHCl₃) (Notes 24 and 25); ¹H NMR (400 MHz, CDCl₃) δ: 1.61–1.73 (m, 8 H), 2.12–2.18 (m, 2 H), 2.29–2.35 (m, 2 H), 2.53 (bs, 8 H), 2.70 (bs, 4 H), 3.57 (d, 2 H, J = 13.2 Hz), 3.66 (s, 8 H), 3.78 (d, 2 H, J = 13.2 Hz), 6.71 (s,

2 H), 10.36 (s, 2 H); ^{13}C NMR (125 MHz CDCl$_3$) δ: 23.2, 23.3, 27.0, 29.3, 53.0, 62.0, 66.7, 117.8, 124.0, 127.6, 128.8, 136.0, 152.1; Anal. calcd. for C$_{30}$H$_{40}$N$_2$O$_4$: C, 73.14; H, 8.18; N, 5.69; found: C, 72.75; H, 8.22; N, 5.63.

23. An Agilent 1110 Series Instrument with a multiple wavelength detector was used for HPLC analysis. A Daicel Chiralcel OD column was eluted with hexanes:isopropanol:triethylamine (95:5:0.1) at 1.0 mL/min. Buffering with triethylamine was required to elute the product, which contains an amine and can stick to the column. The retention times of the R- and S-enantiomers were determined to be as follows: $t_{(R)-1}$ = 15.1 min. and $t_{(S)-1}$ = 31.1 min.

24. The specific rotation was measured on a Perkin-Elmer 241 polarimeter with a sodium lamp (589 nm, D line) using a 1 dm, 2 mL quartz sample cell.

25. The specific rotation value is corrected from the value we previously reported.[7-9]

Safety and Waste Disposal Information

All hazardous materials should be handled and disposed of in accordance with "Prudent Practices in the Laboratory"; National Academy Press; Washington, DC, 1995.

3. Discussion

1,1'-Bi-2-naphthol (BINOL) and its derivatives have been extensively used for asymmetric catalysis.[4] Derivatives of the partially hydrogenated BINOL, (H$_8$BINOL), have also been studied and in many instances have shown improved enantioselectivity over BINOL.[5] This increased stereocontrol is attributed to the increased dihedral angle between the two aromatic rings due to the increased steric interactions of the partially hydrogenated aryl rings. Thus, derivatives of H$_8$BINOL are interesting ligands for asymmetric catalysis.

Furthermore, BINOL derivatives containing both Lewis acidic and Lewis basic functionalities have generated interest as bifunctional chiral catalysts, and have been shown to exhibit good enantioselectivity in an array of reactions.[6] These catalysts are attractive as the bifunctional character enables the activation of both a nucleophile and an electrophile in a chiral environment.

The title compound, (*S*)-**1**, represents such a bifunctional chiral catalyst. The dihydroxy groups yield Lewis acidic centers upon reaction with a metal species, and the nitrogen atoms of the morpholinomethyl groups provide two Lewis basic sites. In addition to its bifunctional character, our catalyst also utilizes the reported advantages of the increased steric bulk of the partially hydrogenated naphthyl rings. Recently, our group has demonstrated the utility of the title compound in the asymmetric synthesis of two widely useful intermediates in organic synthesis, specifically chiral α-substituted benzyl alcohols and chiral propargylic alcohols.[7-9]

In the presence of catalyst (*S*)-**1** (10 mol%), diphenyl zinc addition to aldehydes affords chiral α-substituted benzyl alcohols at room temperature and requires no additives, as shown in Scheme 1.[8-10] The catalyst is widely applicable, as high enantioselectivities are obtained for linear, α-branched, and β-branched aliphatic aldehydes, as well as for α,β-unsaturated aldehydes.[8,9] The ability to control the diphenyl zinc addition to linear aliphatic aldehydes is notable, as this is unique to our catalyst. *para*-Substituted aromatic aldehydes also give high chiral induction, while *ortho*- and *meta*-substituted benzaldehydes show moderate enantioselectivity.[8,9] In addition, (*S*)-**1** can also catalyze the addition of functionalized diarylzincs, generated in situ from aryl iodides, to react with a variety of aldehydes with high enantioselectivity.[11]

Scheme 1. Asymmetric diaryl zinc addition to aldehydes.[8,9,11]

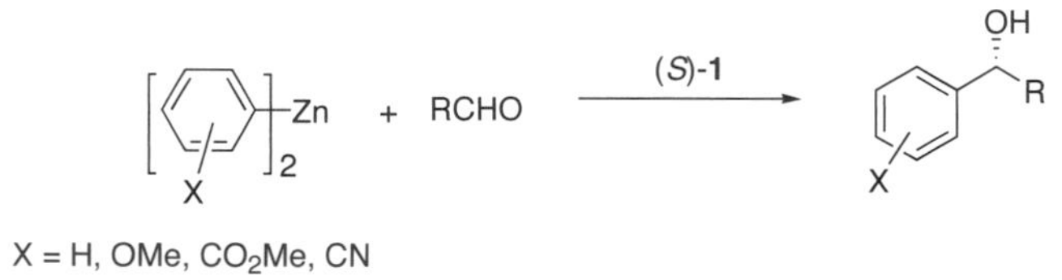

X = H, OMe, CO₂Me, CN

Our lab has also demonstrated the use of (*S*)-**1** in the asymmetric alkyne addition to aldehydes, as illustrated in Scheme 2. In the presence of (*S*)-**1** (10 mol%), ZnEt₂, and Ti(OⁱPr)₄, chiral propargylic alcohols are obtained at room temperature in 4 h. The catalysis proves especially useful for *ortho*-substituted benzaldehydes, affording high enantioselectivities.[7,9,12] In previous studies, H₈BINOL was unable to yield high enantioselectivities in this reaction for *ortho*-substituted benzaldehydes.[5k] The title compound also demonstrated good control of enantioselectivity for *meta*- and *para*-substituted benzaldehydes.[7,9]

64

Scheme 2. Asymmetric alkyne addition to aldehydes.[7,9]

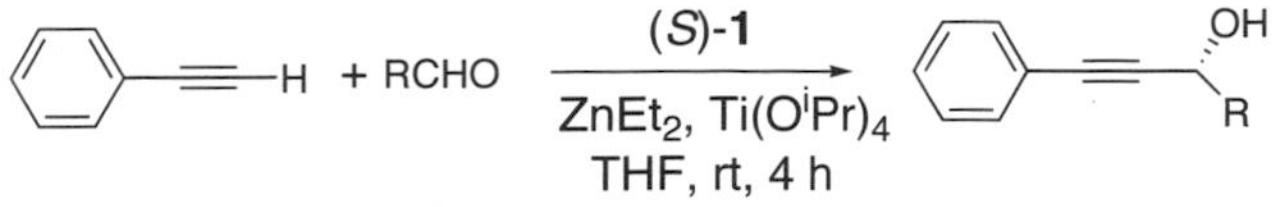

The synthetic procedure of (*S*)-**1** described in this paper provides a convenient route to this functionalized H$_8$BINOL ligand. Unlike other 3,3'-Lewis base substituted BINOL ligands that normally require six or seven synthetic steps from BINOL,[13] the synthesis of (*S*)-**1** involves only one step from the commercially available (*S*)-H$_8$BINOL.

References

1. Department of Chemistry, University of Virginia, Charlottesville, VA 22904-4319, USA.

2. Netscher, T.; Mazzini, F.; Jestin, R. *Eur. J. Org. Chem.* **2007**, 1176–1183.

3. (a) Cram, D. J.; Helgeson, R. C.; Peacock, S. C.; Kaplan, L. J.; Domeier, L. A.; Moreau, P.; Koga, K.; Mayer, J. M.; Chao, Y.; Siegel, M. G.; Hoffman, D. H.; Sogah, G. D. Y. *J. Org. Chem.* **1978**, *43*, 1930–1946. (b) Korostylev, A.; Tararov, V.; Fischer, C.; Monsees, A.; Borner, A. *J. Org. Chem.* **2004**, *69*, 3220-3221.

4. A review: Pu, L. *Chem. Rev.* **1998**, *98*, 2405-2494.

5. (a) Waltz, K. M.; Carroll, P. J.; Walsh, P. J. *Organometallics* **2004**, *23*, 127–134. (b) Reetz, M. T.; Merk, C.; Naberfeld, G.; Rudolph, J.; Griebenow, N.; Goddard, R. *Tetrahedron Lett.* **1997** *38*, 5273–5276. (c) Iida, T.; Yamamoto, N.; Matsunaga, S.; Woo. H. G.; Shibaski, M. *Angew. Chem. Int. Ed.* **1998**, *37*, 2223–2226. (d) Aeilts, S. L.; Cefalo, D. R.; Bonitatebus, P. J.; Houser, J. H.; Hoveyda, A. M.; Schrock, R. R. *Angew. Chem. Int. Ed.* **2001**, *40*, 1452–1456. (e) Schrock. R. R.; Jamieson, J. Y.; Dolman, S. J.; Miller, S. A.; Bonitatebus, P. J.; Hoveyda, A. H. *Organometallics* **2002**, *21*, 409–417. (f) Huang, H.; Liu, X.; Chen, H., Zheng, Z. *Tetrahedron: Asymmetry* **2005**, *16*, 693–697. (g) Zeng, Q. H.; Hu, X. P.; Duan, Z. C.; Liang, X. M.; Zheng, Z. *Tetrahedron: Asymmetry* **2005**, *16*, 1233–1238. (h) Zhang, F. Y.; Chan. A. S. C. *Tetrahedron: Asymmetry* **1997**, *8*, 3651–3655. (i) Long, J.; Hu,

J.; Shen, X.; Ji, B.; Ding, K. *J. Am. Chem. Soc.* **2002**, *124*, 10–11. (j) Wang, B.; Feng, X.; Huang, Y.; Liu, H.; Cui, X.; Jiang, Y. *J. Org. Chem.* **2002**, *67*, 2175–2182. (k) Lu, G.; Li, X.; Chan, W. L.; Chan, A. S. C. *Chem. Commun.* **2002**, 172–173. (l) Chan, A. S. C.; Zhang, F. Y.; Yip, C. W. *J. Am. Chem. Soc.* **1997**, *119*, 4080–4081. (m) Kim, J. G.; Camp, E. H.; Walsh, P. J. *Org. Lett.* **2006**, *8*, 4413–4416. (n) Wu, K. H.; Gau, H. M. *J. Am. Chem. Soc.* **2006**, *128*, 14808–14809. (o) Guo, Q. X.; Liu, H.; Guo, C.; Luo, S. W.; Gu, Y.; Gong, L. Z. *J. Am. Chem. Soc.* **2007**, *129*, 3790–3791.

6.   Reviews of bifunctional chiral catalysts: (a) Rowlands, G. J. *Tetrahedron* **2001**, *57*, 1865–1882. (b) Shibasaki, M.; Kanai, M.; Funabshi, K. *Chem. Commun.* **2002**, 1989–1999. Selected references applicable to BINOL derivatives: (c) Hamashima, Y.; Sawada, D.; Kanai, M.; Shibasaki, M. *J. Am. Chem. Soc.* **1999**, *121*, 2641–2642. (d) DiMauro, E. F.; Kozlowski, M. C. *Org. Lett.* **2001**, *3*, 1641–1644. (e) Casas, J.; Nájera, C.; Sansano, J. M.; Saá, J. M. *Org. Lett.* **2002**, *4*, 2589–2592. (f) Matsui, K.; Takizawa, S.; Sasai, H. *J. Am. Chem. Soc.* **2005**, *127*, 3680–3681.

7.   Liu, L.; Pu, L. *Tetrahedron* **2004**, *60*, 7427–7430.

8.   Qin, Y. C.; Pu, L. *Angew. Chem. Int. Ed.* **2006**, *45*, 273–277.

9.   Qin, Y. C.; Liu, L.; Sabat, M.; Pu, L. *Tetrahedron* **2006**, *62*, 9335–9348.

10. A review on the asymmetric arylzinc additions to carbonyl compounds: Schmidt, F.; Stemmler, R. T.; Rudolph, J.; Bolm, C. *Chem. Soc. Rev.* **2006**, *35*, 454–470.

11. DeBerardinis, A. M.; Turlington, M.; Pu, L. *Org. Lett.* **2008**, *10*, 2709-2712.

12. A review on the asymmetric alkyne addition to carbonyl compounds: Pu, L. *Tetrahedron* **2003**, *59*, 9873-9886.

13. (a) Hamashima, Y.; Sawada, D.; Kanai, M.; Shibasaki, M. *J. Am. Chem. Soc.* **1999**, *121*, 2641–2642. (b) Casas, J.; Nájera, C.; Sansano, J. M.; Saá, J. M. *Org. Lett.* **2002**, *4*, 2589–2592.

Appendix
Chemical Abstracts Nomenclature; (Registry Number)

Paraformaldehyde; (30525-89-4)

66

Morpholine; (110-91-8)

Morpholinomethanol: 4-Morpholinemethanol; (4432-43-3)

(S)-5,5',6,6',7,7',8,8'-Octahydro-1,1'-bi-2-naphthol [(S)-H₈BINOL]: [1,1'-Binaphthalene]-2,2'-diol, 5,5',6,6',7,7',8,8'-octahydro-, (1S)-; (65355-00-2)

(S)-3,3'-Bis-morpholinomethyl-5,5',6,6',7,7',8,8'-octahydro-1,1'-bi-2-naphthol: [1,1'-Binaphthalene]-2,2'-diol, 5,5',6,6',7,7',8,8'-octahydro-3,3'-bis(4-morpholinylmethyl)-, (1S)-; (758698-16-7)

Lin Pu was born in 1965 in Xuyong, Sichuan, China. He received a B.S. degree from Sichuan University in 1984. He then obtained the Doering Fellowship (CGP) and received a Ph.D. degree in 1990 under the supervision of Professor Joseph O'Connor at UC San Diego. From 1991 to 1994, he worked with Professor Henry Taube at Stanford University and Professor Robert Grubbs at California Institute of Technology as a postdoctoral fellow. He was appointed as an assistant professor at North Dakota State University in 1994. He then moved to University of Virginia in 1997 as an associate professor and became a professor in 2003.

Mark Turlington was born in 1984 and grew up in Mills River, North Carolina. In 2002 he attended Furman University and began working in the lab of Professor Moses Lee, synthesizing distamycin analogues capable of molecular recognition of specific DNA sequences. After graduating with a B.S. degree in chemistry in 2006, he is now pursuing his doctoral degree under the guidance of Professor Lin Pu, working on the development of asymmetric catalysts and their application in organic synthesis.

Kyle Kimmel was born in 1985 in Fairfax, VA. He studied as an undergraduate at the University of California, Irvine, where he completed a B.S. degree in Chemistry, working on undergraduate research in the lab of Professor Keith Woerpel. Currently, he is a third year graduate student at University of California, Berkeley, working under the direction of Professor Jonathan A. Ellman. His research is on the development of sulfinyl-based hydrogen-bonding organocatalysts.

CATALYTIC ASYMMETRIC ADDITION OF AN *IN-SITU* PREPARED ARYLZINC TO CYCLOHEXANECARBOXALDEHYDE: (*R*)-(+)-α-CYCLOHEXYL-3-METHOXY-BENZENEMETHANOL

(*S*)-**1**

Submitted by Albert M. DeBerardinis, Mark Turlington, and Lin Pu.[1]
Checked by Kyle L. Kimmel and Jonathan A. Ellman.

1. Procedure

> *CAUTION! Neat diethylzinc may ignite on exposure to air and reacts violently with water. It must be handled and reacted under nitrogen. The reaction solvent must be dried and distilled prior to use and all glassware and syringes must be thoroughly dried.*

An oven-dried 500-mL, 3-necked, round-bottomed flask, stir bar, glass stopper, septum and vacuum adaptor containing a stopcock are taken into a glove box (Note 1). The flask is charged with lithium acetylacetonate (0.771 g, 7.27 mmol, 0.25 equiv) (Note 2), sealed, and brought outside of the glove box and into a fume hood. The vacuum adaptor is connected to a Schlenk line, and the flask is placed under a nitrogen atmosphere (Note 3).

1-Methyl-2-pyrrolidinone (NMP, 48 mL) (Note 4) and 3-iodoanisole (7.62 mL, 64.0 mmol, 2.2 equiv) (Note 5) are added sequentially through the septum via syringe. The reaction flask is cooled to 0 °C with an ice bath, and diethylzinc (3.65 mL, 34.9 mmol, 1.2 equiv) (Note 6) is added via syringe. The pale yellow reaction mixture is stirred at 0 °C for 12 h.

(S)-3,3'-Bis-morpholinomethyl-5,5',6,6',7,7',8,8'-octahydro-1,1'-bi-2-naphthol [(S)-1] (Note 7) is dried prior to use in catalysis. An oven-dried 100-mL Schlenk flask equipped with a stir bar is connected to a Schlenk line, fitted with a glass stopper and cooled under vacuum. Under positive nitrogen pressure, the flask is then charged with (S)-1 (1.43 g, 2.91 mmol, 0.1 equiv) and then 20 mL of tetrahydrofuran (THF) is added (Note 8). After the resulting solution is stirred for 30 min, the solvent is removed under vacuum (Note 9). While maintaining a positive flow of pressure, the glass stopper is exchanged for a rubber septum. To the dried ligand is added 80 mL of THF via syringe.

The THF solution of (S)-1 is transferred via cannula into the reaction flask which is maintained at 0 °C in an ice bath. To ensure that the entire amount of (S)-1 is transferred, an additional 40 mL of THF is added via syringe to the Schlenk flask and then cannula-transferred into the reaction flask. A final 40 mL of THF is added to the round-bottomed flask via syringe, and the reaction mixture is stirred at 0 °C.

After 1 h, the reaction flask is warmed to room temperature for 15 min. Cyclohexanecarboxaldehyde (3.50 mL, 29.1 mmol, 1 equiv) (Note 10) is then added via syringe. The pale yellow reaction mixture is stirred at room temperature.

After 16 h, the reaction is determined to be complete by TLC (Note 11) and quenched with saturated aqueous ammonium chloride (Note 12). The reaction mixture is then transferred to a 1-L round-bottomed flask, and most of the organic solvent is removed via rotary evaporation (20 °C, 10 mmHg). The remaining aqueous layer is transferred to a 500-mL separatory funnel and diluted with 200 mL of diethyl ether and an additional 75 mL of saturated aqueous ammonium chloride. After shaking, the organic layer is collected, and the aqueous layer is extracted with ether (3 x 50 mL). The combined organic layers are then washed with water (3 x 100 mL). The organic layer is dried over sodium sulfate (15 g) (Note 13), filtered, and concentrated by rotary evaporation (20 °C, 10 mmHg to give a non-viscous yellow oil.

This oil is loaded on a coarse-fritted column (5 x 25.4 cm) of SiO$_2$ (170 g) (Note 14). The round-bottomed flask is then rinsed with methylene chloride (10 mL) to ensure that all of the extract is loaded onto the column. The column is eluted as follows: hexanes (1 L), 3% EtOAc/hexanes (1 L),

4% EtOAc/hexanes (500 mL), 5% EtOAc/hexanes (3.5 L). Fraction collection (25 mL fractions) is begun after addition of the second 500 mL portion of 3% EtOAc/hexanes. The product begins to elute at 4% ethyl acetate/hexanes as determined by TLC analysis, and collection is continued until the product is no longer visible by TLC. Due to tailing the product eluted over 3 liters of solvent. Consequently, after 25 fractions of 25 mL that contained product are collected, 500 mL fractions are next collected until no more product is eluted out (Note 15). The desired product is concentrated by rotary evaporation (20 °C, 10 mmHg) and dried overnight under vacuum (5 x 10^{-3} mmHg) to yield 5.91 g (92%) of (*R*)-(+)-α-cyclohexyl-3-methoxybenzenemethanol as a white solid (mp 62–65 °C) (Notes 16 and 17). HPLC analysis demonstrates that the product is obtained with >99% ee and >99% purity (Note 18).

2. Notes

1.   A nitrogen-filled Vacuum Atmospheres inert atmosphere glove box was used.

2.   Lithium acetylacetonate (97%) was used as purchased from Sigma-Aldrich and stored in the glove box.

3.   To place the flask under nitrogen atmosphere the flask is evacuated (5 x 10^{-3} mmHg) and refilled with nitrogen three times. It is crucial that the flask be under a completely inert atmosphere since trace amounts of moisture significantly decrease the yield of product.

4.   1-Methyl-2-pyrrolidinone (99+%, HPLC grade) was purchased from Sigma-Aldrich and refluxed over calcium hydride for 16 h under a nitrogen atmosphere. NMP was then distilled under vacuum (5 x 10^{-3} mmHg) at 70–80 °C. NMP was stored over activated molecular sieves (4 Å) under a nitrogen atmosphere in a dried Schlenk flask (see Note 3) prior to use.

5.   3-Iodoanisole (99%) was purchased from Sigma-Aldrich and was purified by distillation prior to use. Due to the light sensitivity of 3-iodoanisole, the purification is performed with the careful exclusion of light. 3-Iodoanisole was stirred over activated molecular sieves (4 Å) and copper as a stabilizer (used as received from supplier) for 2 h at 40 °C. It was then vacuum distilled (5 x 10^{-3} mmHg) at 70 °C and stored in a dried pear-shaped flask under nitrogen atmosphere in the dark.

6.   Diethylzinc (min. 95%) was used as purchased from Strem Chemicals. Diethylzinc is flammable when it comes into contact with water or oxygen in the air and should be handled with caution.

7.   See the preceding procedure for the synthesis of (S)-1 [(S)-3,3'-Bis-morpholinomethyl-5,5',6,6',7,7',8,8'-octahydro-1,1'-bi-2-naphthol].

8.   THF (Certified) was purchased from Fisher Chemicals and distilled under nitrogen atmosphere from sodium benzophenone ketyl.

9.   Under vacuum (5×10^{-3} mmHg), the flask was warmed with hands and then refilled with nitrogen. This cycle was conducted three times. The flask was then placed under vacuum for 2 h at 40 °C in an oil bath.

10.   Cyclohexanecarboxaldehyde (97%) was purchased from Sigma-Aldrich and purified prior to use. Cyclohexanecarboxaldehyde was distilled under inert atmosphere at reduced pressure (20 mmHg) from activated molecular sieves (4 Å) directly into a Schlenk tube and sealed. By ^{1}H and ^{13}C NMR, the aldehyde remained pure indefinitely when stored in a Schlenk tube under inert atmosphere.

11.   Aluminum-backed silica gel 60 F_{254} TLC plates purchased from EMD Chemicals Inc. were used and visualized with UV light and phosphomolybdic acid (PMA) stain. Hexanes/ethyl acetate (4/1) were used as the eluent. The aldehyde starting material has an $R_f = 0.49$, and the product has an $R_f = 0.24$ and is also visible under UV light.

12.   Quenching is an exothermic process and should be performed carefully. Saturated aqueous ammonium chloride was added via syringe through the septum with needles for venting. Over 25 min, 5 mL of saturated aqueous ammonium chloride was slowly added. An additional 20 mL of saturated aqueous ammonium chloride was added over 20 min, using 1 mL aliquots and stirring for 1 min prior to the next addition.

13.   Sodium sulfate (anhydrous) was purchased from Fisher Chemical.

14.   SiliaFlash P60, 40-63 μm 60 Å silica gel was used as purchased from SiliCycle, Inc.

15.   TLC analysis was performed using alumina backed TLC plates (Note 11), and using hexanes/ethyl acetate (4/1) as eluent. The product was visualized by PMA stain. While the product is UV visible at higher concentrations, PMA stain is necessary to visualize the product at the beginning and end of elution from the column.

16.   (R)-(+)-α-Cyclohexyl-3-methoxy-benzenemethanol   gave   the following analytical data: $[\alpha]_D^{20}$ +24.7 (c 1.30, CHCl$_3$); ^{1}H NMR (500 MHz CDCl$_3$) δ: 0.98–1.27 (m, 5 H), 1.42 (apparent d, J = 13.0 Hz, 1 H), 1.61–

1.68 (m, 3 H), 1.79 (m, 1 H), 1.86 (apparent d, J = 3.5 Hz, 1 H), 2.01 (apparent d, J = 13.0 Hz, 1 H), 3.85 (s, 3 H), 4.37 (dd, J = 3.0, 7.0 Hz, 1 H), 6.84 (m, 1 H), 6.90–6.92 (m, 2 H), 7.27 (t, J = 8.0 Hz, 1 H); [13]C (125 MHz, CDCl$_3$) δ: 26.0, 26.1, 26.4, 28.8, 29.4, 44.9, 55.2, 79.3, 112.2, 112.8, 119.1, 129.2, 145.4, 159.6; Anal. calcd. for C$_{14}$H$_{20}$O$_2$: C, 76.33; H, 9.15; found: C, 76.47; H, 9.49.

17. The specific rotation was measured on a Perkin-Elmer 241 polarimeter with a sodium lamp (589 nm, D line) using a 1 dm, 2 mL quartz sample cell.

18. An Agilent 1100 Series Instrument with a multiple-wavelength detector was used for HPLC analysis for the determination of enantiomeric excess and product purity. A Daicel Chiralcel OD column was eluted with 5% isopropanol in hexanes at 1.0 mL/min. The enantiomeric purity was determined to be >99%, and the retention times of the R- and S-enantiomers were as follows: $t_{(S)-1}$ = 14.6 min, and $t_{(R)-1}$ = 23.8 min. The purity was determined to be >99%. The configuration of the product is assigned to be R as determined by the preparation and NMR analysis of the (R)- and (S)-α-methoxy-α-phenylacetic (MPA) ester derivatives of the alcohol product.[11]

Safety and Waste Disposal Information

All hazardous materials should be handled and disposed of in accordance with "Prudent Practices in the Laboratory"; National Academy Press; Washington, DC, 1995.

3. Discussion

Diaryl- and arylalkyl carbinols (**A**) are useful intermediates for the synthesis of a number of pharmacologically active compounds.[2,3] Two general strategies for the preparation of these secondary chiral alcohols are the asymmetric reduction of prochiral ketones[4] and the asymmetric arylation of aldehydes.[5] Though both strategies can be carried out catalytically, the latter process generates compounds **A** more efficiently because the carbon-carbon bond is formed and the stereocenter is set simultaneously.

A R = alkyl, aryl

The use of diphenylzinc as the arylating agent has been extensively studied.[6] Because diphenylzinc is the only commercially available diarylzinc reagent, other aryl sources have been explored.[7] For example, preparation of diarylzincs from aryl bromides and [n]BuLi at low temperatures and their subsequent asymmetric addition to aldehydes have been studied.[8] Because an alkyllithium is used to generate the arylzincs, the method potentially limits the compatibility of functional groups on the aryl halides.

We have recently reported a catalytic asymmetric method that uses the in situ generation of functionalized arylzincs from aryl iodides for reaction with aldehydes in the presence of the chiral ligand (S)-1 (Scheme 1).[9,10] This ligand not only activates the functionalized arylzinc for reaction with aldehydes but also provides excellent stereocontrol. High enantioselectivity has been achieved in the preparation of various diaryl and arylalkyl carbinols by using this method.[9]

Scheme 1. Preparation of functionalized arylzincs and their addition to aldehydes.

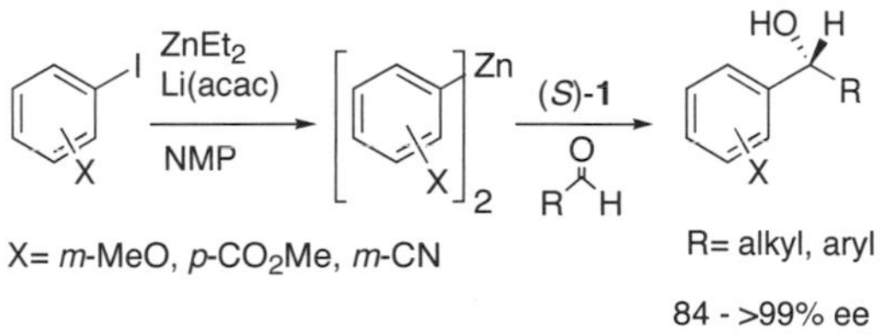

X= m-MeO, p-CO_2Me, m-CN

R= alkyl, aryl

84 - >99% ee

References

1. Department of Chemistry, University of Virginia, Charlottesville, VA 22904-4319.
2. (a) Müller, P.; Nury, P.; Bernardinelli, G. *Eur. J. Org. Chem.* **2001**, *7*, 4137-4147. (b) Torrens, A.; Castrillo, J. A.; Claparols, A.; Redonodo, J. *Synlett.* **1999**, 765-767. (c) Casy, A. F.; Drake, A. F.; Ganellin, C. R.; Mercer, A. D.; Upton, C. *Chirality* **1992**, *4*, 356-366. (d) Ebnother, A.; Weber, H.-P. *Helv. Chim. Acta.* **1976**, *59*, 2462-2468. (e) James, M. N. G.; Williams, G. J. B. *Can. J. Chem.* **1974**, *52*, 1872-1879. (f) Barouh, V.; Dall, H.; Patel, D.; Hite, G. *J. Med. Chem.* **1971**, *14*, 834-836. (h) Shafi'ee, A.; Hite, G. *J. Med. Chem.* **1969**, *12*, 266-270.

3. (a) Bolshan, Y.; Chen, C.-Y.; Chilenski, J. R.; Gosselin, F.; Mathre, D. J.; O'Shea, P. D.; Roy, A.; Tillyer, R. D. *Org. Lett.* **2004**, *6*, 111-114. (b) Astles, P. C.; Brown, T. J.; Halley, F.; Handscombe, C. M.; Harris, N. V.; Majid, T. N.; McCarthy, C.; McLay, I. M.; Morley, A.; Porter, B.; Roach, A. G.; Sargent, C.; Smith, C.; Walsh, R. J. A. *J. Med. Chem.* **2000**, *43*, 900-910. (c) Nilvebrant, L.; Andersson, K.-E.; Gillberg, P.-G.; Stahl, M.; Sparf, B. *Eur. J. Pharmacol.* **1997**, *327*, 195-207. (d) Welch, W. M.; Kraska, A. R.; Sarges, R.; Koe, B. K. *J. Med. Chem.* **1984**, *27*, 1508-1515.

4. (a) Ohkuma, T.; Koizumi, M.; Ikehira, H.; Yokozawa, T.; Noyori, R. *Org. Lett.* **2000**, *2*, 659-662. (b) Corey, E. J.; Helal, C. J. *Tetrahedron Lett.* **1995**, *36*, 9153-9156.

5. Bolm, C.; Hildebrand, J. P.; Muñiz, K.; Hermanns, N. *Angew. Chem. Int. Ed.* **2001**, *40*, 3284-3308.

6. (a) Schmidt, F.; Stemmler, R. T.; Rudolph, J.; Bolm, C. *Chem. Soc. Rev.* **2006**, *35*, 454-470. (b) Qin, Y.-C.; Pu, L. *Angew. Chem. Int. Ed.* **2006**, *45*, 273-277. (c) Huang, W.-S.; Pu, L. *J. Org. Chem.* **1999**, *64*, 4222-4223. (d) Bolm, C.; Muñiz, K. *Chem. Commun.* **1999**, 1295-1296. (e) Dosa, P. I.; Ruble, J. C.; Fu, G. C. *J. Org. Chem.* **1997**, *62*, 444-445.

7. (a) Shannon, J.; Bernier, D.; Rawson, D.; Woodward, S. *Chem. Commun.* **2007**, 3945-3947. (b) Tomita, D.; Wada, R.; Kanai, M.; Shibasaki, M. *J. Am. Chem. Soc.* **2005**, *127*, 4138-4139. (c) Dahmen, S.; Lormann, M. *Org. Lett.* **2005**, *7*, 4597-4600. (d) Braga, A. L.; Lüdtke, D. S.; Vargas, F.; Paixão, M. W. *Chem. Commun.* **2005**, 2512-2514. (e) Prieto, O.; Ramón, D. J.; Yus, M. *Tetrahedron: Asymmetry* **2003**, *14*, 1955-1957. (f) Bolm, C.; Rudolph, J. *J. Am. Chem. Soc.* **2002**, *124*, 14850-14851.

8. Kim, J. G.; Walsh, P. J. *Angew. Chem. Int. Ed.* **2006**, *45*, 4175-4178.

9. DeBerardinis, A. M.; Turlington, M.; Pu, L. *Org. Lett.* **2008**, *10*, 2709-2712.

10. Kneisel, F. F.; Dochnahl, M.; Knochel, P. *Angew. Chem. Int. Ed.* **2004**, *43*, 1017-1021.

11. (a) Trost, B. M.; Belletire, J. L.; Goldleski, P. G.; McDougal, P. G.; Balkovec, J. M.; Baldwin, J. J.; Christy, M.; Ponticello, G.S.; Varga, S. L.; Springer, J. P. *J. Org. Chem.* **1986**, *51*, 2370-2374. (b) Seco, J.; Quinoa, E.; Riguera, R. *Chem. Rev.* **2004**, 104, 17-117.

Appendix
Chemical Abstracts Nomenclature; (Registry Number)

Lithium acetylacetonate: 2,4-Pentanedione, ion(1-), lithium (1:1);
    (18115-70-3)
3-Iodoanisole: Benzene, 1-iodo-3-methoxy-; (766-85-5)
Diethylzinc; (557-20-0)
(S)-3,3'-Bis-morpholinomethyl-5,5',6,6',7,7',8,8'-octahydro-1,1'-bi-2-
    naphthol: [1,1'-Binaphthalene]-2,2'-diol, 5,5',6,6',7,7',8,8'-octahydro-
    3,3'-bis(4-morpholinylmethyl)-, (1S)-; (758698-16-7)
Cyclohexanecarboxaldehyde; (2043-61-0)
(R)-(+)-α-Cyclohexyl-3-methoxy-benzenemethanol; (1036645-45-0)

Lin Pu was born in 1965 in Xuyong, Sichuan, China. He received a B.S. degree from Sichuan University in 1984. He then obtained the Doering Fellowship (CGP) and received a Ph.D. degree in 1990 under the supervision of Professor Joseph O'Connor at UC San Diego. From 1991 to 1994, he worked with Professor Henry Taube at Stanford University and Professor Robert Grubbs at California Institute of Technology as a postdoctoral fellow. He was appointed as an assistant professor at North Dakota State University in 1994. He then moved to University of Virginia in 1997 as an associate professor and became a professor in 2003.

Albert DeBerardinis was born in 1979 and grew up in Aston, Pennsylvania. Encouraged by his entire family, he attended West Chester University from 2000 to 2004, earning a B.S. degree in biochemistry. His undergraduate mentors were invaluable to his interest and further pursuit in chemistry. He was introduced to biochemical research by Professors Blaise Frost and Albert Caffo. Under the direction of Professor Michael Moran, he performed intercalation studies on graphite and prepared polymeric sulfur nitride. After graduating in 2004, he began his doctoral pursuit under the guidance of Professor Lin Pu, working on the development of asymmetric catalysts and their application in organic synthesis.

Mark Turlington was born in 1984 and grew up in Mills River, North Carolina. In 2002 he attended Furman University and began working in the lab of Professor Moses Lee, synthesizing distamycin analogues capable of molecular recognition of specific DNA sequences. After graduating with a B.S. degree in chemistry in 2006, he is now pursuing his doctoral degree under the guidance of Professor Lin Pu, working on the development of asymmetric catalysts and their application in organic synthesis.

Kyle Kimmel was born in 1985 in Fairfax, VA. He studied as an undergraduate at the University of California, Irvine, where he completed a B.S. degree in Chemistry, working on undergraduate research in the lab of Professor Keith Woerpel. Currently, he is a third year graduate student at University of California, Berkeley, working under the direction of Professor Jonathan A. Ellman. His research is on the development of sulfinyl-based hydrogen-bonding organocatalysts.

76

MICROWAVE-ASSISTED SYNTHESIS OF
1,3-DIMESITYLIMIDAZOLINIUM CHLORIDE

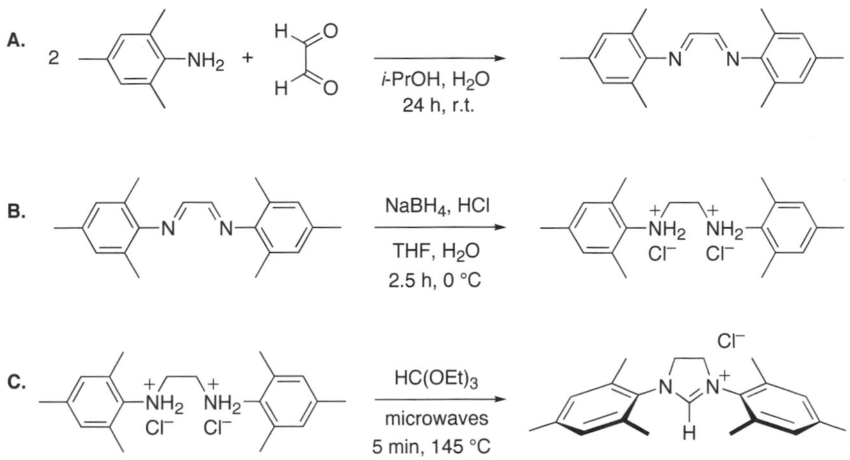

A.

B.

C.

Submitted by Morgan Hans and Lionel Delaude.[1]
Checked by Somenath Chowdhury and Jonathan A. Ellman.

1. Procedure

A.   N,N'-Dimesitylethylenediimine. A 500-mL, one-necked, round-bottomed flask equipped with a magnetic stirring bar and a stopper is charged with 2,4,6-trimethylaniline (40.56 g, 0.300 mol, 2.0 equiv) and isopropyl alcohol (150 mL) (Notes 1, 2). Glyoxal (21.76 g of a 40% aqueous solution, 0.150 mol, 1.0 equiv) is weighed in a 250-mL Erlenmeyer flask (Note 3) and further diluted with deionized water (50 mL) and isopropyl alcohol (50 mL) (Note 2). This colorless mixture is briefly swirled. It is then poured into the aniline solution. The resulting solution is stirred for 24 h at room temperature. A bright yellow precipitate appears within a few minutes and progressively builds up. After 24 h, the suspension is filtered through a Büchner funnel, and the precipitate is washed with deionized water (2 × 100 mL) (Note 4). It is spread out in a 20-cm diameter evaporating dish and dried in an oven at 40 °C under 30 mmHg vacuum until constant weight (Note 5) to afford 40.09 g (91%) of *N,N'*-dimesitylethylenediimine as a bright yellow solid (Note 6).

Org. Synth. **2010**, *87*, 77-87
Published on the Web 1/8/2010

B. N,N'-Dimesitylethylenediamine Dihydrochloride. A 500-mL, two-necked, round-bottomed flask equipped with a magnetic stirring bar, a thermometer, and a solid addition funnel is charged with *N,N'*-dimesityl-ethylenediimine (36.55 g, 0.125 mol, 1.0 equiv) and tetrahydrofuran (300 mL) (Note 7). The bright yellow suspension is cooled to 0 °C internal temperature using an ice-water bath before sodium borohydride (18.92 g, 0.500 mol, 4.0 equiv) is added in one portion (Notes 8, 9). Next, concentrated hydrochloric acid (21 mL, 0.25 mol, 2.0 equiv) is added dropwise at a rate such that the internal temperature remains between 12–15 °C, which took 40 min (Notes 10, 11). The resulting white suspension is further stirred for 30 min at 0 °C. It is then transferred into a 2-L, two-necked, round-bottomed flask before 1 L of cold 3 M aqueous hydrochloric acid is added dropwise over a 30 min period of time (Note 12). The cooling bath is removed, and the off-white suspension is stirred for 1 h at room temperature. It is then filtered through a Büchner funnel. The white precipitate is washed with deionized water (3 × 200 mL) (Note 13). It is spread out in a 20-cm diameter evaporating dish and dried in an oven at 40 °C under 30 mmHg vacuum until constant weight (Note 5) to afford 37.4 g (81%) of *N,N'*-dimesitylethylenediamine dihydrochloride as a white powder (Note 14).

C. 1,3-Dimesitylimidazolinium Chloride. Four 20-mL pressure vials each equipped with a magnetic stirring bar are individually charged with *N,N'*-dimesitylethylenediamine dihydrochloride (4.62 g, 0.0125 mol, 1.0 equiv) and triethyl orthoformate (6.25 mL, 0.0375 mol, 3.0 equiv) (Note 15). The mixtures are mixed with a spatula until a homogeneous thick white paste is obtained. The vials are closed with screw caps equipped with a pressure sensor and a glass inlet tube ended by a sapphire window containing an optical fiber temperature sensor. The reaction mixtures are heated for 5 min at 145 °C in a monomodal microwave reactor (Note 16). After cooling to room temperature, the resulting beige-yellow suspensions are combined in a 250-mL beaker placed in an ice bath and each reaction vessel is rinsed with diethyl ether (2 × 12.5 mL), which is also added to the beaker (Note 17). The beige suspension is filtered through a Büchner funnel. The white precipitate is washed with diethyl ether (3 × 50 mL) (Note 17). The white precipitate is further purified by dissolution into hot acetonitrile (50 mL) (Note 18). The solution is allowed to cool to room temperature and then diethyl ether is added (200 mL) (Note 17). The resulting pale yellow suspension is filtered through a Büchner funnel. The white precipitate is

78

washed with diethyl ether (50 mL) (Note 17). The precipitate is spread out in a 20-cm diameter evaporating dish and dried in an oven at 40 °C under 30 mmHg vacuum until constant weight (Note 5) to afford 13.5 g (79%) of 1,3-dimesitylimidazolinium chloride as white microcrystals (Note 19).

2. Notes

1.   2,4,6-Trimethylaniline ($\geq$98%) was purchased from Aldrich and used as received.

2.   Reagent grade isopropyl alcohol was obtained from EMD Chemical Inc. and used without any further purification.

3.   Glyoxal (ca. 40% aqueous solution) was purchased from Aldrich and stored in a refrigerator at + 6 °C.

4.   A second crop of yellow precipitate formed in the filtrate upon standing at room temperature for a few hours. It was filtered through a Büchner funnel and washed with deionized water (2 × 50 mL) to afford ca. 0.75 g of N,N'-dimesitylethylenediimine (mp 156–157 °C) that was added to the first crop. Upon resting overnight under the fume cupboard, the filtrate afforded a third crop of dark yellow solid (ca. 2.3 g, mp 149–152 °C) that was discarded.

5.   A vacuum oven (NAPCO, Model 5831 from Cascade TEK, Oregon) was used at 40 ºC under 30 mmHg vacuum. The submitters dried the product in open air under an IR lamp.

6.   Spectral and analytical data for N,N'-dimesitylethylenediimine are as follows: mp 157 °C; IR (ATR) 2911, 1723, 1615, 1475, 1373, 1201, 1139, 1031, 1012, 928, 850 cm^{-1}; ^{1}H NMR (400 MHz, CDCl$_3$) δ: 2.16 (s, 12 H), 2.29 (s, 6 H), 6.91 (s, 4 H), 8.10 (s, 2 H); ^{13}C NMR (100 MHz, CDCl$_3$) δ: 18.3, 20.8, 126.6, 129.0, 134.3, 147.5, 163.5; HRMS (ESI) m/z calcd. for C$_{20}$H$_{25}$N$_2^+$ ([M+H]$^+$) 293.2012; found 293.2018; Anal. calcd for C$_{20}$H$_{24}$N$_2$: C, 82.15; H, 8.27; N, 9.58; found: C, 81.98; H, 8.07; N, 9.54.

7.   Tetrahydrofuran (99+%, stabilized) was purchased from Acros and used without any further purification.

8.   Sodium borohydride (powder, 98%) was purchased from Aldrich and used as received.

9.   The submitters note that the use of a smaller excess of sodium borohydride compared to N,N'-dimesitylethylenediimine led to significantly lower yields of N,N'-dimesitylethylenediamine dihydrochloride tainted with colored impurities

10. Hydrochloric acid for analysis (ca. 37% solution in water) was purchased from Acros and used as received.

11. A strongly exothermic reaction takes place. During the addition of HCl, the reaction temperature slowly increases to 15 °C within 5 min and then is maintained between 12–15 °C by controlling the addition rate.

12. The 3 M hydrochloric acid was cooled in an ice bath to 0 °C. The cold hydrochloric acid solution was then transferred to the dropping funnel in 50 mL batches and was added dropwise at such a rate that the temperature is maintained between 8–10 °C. During the addition, hydrogen is evolved and frothing occurs.

13. A second crop of white precipitate formed in the filtrate upon standing at room temperature for a few hours. It was filtered through a Büchner funnel and washed with deionized water (2 × 50 mL) to afford ca. 0.6 g of *N,N'*-dimesitylethylenediamine dihydrochoride (mp 224–225 °C (dec)) that was added to the first crop.

14. Spectral and analytical data for *N,N'*-dimesitylethylenediamine dihydrochloride are as follows: mp 224–225 °C (dec); IR (ATR) 3424, 3254, 2961, 2917, 2660, 2470, 2355, 1668, 1607, 1563, 1470, 1388, 1304, 1219, 1200, 1190, 1136, 1034, 980, 900, 875, 851 cm^{-1}; ^{1}H NMR (400 MHz, DMSO-d_6) δ: 2.25 (s, 6 H), 2.48 (s, 12 H), 3.71 (s, 4 H), 6.99 (s, 4 H); ^{13}C NMR (100 MHz, DMSO-d_6) δ: 17.9, 20.1, 45.8, 130.1, 131.4, 132.5, 137.0; HRMS (ESI) *m/z* calcd. for $C_{20}H_{29}N_2^+$ ([M–H–2Cl]$^+$) 297.2325; found 297.2331; Anal. calcd. for $C_{20}H_{30}Cl_2N_2$: C, 65.03; H, 8.19; N, 7.58; found: C, 65.94; H, 8.69; N, 7.63.

15. Triethyl orthoformate (98%) was purchased from Aldrich and used as received.

16. An INITIATOR @Biotage, Serial No. 10662-17U Microwave reactor was employed. The maximum microwave power was set at 120 W and the pressure limit at 250 psi (17 bar). No ramp and no simultaneous cooling were applied. The submitters performed the microwave reaction in a single 80-mL pressure vial equipped with a magnetic stirring bar using a Discover® LabMate instrument (CEM Corp., Matthews, NC). The maximum microwave power was set at 120 W and the pressure limit at 250 psi (17 bar). No ramp and no simultaneous cooling were applied.

17. Reagent grade diethyl ether was obtained from Fisher Chemical and used without any further purification.

18. HPLC grade acetonitrile was obtained from EMD Chemicals Inc. and used without any further purification.

19. Spectral and analytical data for 1,3-dimesitylimidazolinium chloride: mp 279–281 °C (dec); IR (ATR) 3642, 3348, 3213, 2848, 1623, 1484, 1448, 1383, 1274, 1216, 1038, 869, 846 cm^{-1}; ^{1}H NMR (400 MHz, CDCl$_3$) δ: 2.24 (s, 6 H), 2.32 (s, 12 H), 4.47 (s, 4 H), 6.86 (s, 4 H), 9.69 (s, 1 H); ^{13}C NMR (100 MHz, CDCl$_3$) δ: 17.9, 20.9, 51.7, 129.7, 130.3, 134.9, 139.9, 160.1; HRMS (ESI) m/z calcd. for C$_{21}$H$_{27}$N$_2$$^+$ ([M–Cl]$^+$) 307.2169; found 307.2177; Anal. calcd. for C$_{21}$H$_{27}$ClN$_2$: C, 73.56; H, 7.94; N, 8.17; found: C, 71.68; H, 8.19; N, 7.92.

Safety and Waste Disposal Information

All hazardous materials should be handled and disposed of in accordance with "Prudent Practices in the Laboratory"; National Academy Press; Washington, DC, 1995.

3. Discussion

Imidazolinium salts are the most common synthetic precursors for generating saturated N-heterocyclic carbenes (NHCs), a new class of stable divalent carbon species that has already afforded a wealth of nucleophilic reagents, organocatalysts, and organometallic complexes, including optically active compounds.[2] In particular, deprotonation of 1,3-dimesitylimidazolinium chloride with a non-nucleophilic strong base gives direct access to the important NHC nicknamed SIMes or H$_2$IMes that serves as an ancillary ligand in the second generation Grubbs and Hoveyda–Grubbs metathesis catalysts, among other uses.[3]

The preparation of imidazolinium salts is usually achieved via condensation of an N,N'-disubstituted ethane-1,2-diamine and an inorganic ammonium salt with a triethyl orthoester in the presence of a catalytic amount of formic acid.[4] Alternatively, a suitable ethane-1,2-diammonium salt may be used as a single starting material for the heterocyclic cation and its counteranion. In all cases, the orthoester serves both as a solvent and a reagent. Numerous variations of this experimental procedure have been reported in the literature.[5] All of these reactions require prolonged heating under reflux conditions in order to reach satisfactory conversions. Thus, reaction times ranging between a few hours and a few days are commonly encountered, unless ethanol is distilled off the reaction medium to drive the cyclization more rapidly to completion.[6]

Microwave-assisted organic synthesis (MAOS) has received increasing attention in recent years as a valuable technique for accelerating chemical reactions.[7] The development of safe and reliable mono- or multimodal microwave reactors specifically designed for chemical applications has significantly reinvigorated time-honored laboratory practices. Condensation reactions leading to heterocyclic products are particularly effective under microwave irradiation conditions.[8]

In 2006, we disclosed a very simple and efficient procedure for the microwave-assisted synthesis of 1,3-diarylimidazolinium chlorides by cyclocondensation of N,N'-diarylethane-1,2-diamine dihydrochlorides with triethyl orthoformate (Table 1).[9] The methodology was further extended to prepare a wide range of imidazolinium and tetrahydropyrimidinium salts bearing diverse aromatic and aliphatic substituents on the nitrogen and carbon atoms of the cyclic amidinium function. In these experiments, recourse to an amine free base together with an inorganic ammonium salt as starting materials provided an easy way to vary the nature of the counterion (Table 2).[10] Indeed, bulky polyatomic anions such as BF_4^-, PF_6^-, TfO^-, or Tf_2N^- could be introduced directly using readily available ammonium salts, thereby eliminating the need for subsequent ion exchange.

The procedure described above is a preparative-scale application of our methodology starting from simple, commercially available reagents. Prior to the microwave-assisted cyclization, it involves the formation of N,N'-di-mesitylethane-1,2-diamine dihydrochloride via condensation of glyoxal with two equivalents of mesitylamine, followed by reduction of the intermediate Schiff base with sodium borohydride under acidic conditions. All three steps proceed readily under normal atmosphere. Laboratory grade solvents and reagents taken straight from the bottles do not require any additional purification. The two intermediates and the final product are isolated in high yield and purity by simple filtration and washing and may be used without any further purification for most applications.

Table 1. Microwave-assisted synthesis of 1,3-diarylimidazolinium chlorides.[9]

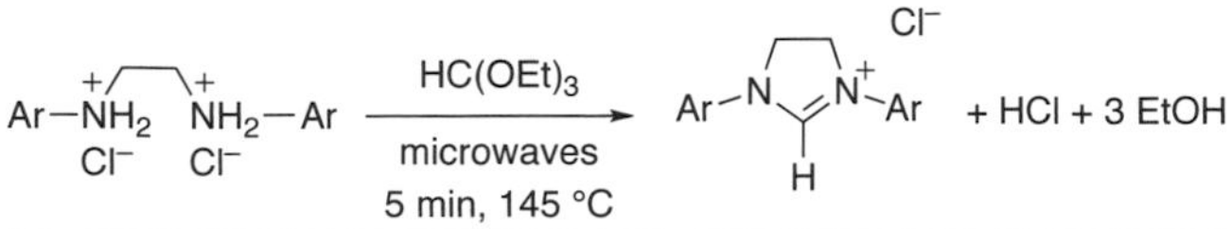

Aryl Group	Isolated Yield (%)
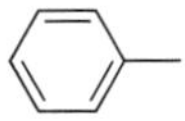	62
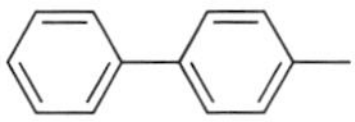	82
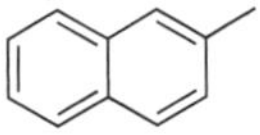	91
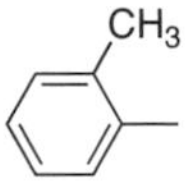	49
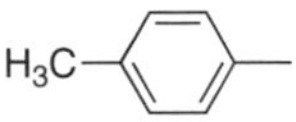	96
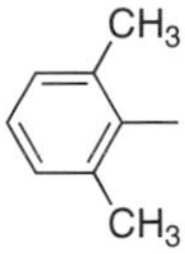	90
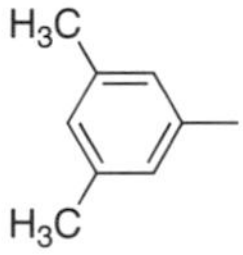	93
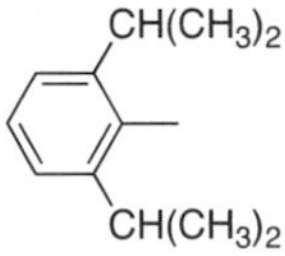	72
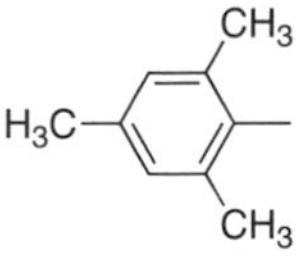	94
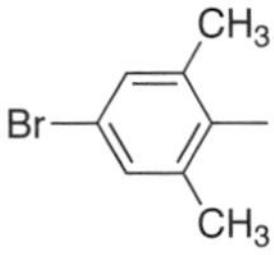	98

Table 2. Microwave-assisted synthesis of cyclic amidinium salts.[10]

$$\begin{array}{c}(CH_2)_n\\R^1\text{-NH}\quad HN\text{-}R^1\end{array} + NH_4X \xrightarrow[\substack{\text{microwaves}\\ \text{5 min, 145 °C}}]{(EtO)_3CR^2} \begin{array}{c}(CH_2)_n\quad X^-\\R^1\text{-N}\diagdown_{\diagup}N\text{-}R^1\\R^2\end{array} + NH_3 + 3\ EtOH$$

R^1	R^2	n	X^-	Isolated Yield (%)
Me	H	2	BF_4^-	99
Me	Me	2	BF_4^-	87
Me	Et	2	BF_4^-	86
Me	Ph	2	BF_4^-	91
Me	Ph	2	TfO^-	85
Me	Ph	2	Tf_2N^-	54
Et	H	2	BF_4^-	92
2-Hydroxyethyl	H	2	BF_4^-	93
i-Pr	H	2	BF_4^-	78
t-Bu	H	2	BF_4^-	84
Mes	H	2	F^-	82[a]
Mes	H	2	Cl^-	68
Mes	H	2	Br^-	36
Mes	H	2	I^-	38
Mes	H	2	NO_3^-	66
Mes	H	2	SCN^-	82
Mes	H	2	BF_4^-	91
Mes	H	2	PF_6^-	93
Mes	H	2	TfO^-	40
Mes	H	2	Tf_2N^-	58
Me	Me	3	BF_4^-	94
Me	Et	3	BF_4^-	92
Me	Ph	3	BF_4^-	49
Me	Ph	3	TfO^-	95
Me	Ph	3	Tf_2N^-	77
Me	H	3	PF_6^-	93
Et	H	3	PF_6^-	93
i-Pr	H	3	PF_6^-	92
t-Bu	H	3	PF_6^-	93
Cyclohexyl	H	3	PF_6^-	90

[a] Starting from *N,N'*-dimesityl-1,2-ethanediamine dihydrofluoride.

1. Department of Chemistry, University of Liège, Sart-Tilman par 4000 Liège, Belgium (email: l.delaude@ulg.ac.be).

2. (a) *Carbene Chemistry: From Fleeting Intermediates to Powerful Reagents*; Bertrand, G., Ed.; Marcel Dekker: New York, 2002. (b) *N-Heterocyclic Carbenes in Synthesis*; Nolan, S. P., Ed.; Wiley-VCH: Weinheim, 2006. (c) *N-Heterocyclic Carbenes in Transition Metal Catalysis*; Topics in Organometallic Chemistry, Vol. 21; Glorius, F., Ed.; Springer: Berlin, 2007.

3. (a) Scholl, M.; Ding, S.; Lee, C. W.; Grubbs, R. H. *Org. Lett.* **1999**, *1*, 953–956. (b) Garber, S. B.; Kingsbury, J. S.; Gray, B. L.; Hoveyda, A. H. *J. Am. Chem. Soc.* **2000**, *122*, 8168–8179.

4. Saba, S.; Brescia, A.; Kaloustian, M. S. *Tetrahedron Lett.* **1991**, *32*, 5031–5034.

5. See, for example: (a) Arduengo, A. J., III; Krafczyk, R.; Schmutzler, R.; Craig, H. A.; Goerlich, J. R.; Marshall, W. J.; Unverzagt, M. *Tetrahedron* **1999**, *55*, 14523–14534. (b) Delaude, L.; Szypa, M.; Demonceau, A.; Noels, A. F. *Adv. Synth. Catal.* **2002**, *344*, 749–759. (c) Ma, Y.; Song, C.; Jiang, W.; Wu, Q.; Wang, Y.; Liu, X.; Andrus, M. B. *Org. Lett.* **2003**, *5*, 3317–3319. (d) Funk, T. W.; Berlin, J. M.; Grubbs, R. H. *J. Am. Chem. Soc.* **2006**, *128*, 1840–1846.

6. See, for example: (a) Alder, R. W.; Blake, M. E.; Bufali, S.; Butts, C. P.; Orpen, A. G.; Schütz, J.; Williams, S. J. *J. Chem. Soc., Perkin Trans. 1* **2001**, 1586–1593. (b) Van Veldhuizen, J. J.; Gillingham, D. G.; Garber, S. B.; Kataoka, O.; Hoveyda, A. H. *J. Am. Chem. Soc.* **2003**, *125*, 12502–12508.

7. (a) *Microwaves in Organic and Medicinal Chemistry*; Kappe, C. O.; Stadler, A., Eds.; Methods and Principles in Medicinal Chemistry, Vol. 25; Wiley-VCH: Weinheim, 2005. (b) *Microwave Assisted Organic Synthesis*; Tierney, J. P.; Lidström, P., Eds.; Blackwell: Oxford, 2005. (c) *Microwaves in Organic Synthesis*; 2nd ed.; Loupy, A., Ed.; Wiley-VCH: Weinheim, 2006.

8. For a review, see: Xu, Y.; Guo, Q.-X. *Heterocycles* **2004**, *63*, 903–974.

9. Aidouni, A.; Demonceau, A.; Delaude, L. *Synlett* **2006**, 493–495.

10. Aidouni, A.; Bendahou, S.; Demonceau, A.; Delaude, L. *J. Comb. Chem.* **2008**, *10*, 886–892.

Appendix
Chemical Abstracts Nomenclature; (Registry Number)

Mesitylamine: Benzenamine, 2,4,6-trimethyl-; (88-05-1)

Glyoxal: Ethanedial; (107-22-2)

N,N'-Dimesitylethylenediimine: Benzenamine, N,N'-1,2-ethane-
diylidenebis[2,4,6-trimethyl-; (56222-36-7)

Sodium borohydride: Borate(1-), tetrahydro-, sodium; (16940-66-2)

N,N'-dimesitylethylenediamine dihydrochloride: 1,2-Ethanediamine,
N,N'-bis(2,4,6-trimethylphenyl)-, dihydrochloride; (258278-23-8)

Triethyl orthoformate: Ethane, 1,1',1"-[methylidynetris(oxy)]tris- ;
(122-51-0)

1,3-Dimesitylimidazolinium Chloride: 1H-Imidazolium, 4,5-dihydro-
1,3-bis(2,4,6-trimethylphenyl)-, chloride (1:1); (173035-10-4)

Lionel Delaude was born in 1966 in Huy, Belgium. He received his Ph.D. from the University of Liège, Belgium under the supervision of Pierre Laszlo. After postdoctoral research at the University of Ottawa, Canada with Howard Alper and at Columbia University in the City of New York with Ronald Breslow, he returned to Liège where he is presently appointed as a Senior Lecturer within the Center for Education and Research on Macromolecules (CERM). His current research interests focus on the design of new ruthenium–N-heterocyclic carbene complexes for catalytic applications.

Morgan Hans was born in 1985 in Malmedy, Belgium. He received his M.Sc. degree in chemistry in 2009 from the University of Liège and is currently pursuing a Ph.D. in organometallic chemistry at the University of Liège under the supervision of Lionel Delaude, working on the synthesis and catalytic applications of new zwitterions derived from N-heterocyclic carbenes.

Somenath Chowdhury grew up in Majirdanga, India. He obtained his M.Sc. degree in Chemistry from the Indian Institute of Technology, Kharagpur, India and his Ph.D. degree from the University of Saskatchewan, Canada. In his Ph.D. research he worked on the design and synthesis of peptide beta sheet mimics under the supervision of Professor Heinz-Bernhard Kraatz. After finishing his Ph.D. he carried out research with Professor Giuseppe Melacini's group at McMaster University, Canada for one and a half years where he studied protein ligand interactions by NMR spectroscopy. He then joined Professor Jonathan Ellman's lab at the University of California, Berkeley in Sept 2008 as an NSERC postdoctoral fellow. His current research focuses on developing small molecule protease inhibitors.

SYNTHESIS OF (3-CHLOROBUTYL)BENZENE BY THE COBALT-CATALYZED HYDROCHLORINATION OF 4-PHENYL-1-BUTENE

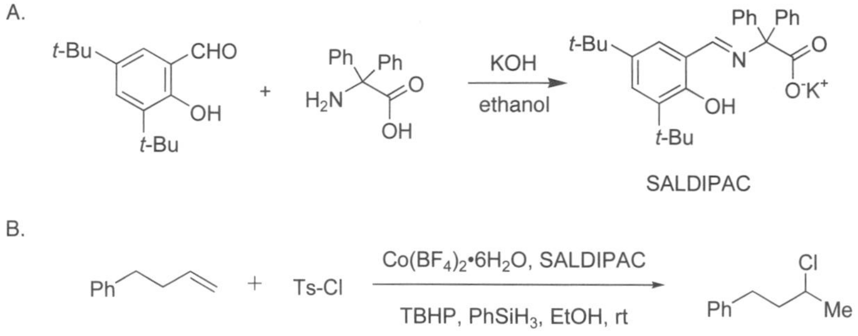

A.

SALDIPAC

B.

Submitted by Boris Gaspar, Jerome Waser, and Erick M. Carreira.[1]
Checked by David Hughes.

1. Procedure

In-situ preparation of 2-(3,5-di-tert-butyl-2-hydroxybenzylideneamino)-2,2-diphenylacetic acid potassium salt (SALDIPAC). A 500-mL, three-necked, round-bottomed flask is equipped with a rubber septum pierced with a thermocouple thermometer probe (Note 1), a gas adapter connected to a nitrogen line and oil bubbler, and a ground-glass stopper. A 3-cm Teflon-coated oval stir bar is added to the flask. The flask is charged with α,α-diphenylglycine (0.94 g, 4.1 mmol) and absolute ethanol (40 mL) (Note 2). The suspension is stirred at 22 °C and 0.5 M KOH in ethanol (10.0 mL, 5.0 mmol) is added (Note 3). After stirring for 10 min, most of the solids are dissolved. 3,5-Di-*t*-butyl-2-hydroxybenzaldehyde (1.00 g, 4.27 mmol) is added, forming a bright yellow solution. The solution is stirred at 22 °C for 16 h to provide 2-(3,5-di-*tert*-butyl-2-hydroxybenzylideneamino)-2,2-diphenylacetic acid potassium salt as a solution in ethanol (Note 4).

(3-Chlorobutyl)benzene. To the ligand solution prepared above is added absolute ethanol (140 mL) and Co(BF₄)₂•6H₂O (1.40 g, 4.11 mmol, 8 mol%). The resulting dark red-brown solution is stirred at 22 °C for 10 min. To the vigorously stirred mixture (Note 5) is added 4-phenyl-1-butene (6.69 g, 50.6 mmol, 1.00 equiv) in one portion by syringe followed

88

by *p*-toluenesulfonyl chloride (11.80 g, 61.89 mmol, 1.2 equiv). Then *t*-butyl hydroperoxide (2.8 mL, 14-17 mmol, 0.3 equiv) is added, followed by phenylsilane (PhSiH$_3$) (6.23 g, 57.6 mmol, 1.1 equiv); both are added by syringe in one portion. The temperature slowly rises to 38 °C over 10 min as the color changes to dark green and weak gas evolution occurs (Notes 6 and 7). The reaction mixture slowly cools to 22 °C over one h and is vigorously stirred at 22 °C for an additional 3 h (Note 8). The mixture is then transferred into a 1-L flask and the solvent is removed under reduced pressure (20 mmHg, 40 °C bath temperature) by rotary evaporation (Notes 9 and 10) to afford 30 g of a blue-green gum. Hexanes (200 mL) are added and the mixture is sonicated (Note 11) for 5 min to form a suspension, which is then filtered through a pad of Celite (50 g) in a 350-mL medium-porosity sintered glass funnel. The flask and the Celite pad are washed with hexanes (2 x 100 mL) and the combined filtrate is concentrated by rotary evaporation (20 mmHg, 40 °C bath temperature) to afford 17.5 g of crude product. The crude product is purified by chromatography on SiO$_2$ (Note 12) to afford (3-chlorobutyl)benzene (7.16 g, 84 %) as a colorless oil (Notes 13 and 14).

2. Notes

1.   The internal temperature is monitored using a J-Kem Gemini digital thermometer with a Teflon-coated T-Type thermocouple probe (12-inch length, 1/8 inch outer diameter, temperature range -200 to +250 °C)
2.   The following reagents and solvents used in this preparation were obtained from Sigma-Aldrich and used without further purification: 3,5-di-*t*-butyl-2-hydroxybenzaldehyde (99%), 4-phenyl-1-butene (99%), cobalt(II) tetrafluoroborate hexahydrate (99%), *p*-toluenesulfonyl chloride (99%), *t*-butyl hydroperoxide (5 – 6 M in decane), hexanes (ACS reagent, >98.5%), dichloromethane (ACS reagent, >99.5%), Celite 545, and silica gel (200-400 mesh, 60 Å).   The following reagents were obtained from Acros and used without further purification: α,α-diphenylglylcine (98%), KOH (powdered, >85%) and phenylsilane (97%). Absolute ethanol was obtained from Pharmaco. Deionized tap water was used throughout the procedure.
3.   A 0.5 M solution of KOH in ethanol is prepared by adding 3.2 g of 85% powdered KOH to 100 mL of absolute ethanol and sonicating for 5 minutes. (Note 11) The resulting hazy solution is allowed to settle overnight

to provide a clear solution with residual powdered solids in the bottom of the flask.

4. The reaction was monitored by ^{1}H NMR as follows. An approx. 0.05 mL aliquot of the reaction mixture was diluted with CD$_3$OD for NMR analysis. The resonances at 9.9, 7.7, and 7.6 ppm of the salicylaldehyde were readily observable and were integrated vs the imine resonances at 8.1 and 6.9 ppm. The reaction typically proceeded to 90% conversion of the aldehyde.

5. The mixture is stirred at 500 rpm on a magnetic stirring plate throughout the reaction.

6. For larger scale preparations, external cooling is recommended.

7. Slow bubbling started after *t*-butyl hydroperoxide addition and was most probably hydrogen gas. Hydrogen chloride can be excluded as a wet Riedel-de-Haen pH paper placed in the neck of the flask gave a negative test for acid.

8. The reaction was monitored by TLC on Merck silica gel 60 F$_{254}$ TLC glass plates and visualized with UV light and permanganate stain. The R$_f$ values in hexane:CH$_2$Cl$_2$ (7:1) are 0.65 for 4-phenylbutene and 0.56 for (3-chlorobutyl)benzene. The reaction was also followed using ^{1}H NMR as follows: One drop of the reaction mixture was added to 1 mL of CDCl$_3$, then filtered through a plug of Celite. The multiplet at 6 ppm of the starting material was monitored to assess reaction completeness vs. the doublet at 1.6 ppm of the product. *p*-TsCl (doublet at 7.9 ppm) and phenylsilane (singlet at 4.2 ppm) could also be monitored. The reaction was >95% complete within the first hour. More concentrated samples caused line broadening.

9. Solids in the flask require slow lowering of the pressure during the concentration to prevent bumping.

10. TLC and ^{1}H NMR analysis of the ethanol distillate from the concentration indicated the presence of a small amount of product. In one run, the ethanol distillate was added to 200 mL of water then extracted with 200 mL of hexanes. The hexanes extract was washed with water (2 x 200 mL), then concentrated to 1.2 g which was purified by chromatography on 15 g of SiO$_2$ using 7:1 hexanes: dichloromethane to afford 0.30 g of product (3.5% yield).

11. Sonication is carried out using a Fisher Scientific Ultrasonic Cleaner, Model FS20, having a capacity of 2.8L and power of 143 watts.

12. Chromatography conditions: 270 g of SiO$_2$ packed and eluted with hexanes:CH$_2$Cl$_2$ (7:1), column diameter 5 cm, fraction volume 40 mL. After loading the crude product on the column, 150 mL of eluent is collected before fractions are collected. The product appears in fractions 13-23.

13. The physical and spectroscopic properties of (3-chlorobutyl)benzene are as follows: IR (film) 3064 (w), 3028 (w), 2971 (w), 2927 (w), 2863 (w), 2361 (w), 1604 (w), 1496 (w), 1454 (m), 1379 (w), 1275 (w), 1117 (w), 1030 (w), 820 (w), 748 (m), 699 (s), 612 (w), 574 (w), 506 (w), 454 (w) cm^{-1}; ^{1}H NMR (400 MHz, CDCl$_3$) δ: 1.55 (d, 3 H, J=6.5 Hz), 2.01–2.08 (m, 2 H), 2.73–2.91 (m, 2 H), 3.97–4.06 (m, 1 H), 7.20–7.33 (m, 5 H); ^{13}C NMR (100 MHz, CDCl$_3$) δ: 25.6, 33.1, 42.1, 58.1, 126.3, 128.68, 128.72, 141.3; HRMS (EI) m/z calcd. for C$_{10}$H$_{13}$Cl [M]$^+$ 168.0700; found 168.0700; Anal. calcd. for C$_{10}$H$_{13}$Cl: C, 71.21, H, 7.77; found: C, 70.91, H, 7.68.

14. An analytically pure sample was prepared by dissolving 200 mg of the chromatographed product in 5 mL of pentane, filtering the solution through a 0.45 micron Teflon filter, and concentrating to dryness by rotary evaporation, then further removing residual solvent under vacuum at room temperature (20 mm Hg) for 3 h.

Waste Disposal Information

All hazardous materials should be handled and disposed of in accordance with "Prudent Practices in the Laboratory"; National Academy Press; Washington, DC, 1995.

3. Discussion

Olefins are inexpensive and readily available starting materials for organic synthesis. For this reason, the direct heterofunctionalization of the C=C has been of interest for many years, especially in a regioselective manner.[2] Among these, the hydrochlorination reaction belongs to one of the first fundamental reactions discussed in introductory organic chemistry. However, the process is very limited in scope, as the addition at useful rates occurs only to highly substituted or strained olefins[3] and to styrene-like substrates.[4] Lewis acid or surface mediated reactions of HCl were reported for simple olefins such as cyclohexene and cycloheptene.[5] In attempts to avoid strongly acidic conditions different precursors were recognized to form HCl in small amounts *in situ*, however these methods are still limited to polysubstituted or activated alkenes and acid sensitive functional groups are not tolerated.[6]

The cobalt-catalyzed hydrochlorination described above is applicable to a range of unactivated alkenes and tolerates a variety of functional groups.[7] Importantly, the reaction displays complete Markovnikov selectivity and operates under very mild conditions (EtOH as solvent, room temperature). Furthermore, all the reaction components are commercially available. In a broader sense, the role of *p*-TsCl as a Cl-transfer reagent is intriguing and may have additional applications in other processes.[8,9]

1. Laboratory of Organic Chemistry, ETH Zürich, Wolfgang-Pauli-Strasse 10, 8093, Zürich.
2. Beller, M.; Seayad, J.; Tillack, A.; Jiao, H. *Angew. Chem. Int. Ed.* **2004**, *43*, 3368-3398.
3. (a) Whitmore, F. C.; Johnston, F. *J. Am. Chem. Soc.* **1933**, *55*, 5020–5022; (b) Schmerling, L. *J. Am. Chem. Soc.* **1946**, *68*, 195–196; (c) Stille, J. K.; Sonnenberg, F. M.; Kinstle, T. H. *J. Am. Chem. Soc.* **1966**, *88*, 4922–4925; (d) Fahey, R. C.; McPherson, C. A. *J. Am. Chem. Soc.* **1971**, *93*, 2445–2453; (e) Becker, K. B.; Grob, C. A. *Synthesis* **1973**, *12*, 789–790; (f) Becker, K. B.; Grob, C. A. *Helv. Chim. Acta* **1973**, *56*, 2723–2732.
4. (a) Dewar, M. J. S.; Fahey, R. C. *J. Am. Chem. Soc.* **1963**, *85*, 2245–2248; (b) Brown, H. C.; Rei, M.-H. *J. Org. Chem.* **1965**, *31*, 1090–1093.
5. (a) Kennedy, J. P.; Sivaram, S. *J. Org. Chem.* **1973**, *38*, 2262–2264; (b) Kropp, P. J.; Daus, K. A.; Crawford, S. D.; Tubergen, M. W.; Kepler, K. D.; Craig, S. L.; Wilson, V. P. *J. Am. Chem. Soc.* **1990**, *112*, 7433–7434; (c) Alper, H.; Huang, Y. *Organometallics* **1991**, *10*, 1665–1671.
6. (a) Kropp, P. J.; Daus, K. A.; Tubergen, M. W.; Kepler, K. D.; Wilson, V. P.; Craig, S. L.; Baillargeon, M. M.; Breton, G. W. *J. Am. Chem. Soc.* **1993**, *115*, 3071–3079; (b) Boudjouk, P.; Kim, B.-K.; Han, B.-H. *Synth. Commun.* **1996**, *26*, 3479–3484; (c) Yadav, V. K.; Babu, K. G. *Eur. J. Org. Chem.* **2005**, 452–456.
7. Gaspar, B.; Carreira E. M. *Angew. Chem. Int. Ed.* **2008**, *47*, 5758–5760.

8. For a recent example of Pd-catalyzed chlorination with TsCl see: Zhao, X.; Dimitrijevic, E.; Dong V. M. *J. Am. Chem. Soc.* **2009**, *131*, 3466–3467.

9. For the use of related cobalt catalysts in a wide range of other olefin functionalization reactions, see: (a) Waser, J.; Carreira, E. M. *J. Am. Chem. Soc.* **2004**, *126*, 5676–5677; (b) Waser, J.; Carreira, E. M. *Angew. Chem. Int. Ed.* **2004**, *43*, 4099–4102; (c) Waser, J.; Nambu, H.; Carreira, E. M. *J. Am. Chem. Soc.* **2005**, *127*, 8294–8295; (d) Waser, J.; González-Gómez, J. C.; Nambu, H.; Huber, P.; Carreira, E. M. *Org. Lett.* **2005**, *7*, 4249–4252; (e) Waser, J.; Gaspar, B.; Nambu, H.; Carreira, E. M. *J. Am. Chem. Soc.* **2006**, *128*, 11693–11712; (f) Gaspar, B.; Waser, J.; Carreira, E. M. *Synthesis* **2007**, *24*, 3839–3845.

Appendix
Chemical Abstracts Nomenclature; (Registry Number)

(3-Chlorobutyl)benzene; (4830-94-8)

2-(3,5-Di-*tert*-butyl-2-hydroxybenzylideneamino)-2,2-diphenylacetic acid potassium salt, (SALDIPAC); (858344-69-1)

α,α-Diphenylglycine: Benzeneacetic acid, α-amino-α-phenyl-; (3060-50-2)

3,5-Di-*t*-butyl-2-hydroxybenzaldehyde: Benzaldehyde, 3,5-bis(1,1-dimethylethyl)-2-hydroxy-; (37942-07-7)

Cobalt(II) tetrafluoroborate hexahydrate; (15684-35-2)

p-Toluenesulfonyl chloride: Benzenesulfonyl chloride, 4-methyl-; (98-59-9)

t-Butyl hydroperoxide: Hydroperoxide, 1,1-dimethylethyl; (75-91-2)

Phenylsilane: Benzene, silyl-; (694-53-1)

4-Phenyl-1-butene: Benzene, 3-buten-1-yl-; (768-56-69)

Prof. Erick M. Carreira obtained a B.S. degree in 1984 from the University of Illinois at Urbana-Champaign and a Ph.D. degree in 1990 from Harvard University. After carrying out postdoctoral work with Peter Dervan at the California Institute of Technology through late 1992, he joined the faculty at the same institution as an assistant professor of chemistry and subsequently was promoted to the rank of full professor. Since September 1998, he has been professor of chemistry at the ETH Zürich. Most recently, he is the recipient of the Tetrahedron Chair Award, Thieme Prize, the Springer Award, American Chemical Society Award in Pure Chemistry, Nobel Laureate Signature Award, Young Investigator Awards from Merck, Novartis, Pfizer, Eli Lilly, as well as Astra Zeneca, and a recipient of the David and Lucile Packard foundation Fellowship in Science and Engineering.

Boris Gaspar was born in 1982 in Nove Zamky, Slovakia. He completed his undergraduate studies in chemistry at the Comenius University in Bratislava while working in the group of Associate Professor M. Salisova. During his undergraduate studies, he also worked with Professor A. Solladiè-Cavallo, (Université Louis Pasteur, Strasbourg, France) as a Socrates-Erasmus exchange fellow and carried out an internship at Syngenta (Basel, Switzerland). He recently completed his Ph.D. studies at the ETH Zürich in the group of Professor E. M. Carreira where he was involved in the development of metal-catalyzed functionalizations of olefins.

Jérôme Waser was born in Sierre, Valais, Switzerland in 1977. He studied chemistry at ETH Zurich and obtained his Diploma in 2001. In 2002, he started his Ph.D. studies at ETH Zurich with Prof. Erick M. Carreira, working on the development of metal-catalyzed amination reactions of olefins. In 2006, he joined Prof. Barry M. Trost at Stanford University and accomplished the total synthesis of Pseudolaric Acid B, a diterpene natural product. Since October 2007, he is working as tenure-track assistant professor at EPF Lausanne, focusing on the development and application of catalytic methods for the synthesis of bioactive compounds.

94

SYNTHESIS OF SUBSTITUTED INDAZOLES VIA [3+2] CYCLOADDITION OF BENZYNE AND DIAZO COMPOUNDS
[1H-indazole-3-carboxylic acid, ethyl ester]

Submitted by Feng Shi and Richard C. Larock.[1]
Checked by Alistair Boyer and Mark Lautens.

1. Procedure

> *CAUTION! Due to the potential explosive nature of diazo compounds, the reaction should be conducted with proper precautions. A safety shield in a closed fume hood is recommended.*

1H-Indazole-3-carboxylic acid, ethyl ester. A 1-L, flame-dried (Note 1), round-bottomed, three-necked flask (the central neck is sealed with a rubber septum, the other necks are sealed with a ground glass stopper and a tap adaptor attached to a nitrogen line) equipped with a large, egg-shaped stirring bar is charged with 2-(trimethylsilyl)phenyl trifluoromethanesulfonate (9.84 g, 32.0 mmol, 1.0 equiv) (Note 2) and ethyl diazoacetate (5.51 mL, 48.0 mmol, 1.5 equiv) (Note 3). Tetrahydrofuran (THF, Note 2) (360 mL) is introduced by cannula into the flask and the reaction mixture is cooled to –78 °C (bath temperature) in an acetone/dry ice bath. Upon vigorous stirring, a TBAF solution (57.6 mL of 1 M THF solution, 57.6 mmol, 1.8 equiv) (Note 4) is added dropwise via syringe over ca. 40 min (Note 5). After complete addition, the reaction mixture is stirred at –78 °C in the acetone/dry ice bath for 1.5 h. The flask is then transferred to a cold-acetone bath (ca. −65 °C, bath temperature) and allowed to warm to room temperature overnight, over which time the reaction mixture becomes orange. After a further 12 h stirring at room temperature, the reaction is judged complete (Note 6). The reaction mixture is concentrated at room temperature by rotary evaporation (35 °C, 40 mmHg) to ca. 100 mL and poured into a 1-L separatory funnel containing EtOAc (150 mL) and saturated aq. NaHCO$_3$ (200 mL). The flask is rinsed with EtOAc (20 mL), which is added to the separatory funnel. The layers are separated, and the aqueous layer is extracted twice (2 x 50 mL) with EtOAc. The combined EtOAc extracts are dried over $MgSO_4$ (ca. 20 g for 30 min), filtered through a fritted glass funnel, rinsing twice with EtOAc (20 mL), and evaporated to give an orange oil (ca. 27.6 g). The residue is purified by chromatography on SiO_2 (Note 7) to afford 4.98 g (26.2 mmol, 82%) of *1H-indazole-3-carboxylic acid, ethyl ester* as an off-white solid (Notes 8, 9).

2. Notes

1. Although benzyne is highly reactive, the reaction summarized in this procedure does not exhibit air sensitivity. Thus, all weighing and transferring procedures can be carried out in air without a problem. The reaction does exhibit some, although not significant, moisture sensitivity, so drying the glassware and preventing the reaction mixture from absorbing moisture from outside air are desirable. It should be noted, however, that commercial TBAF solution contains about 5 wt% of water.

2. The following chemicals were obtained from Aldrich and were used as received: ethyl diazoacetate; 2-(trimethylsilyl)phenyl trifluoromethanesulfonate (97%); and tetrabutylammonium fluoride (1.0 M solution in THF). The THF solvent (reagent grade, ~0.05% H_2O, obtained from Caledon Laboratories Ltd.) was dried over KOH then distilled from Na/benzophenone ketyl. All solvents used in the work-up procedures were standard reagent grade. Aluminium-backed TLC plates (silica gel, 60 F_{254}) were purchased from EMD Chemicals and 60 Å, 40-63 μm silica gel was purchased from Silicycle (submitters used 0.25 mm thick glass-backed TLC plates and 60 Å, 230-400 mesh silica gel purchased from Sorbent Technology).

3. Commercial ethyl diazoacetate contains a small quantity of dichloromethane. The presence of dichloromethane does not hamper the reaction of ethyl diazoacetate with benzyne, but the added weight of dichloromethane should be taken into account. A ^1H NMR spectrum should be taken prior to the use of this compound to determine the molar ratio of ethyl diazoacetate to dichloromethane. For both checkers and submitters, this ratio was 100:13. Thus, per 1.00 g of commercial ethyl diazoacetate (corresponding to 0.91 mL, based on d = 1.10 g/mL), 0.91 g of pure ethyl diazoacetate was present.

4. The quality of commercial TBAF solutions can be variable. The

96

checkers required 1.8 equiv TBAF whereas the submitters needed only 1.2 equiv. However, if incomplete conversion is observed, the reaction mixture can be re-cooled to −78 °C and the protocol repeated using supplementary TBAF without any decrease of overall yield.

5.   The needle should be positioned a sufficient distance above the reaction mixture to prevent freezing of the solution in the needle and such that the TBAF solution is added directly to the reaction mixture to prevent freezing of the solution on the cold flask wall.

6.   A small aliquot (ca. 0.1 mL) of the reaction mixture was removed and concentrated *in vacuo*. Direct ^{1}H NMR (CDCl$_3$) analysis of this sample revealed the reaction mixture composition. Alternatively, a TLC analysis can be performed using 4:1 hexanes/EtOAc as the eluent. Unreacted starting material appears at R$_f$ 0.65, a by-product (*1H-indazole-3-carboxylic acid, 1-phenyl, ethyl ester,* blue visualized by short-wave UV) appears at R$_f$ 0.59, and the desired product appears at R$_f$ 0.10.

7.   Column chromatography is performed on 400 mL of SiO$_2$ (~160 g) packed into a 5.5 cm × 18 cm column with 3:1 hexanes/EtOAc. The crude material is loaded directly onto the packed SiO$_2$, washing with 50 mL of the eluent. Elute with 1.4 L of 3:1 hexanes/EtOAc, then change to 0.8 L of 1:3 hexanes/EtOAc, and collect 70 mL fractions throughout. The high running spots eluting together (1.78 g, fractions 7–16) contain a mixture of unreacted ethyl diazoacetate (1.55 g, 13.6 mmol), unreacted 2-(trimethylsilyl)phenyl trifluoromethanesulfonate (0.12 g, 0.40 mmol, 1%) and the by-product: 1*H*-indazole-3-carboxylic acid, 1-phenyl, ethyl ester (0.11 g, 0.41 mmol, 3%). The low-running spot is the desired product (fractions 17-31), these fractions are concentrated by rotary evaporation (35 °C, 100 mmHg).

8.   This material is analytically pure, however, the color can be removed by crystallization. Toluene (20 mL) is added to the colored product (4.98 g) in a single-necked, round-bottomed flask. The flask is fitted with a reflux condenser and the mixture is heated under reflux until the solid has dissolved (ca. 5 min). The flask is allowed to cool to ambient temperature over 1 h. The flask is transferred to a freezer (temperature −19 °C) for 10 h and the product is collected by filtration, washing with ice-cold toluene (2 × 5 mL) and dried *in vacuo* to give a white crystalline solid (4.71 g). ^{1}H NMR analysis of the concentrated mother liquor reveals only the product and trace impurities.

9.   Data for product: R$_f$ 0.30 (1.5:1 hexanes/EtOAc); mp 133–134 °C (lit.[9] 130 °C); IR (neat) 3292 (s), 1713 (s), 1479 (m), 1421 (m), 1273 (s),

1231 (s), 1140 (m) cm^{-1}; ^{1}H NMR (400 MHz, CDCl$_3$) δ: 1.48 (3 H, t, J = 7.1 Hz, CH$_3$), 4.56 (2 H, q, J = 7.1 Hz, CH$_2$), 7.34 (1 H, t, J = 7.5 Hz, Ar), 7.47 (1 H, t, J = 7.6 Hz, Ar), 7.75 (1 H, d, J = 8.5 Hz, Ar), 8.23 (1 H, d, J = 8.2 Hz, Ar), 12.14 (1 H, br s, NH); ^{13}C NMR (100 MHz, CDCl$_3$) δ: 14.4 (CH$_3$), 61.1 (CH$_2$), 111.1 (Ar-H), 121.8 (Ar-H), 122.4 (Ar), 123.2 (Ar-H), 127.3 (Ar), 127.3 (Ar-H), 141.4 (Ar), 163.0 (C=O); LRMS (EI) 88 (21), 90 (26), 118 (100), 145 (100), 162 (22), 190 (M$^+$, 66); HRMS (ESI$^+$) m/z calcd. for C$_{10}$H$_{11}$N$_2$O$_2$ ([M+H]$^+$) 191.0815; found 191.0819; Anal. calcd. for C$_{10}$H$_{10}$N$_2$O$_2$: C, 63.15; H, 5.30; N, 14.73; found: C, 62.92; H, 5.11; N, 14.76.

Waste Disposal Information

All toxic materials were disposed of in accordance with "Prudent Practices in the Laboratory"; National Academy Press; Washington, DC, 1995.

3. Discussion

Benzyne is a highly reactive intermediate that has attracted wide attention from synthetic organic chemists recently.[2] Due to the high reactivity of benzyne, it is generated *in situ* from various precursors. Two widely used precursors of benzyne are 2-(trimethylsilyl)phenyl trifluoromethanesulfonate[3] and (phenyl)[2-(trimethylsilyl)phenyl]iodonium trifluoromethanesulfonate.[4] The first compound and a few derivatives are now commercially available. These benzyne precursors undergo fluoride-promoted *ortho*-elimination to generate benzyne under mild reaction conditions with a range of organic functional groups tolerated.

Benzyne has been shown to react with 1,3-dipoles under mild conditions to afford various important heterocycles.[5-7] Among these, reactions with diazo compounds afford substituted indazoles efficiently. It has been shown that an excess of a monosubstituted diazo compound reacts with benzyne as the limiting reagent to afford a 1*H*-indazole after a hydrogen shift, as seen in the title reaction. Reaction with an excess of benzyne furnishes the product of *N*-arylation. Thus, the title reaction can be performed on a 2.2:1 stoichiometry using cesium fluoride as the fluoride source in anhydrous acetonitrile at room temperature for 1 day to afford *N*-arylated product (1*H*-indazole-3-carboxylic acid, 1-phenyl, ethyl ester) in high yield (Eq 1).

$$\text{TMS} / \text{OTf} \quad + \quad \underset{H}{\overset{N_2}{\|}} CO_2Et \quad \xrightarrow[\text{MeCN}]{\text{CsF}} \quad \text{indazole-}CO_2Et, N\text{-}Ph \qquad (1)$$

On the other hand, disubstituted diazo compounds can react with benzyne to afford 1H-indazoles and/or 3H-indazoles, depending upon the nature of the diazo compound and benzyne precursor. The product can be hard to predict.[5b] Specifically, diazo compounds containing an acyl group may undergo acyl migration to afford 1H-indazoles, as shown in Eq 2.

$$\text{TMS} / \text{OTf} \quad + \quad EtO_2C \overset{N_2}{\diagup} CO_2Et \quad \xrightarrow[\text{MeCN}]{\text{CsF}} \quad \text{indazole-}CO_2Et, N\text{-}CO_2Et \qquad (2)$$

It has been established that acyclic ketone groups migrate preferentially and acyclic ester groups only migrate occasionally. Amide groups and aryl groups have not been observed to migrate. Cyclic diazo compounds containing acyl groups do not seem to react with benzynes under these conditions (Table 1).

It should be noted that these reactions perform much better on a small scale, often giving near quantitative yields. Once scaled up, these reactions tend to become dirty and give lower yields. In the reaction of Eq 2 mentioned above, we have observed an 85-90% yield on a 0.3-4.0 mmol scale; however, on a 20 mmol scale, this reaction gives incomplete conversion and only a 70% yield. A significant by-product is **III**, the product of Eq 1, apparently formed by N-deacylation, followed by N-arylation. We reason that the acyl migration step may be responsible for this observation. The acyl migration has been suggested to proceed by a dissociation-ion pair intermediate-recombination process (Scheme 1).[8] In such a process, the [3+2] cycloaddition reaction should directly form intermediate **I**, which dissociates to ion pair **IIa/b**, which can recombine to form the desired product. The formation of the by-product **III** is likely due to the reaction of the ion pair **IIb** with free benzyne. However, the N-deacylation of **IV** by either trace amounts of moisture or the equilibrium between **IV** and **IIb**, followed by reaction with benzyne, cannot be ruled out. A control experiment has confirmed that isolated **IV** reacts with benzyne under the same conditions to form a significant amount of **III**. A similar observation has been made during the reaction of triethyl diazophosphonoacetate with benzyne, in which case the phosphonate group is lost and the product formed is again **III**.[5b]

Table 1. Reaction of disubstituted diazo compounds with benzynes

entry[a,b]	substrate	product	% yield
1			90
2			83
3			N. R.
4			87
5			92
6			72+25
7			55
8			44

[a] See ref 5b. [b] 0.3 mmol scale.

Scheme 1. Proposed mechanism for acyl migration and by-product formation.

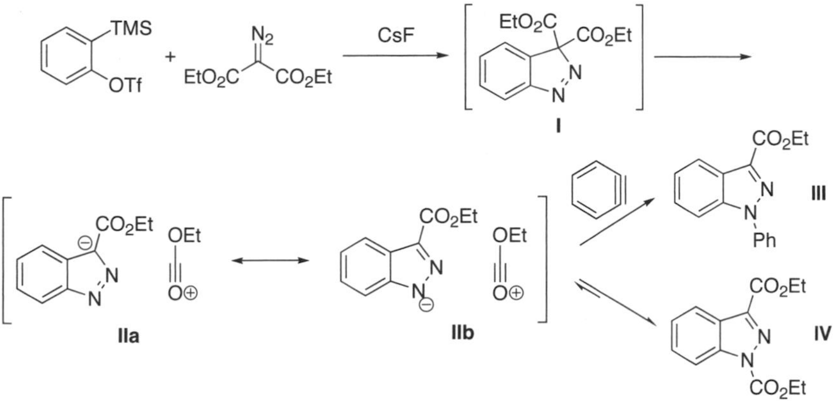

1. Department of Chemistry, Iowa State University, Ames, IA 50011; email: larock@iastate.edu.

2. For a recent comprehensive review of aryne chemistry, see Chen, Y.; Larock, R. C. In *Modern Arylation Methods*; Ackermann, L., Ed.; Wiley-VCH: Weinheim, 2009; pp 401-473.

3. Himeshima, Y.; Sonoda, T.; Kobayashi, H. *Chem. Lett.* **1983**, 1211-1214; Peña, D.; Cobas, A; Pérez, D; Guitián, E. *Synthesis* **2002**, 1454-1458; Wu, Q.-C.; Li, B.-S.; Shi, C.-Q.; Chen, Y.-X. *Chin. J. Synth. Chem.* **2007**, *15*, 111-113; Bronner, S. M.; Garg, N. K. *J. Org. Chem.* **2009**, *74*, 8842–8843.

4. For a preparation and applications of this precursor, see: Kitamuraa, T.; Todaka, M.; Fujiwar, Y. *Org. Synth.* **2002**, *78*, 104-108.

5. For reactions with diazo compounds, see (a) Jin, T.; Yamamoto, Y. *Angew. Chem., Int. Ed.* **2007**, *46*, 3323-3325. (b) Liu, Z.; Shi, F.; Martinez, P. D. G.; Raminelli, C.; Larock, R. C. *J. Org. Chem.* **2008**, *73*, 219-226.

6. For reactions with azides, see: (a) Shi, F.; Waldo, J. P.; Chen, Y.; Larock, R. C. *Org. Lett.* **2008**, *10*, 2409-2412. (b) Zhang, F.; Moses, J. E. *Org. Lett.* **2009**, *11*, 1587-1590. (c) Chandrasekhar, S.; Seenaiah, M.; Rao, C. L.; Reddy, C. R. *Tetrahedron* **2008**, *64*, 11325-11327. (d)

Bronner, S. M.; Bahnck, K. B.; Garg, N. K. *Org. Lett.* **2009**, *11*, 1007-1010. (e) Campbell-Verduyn, L.; Elsinga, P. H.; Mirfeizi, L.; Dierckx, R. A.; Feringa, B. L. *Org. Biomol. Chem.* **2008**, *6*, 3461-3463.

7. For reactions with azomethine imines, see: Shi, F.; Mancuso, R.; Larock, R. C. *Tetrahedron Lett.* **2009**, *50*, 4067-4070.

8. Yamazaki, T.; Baum, G.; Shechter, H. *Tetrahedron Lett.* **1974**, *15*, 4421-4424.

9. Schmidt, A.; Merkel, L.; Eisfeld, W. *Eur. J. Org. Chem.* **2005**, 2124-2130.

10. Croce, P. D.; La Rosa, C. *Synthesis* **1984**, 982-983.

Appendix
Chemical Abstracts Nomenclature (Registry Number)

Ethyl diazoacetate; (623-73-4)
2-(Trimethylsilyl)phenyl trifluoromethanesulfonate; (88284-48-4)
Tetrabutylammonium fluoride; (429-41-4)

Richard C. Larock received his B.Sc. at the University of California, Davis in 1967. He then joined the group of Prof. Herbert C. Brown at Purdue University, where he received his Ph.D. in 1972. He worked as an NSF Postdoctoral Fellow at Harvard University in Prof. E. J. Corey's group and joined the Iowa State University faculty in 1972. His current research interests cover aryne chemistry, electrophilic cyclization, palladium catalysis, and polymer chemistry based on biorenewable resources.

Feng Shi was born in 1978 in China. He obtained his B.Sc. at Beijing University, China in 2001. Afterwards he moved to Michigan State University under the supervision of Prof. Robert E. Maleczka, Jr., where he received his Ph.D. in 2007 on iridium-catalyzed C–H activation chemistry. He joined Prof. Richard C. Larock's group in 2007 working on benzyne annulation chemistry. He is currently an associate professor at Henan University, China.

Alistair Boyer was born in 1982 in Warrington, UK. He obtained his M. Sci. at the University of Cambridge in 2004, completing his master project with Prof. Andrew B. Holmes. He stayed at Cambridge to perform his Ph.D. studies under the supervision of Prof. Steven Ley working on the synthesis of azadirachtin. In 2009, he moved to Toronto, Canada to become a post-doctoral research associate in the group of Prof. Mark Lautens, investigating novel rhodium-catalyzed reactions.

PALLADIUM-CATALYZED CROSS-COUPLING USING AN AIR-STABLE TRIMETHYLALUMINUM SOURCE. PREPARATION OF ETHYL 4-METHYLBENZOATE

A.

DABAL-Me$_3$ **1**

B.

Submitted by Andrej Vinogradov and Simon Woodward.[1]
Checked by Denise Rageot and Andreas Pfaltz.

1. Procedure

Caution! Trialkylaluminum compounds are pyrophoric and must not be allowed to come into contact with air or moisture. These compounds should only be handled by individuals trained in their proper and safe use.

A. *[m-(1,4-Diazabicyclo[2.2.2]octane-kN1:kN4)]hexamethyldialuminum DABAL-Me$_3$* (**1**) (Note 1). A 250-mL, three-necked, round-bottomed flask is equipped with an egg-shaped, teflon-coated, magnetic stir bar (*ca.* 25 mm), a glass stopper, a reflux condenser and a 100-mL pressure-equalizing dropping funnel (Note 2). The reflux condenser is fitted at the top with a two-tap Schlenk adaptor connected to a bubbler and an argon/vacuum manifold (Note 3). A modest flow of argon is applied as judged by the exit bubbler (1-2 bubbles per second) and the reaction is kept under argon during the entire procedure. The glass stopper is removed and dry DABCO (5.40 g, 48.1 mmol, 1.00 equiv) (Note 4) is added in one portion using a powder funnel (Note 5). The open joint is sealed with a rubber septum and dry deoxygenated toluene (30 mL) (Note 6) is added via syringe. The solution is stirred for 10 min until all DABCO has dissolved (Note 7) and the rubber septum is replaced by a 0–150 °C thermometer in a gas-tight adaptor. The dropping funnel is charged with 2 M

Org. Synth. **2010**, *87*, 104-114
Published on the Web 2/2/2010

trimethylaluminum in toluene (47.0 mL, 94.0 mmol, 1.95 equiv) (Note 8) using a syringe. The trimethylaluminum solution is then added to the DABCO solution over 25 min at room temperature (Note 9) causing the formation of a white precipitate after *ca.* 10 min. Residual trimethylaluminum in the dropping funnel is rinsed with additional dry deoxygenated toluene (3 mL) (Note 6). The solution is stirred for 30 min at room temperature before DABAL-Me$_3$ (**1**) is allowed to settle for ca. 10 min. The dropping funnel is replaced by a rubber septum (Note 10, 11) and the toluene is removed by cannular filtration (Note 12). The colorless DABAL-Me$_3$ (**1**) is washed by injection of dry deoxygenated diethyl ether (40 mL) (Note 13), brief stirring and removal of the solvent by cannular filtration. This procedure is repeated twice (2 x 40 mL of dry deoxygenated diethyl ether). The product is dried under high vacuum (0.05 mmHg) at room temperature to afford DABAL-Me$_3$ (**1**) (9.97 g, 38.9 mmol, 81%) (Notes 14, 15) as a colorless solid.

B. *Ethyl 4-methylbenzoate* (**2**). A 500-mL, two-necked, round-bottomed flask is equipped with an egg-shaped, teflon-coated, magnetic stir bar (*ca.* 25 mm), a rubber septum and a reflux condenser fitted at the top with a two-tap Schlenk adaptor connected to a bubbler and an argon manifold (Note 2, 4). The flow of argon is reduced to a modest level as judged by the exit bubbler (1-2 bubbles per second), the rubber septum is removed and solid Pd$_2$(dba)$_3$ (550 mg, 0.600 mmol, 1.50 mol%) (Note 16) and X-Phos (573 mg, 1.20 mmol, 3.00 mol%) (Note 17) are added. The joint is sealed again with the rubber septum and dry THF (200 mL) (Note 18) and ethyl 4-bromobenzoate (9.16 g, 40.0 mmol, 1.00 equiv) (Note 19) are added by syringe. The joint is opened again to add DABAL-Me$_3$ (**1**) (8.31 g, 32.4 mmol, 0.801 equiv) using a powder funnel. The open socket is sealed with a glass stopper and the dark purple reaction mixture is heated at reflux for 4 h in an oil bath (90 °C). Completion of the reaction can by confirmed by working up an aliquot (1 mL) of the reaction mixture, as described below, and recording its ^{1}H NMR spectrum (Note 20). The argon flow is slightly increased (3-4 bubbles per second), heating is stopped and the reaction mixture is allowed to cool to room temperature over an hour. The flask is further cooled in an ice bath for 15 min and the reaction is quenched by cautious, portionwise, addition of 2 M aqueous HCl (160 mL) (Note 21). The resulting biphasic mixture is transferred into a separatory funnel (1 L) with diethyl ether (2 x 100 mL) (Note 22). The layers are separated, retained separately and the aqueous fraction is re-extracted with diethyl ether

(3 x 150 mL). The organic layers are combined, dried over MgSO$_4$ (30 g) and filtered though a plug of activated charcoal (5 g) (Note 23) layered on the top of SiO$_2$ (50 g) (Note 24). The plug is further washed with diethyl ether (1 L). Removal of the solvent by rotatory evaporation (25 °C, 150 mmHg) affords the crude product as a clear brown oil in quantitative yield. Short-path distillation under reduced pressure (118–120 °C at 0.1 mmHg) gives the product **2** (6.11 g, 37.2 mmol, 93%) as a colorless oil (Note 25).

2. Notes

1. DABAL-Me$_3$ (**1**) is available from Aldrich (Catalog No. 682101) which may alternatively be used as received in part B. The preparation of **1** presented here requires the use of *pyrophoric* AlMe$_3$ solutions; the synthesis should be conducted in a fumehood.

2. The glassware was dried in a >120 °C oven overnight, assembled hot and a brisk flow of argon was applied until residual air had been swept out of the apparatus and through the dropping funnel (*ca.* 5 min).

3. A two-tap Schlenk adaptor connected to a bubbler and an argon/vacuum manifold is illustrated in Yu, J.; Truc. V.; Riebel, P.; Hierl, E.; Mudryk, B., *Org. Synth.* **2008**, *85*, 64-71.

4. 1,4-Diazabicyclo[2.2.2]octane (DABCO) was purchased from Aldrich (98% grade). This slightly hygroscopic amine was sublimed (60–90 °C at 0.1 mm Hg) in a Kugelrohr distillation apparatus by cooling the collecting flask with dry ice. The checkers stored the freshly sublimed DABCO in a desiccator over P$_2$O$_5$ overnight.

5. The flow of argon should be moderate to facilitate addition without blowing the solid out of the addition funnel.

6. The submitters purchased toluene (> 99%) from Fisher Scientific. The solvent was dried over sodium overnight and deoxygenated by aerating with argon for 10 min. Alternatively, toluene can be distilled from sodium-benzophenone under argon. The checkers purchased toluene (>99%, over 4 Å molecular sieves from Fluka. It was deoxygenated by aerating with argon for 15 min and used without further drying.

7. The submitters warmed the mixture at 40 °C until the DABCO was completely dissolved. If necessary, to facilitate final dissolution of DABCO, submitters reported the possibility to add an additional quantity of toluene (5 mL). The checkers did not observe any need to help the dissolution of DABCO; it completely dissolved upon stirring.

8. Trimethylaluminum solutions in toluene were purchased from Aldrich (2 M) and used as received. Such solutions should be treated as *pyrophoric* and transferred by syringe under an argon atmosphere. Re-use of a 20 mL syringe proved to be the most effective method.

9. The temperature remained at <30 °C during addition. The submitters added the trimethylaluminum solution over 15 min and observed a *ca.* 20-30 °C rise in the temperature of the reaction mixture and formation of a white precipitate after 5-10 min. If the internal temperature rises above *ca.* 60 °C, the submitters reported the possibility to control the temperature by using a water bath.

10. Occasionally, slight smoking at the end of the addition funnel could be seen but ceased readily. To maximize safety, the discarded dropping funnel was opened and placed at the back of the fumehood overnight.

11. The set-up should be left under a slight positive pressure of argon to avoid ingress of air.

12. The cannular filtration was conducted under an argon atmosphere using Teflon tubing (800 mm x 2 mm). One side of the tubing was covered with filter paper, secured with a Teflon band, and fitted to the reaction flask with a rubber septum. The cannular filtration is described in detail in ref. 2. To the collected solvents and washings was added dropwise *iso*-propanol (*ca.* 5 mL) before storing them at the rear of the fumehood overnight and later subjected to appropriate disposal.

13. The submitters purchased diethyl ether (>99%) from Sigma-Aldrich. The solvent was distilled from sodium-benzophenone under an argon atmosphere and collected by syringe. The checkers used diethyl ether (VWR, HPLC grade) dried and degassed using a Pure-Solve™ system.

14. The submitters obtained 9.20-10.0 g (74-81%) of DABAL-Me$_3$ (**1**). When carried out at half of the scale reported in the procedure, the checkers obtained 4.87 g (79%) of **1**.

15. Combustion of neat DABAL-Me$_3$ (**1**) can be induced by water, aqueous acids and other strong proton sources. It should be regarded as incompatible with strong oxidizing agents. Samples can be stored in air-tight containers under argon or nitrogen at room temperature (storage lifetime under these conditions is at least one year). The compound can be handled in air (15 min to 4 h, depending on the moisture content of the laboratory air). Longer exposure results in slow controlled decomposition to aluminum hydroxides/oxides. Unwanted samples of DABAL-Me$_3$ (**1**) can be disposed of by cautious, slow hydrolysis with alcohols or for small amounts even

ice/ice-water mixtures. The reagent should be considered as harmful by ingestion and strongly irritating to the eyes and mucous membranes. DABAL-Me$_3$ (**1**) has the following properties: mp 230 °C (dec., in air chars from 160 °C); it is soluble in THF, CH$_2$Cl$_2$, C$_6$H$_6$ and CHCl$_3$, but only sparingly soluble in Et$_2$O or in toluene at room temperature. Its physical properties are as follows: ^{1}H NMR (400 MHz, C$_6$D$_6$, 25 °C) δ: –0.64 (s, 18 H), 1.97 (s, 12 H) ppm; ^{13}C NMR (101 MHz, C$_6$D$_6$, 25 °C) δ: –9.8, 43.6 ppm. NMR samples need to be prepared in dry, non-protic solvents. Recrystallization of **1** from dry benzene under argon (crude **1** (1.00 g) dissolved in hot C$_6$H$_6$ (8-10 mL) provides 0.55-0.60 g of colorless rhomboidal crystals (unit cell parameters and X-ray structure identical to literature values[5]).

16. Pd$_2$(dba)$_3$ (dba = dibenzylideneacetone) was purchased from Alpha Aesar by the submitters and used as received. The checkers obtained Pd$_2$(dba)$_3$ from Sigma-Aldrich and used it without further purification. Literature preparations of this compound are available in ref. 3.

17. X-Phos was purchased from Alfa Aesar by the submitters and used as received. The checkers obtained X-Phos (97%) from Sigma-Aldrich.

18. The submitters obtained tetrahydrofuran (THF, >99%) from Sigma-Aldrich and distilled it from sodium-benzophenone under an argon atmosphere. The checkers used tetrahydrofuran (VWR, HPLC-grade) dried using a Pure-Solve^{TM} system.

19. Ethyl 4-bromobenzoate (Aldrich, 98%) was dried overnight using activated 4Å molecular sieves before use.

20. Complete consumption of ethyl 4-bromobenzoate is easily identified by the absence of the two apparent doublets of the phenylene group in its ^{1}H NMR spectrum. Its physical properties are as follows: ^{1}H-NMR (400 MHz, CDCl$_3$, 25 °C) δ: 1.42 (t, J = 7.1 Hz, 3 H), 4.40 (q, J = 7.1 Hz, 2 H), 7.60 (app. d, J = 8.5 Hz, 2 H), 7.92 (app. d, J = 8.5 Hz, 2 H) ppm. Alternatively, the submitters reported that GC analysis can be used to confirm completion of the reaction: 15 m factorFOUR column, column flow: He, 1.5 mL·min^{-1}, run isothermally at 120 °C, t_R = 2.75 min (**2**), t_R = 4.95 min (ethyl 4-bromobenzoate).

21. The quenching procedure releases methane. Care should be taken to add the acid at such rate that excessive foaming is avoided. If the flask is re-stoppered the rate of the hydrolysis reaction can be judged by the rate of out-gassing from the exit bubbler. Typically, addition of the 2 M HCl over 10-15 min is required.

22. The submitters purchased diethyl ether (>99%) from Sigma-Aldrich and used it as supplied. The checkers obtained diethyl ether (>99%) from J. T. Baker and used it as received.

23. The checkers purchased activated charcoal from Fluka.

24. Silica gel 60 (220 240 mesh) supplied by Fluka was used as received. The SiO_2 was slurried up in diethyl ether, before layering it with activated charcoal and filtering the organic phase.

25. The submitters obtained 5.98-6.31 g (91-96%) of **2**. When carried out at half of the scale reported in the procedure the checkers obtained 2.95 g (90%) of **2**. The compound has literature properties[4] and is stable indefinitely. Its physical properties are as follows: ^{1}H NMR (400 MHz, CDCl$_3$, 25 °C) δ: 1.38 (t, J = 7.1 Hz, 3 H), 2.40 (s, 3 H), 4.36 (q, J = 7.1 Hz, 2 H), 7.22 (app. d, J = 8.3 Hz, 2 H), 7.94 (app. d, J = 8.3 Hz, 2 H); ^{13}C NMR (101 MHz, CDCl$_3$, 25 °C) δ: 14.4, 21.7, 60.8, 127.9, 129.1, 129.7, 143.5, 166.8; Anal. calcd. for $C_{10}H_{12}O_2$: C, 73.15%; H, 7.37%; found: 72.90%; H, 7.38%. The purity of the compound was checked by GC analysis; Rtx-1701 column (30 m x 0.25 mm x 0.25 μm), 60 kPa He, 100 °C, 2 min isotherm, 7 °C·min^{-1}, 250 °C, 10 min isotherm, t_R = 13.6 min (**2**), t_R = 16.5 (ethyl 4-bromobenzoate).

Safety and Waste Disposal Information

All hazardous materials should be handled and disposed of in accordance with "Prudent Practices in the Laboratory": National Academy Press: Washington. DC. 1995.

3. Discussion

The abnormally high air stability of the DABCO *bis*-AlMe$_3$ adduct (**1**) was first noted by Bradley in the early 1990s.[5] However, synthetic applications of the material only recently became popular – especially since the commercialization of this reagent. Originally prepared in Et$_2$O, the combination of pyrophoric trimethylaluminium and a very low flash point of ether is less desirable on large scales for safety reasons, even though somewhat higher yields can be realized. The toluene-based procedure described here is effective and safe at scales up to at least 10 g. In combination with Pd$_2$(dba)$_3$ and Buchwald's X-Phos,[6] DABAL-Me$_3$ (**1**) is effective for the methylation of aryl halides ArX (X = Cl, Br, I) and triflates

(X = OTf).[7] A wide range of other functional groups is tolerated under these reaction conditions (F, CF$_3$, vinyl, CO$_2$R, CN, CHO, OTs, NO$_2$, OMe, and OAc) (Table 1). Complete conversion is attained within 4 h at THF reflux (with 0.8 equiv of DABAL-Me$_3$), avoiding the often problematic post-reaction chromatographic separation of residual halide starting materials.

Chemoselective methylation (>95% selectivity) of C-Br over C-Cl bonds can be attained through use of Cy-JohnPhos.[8] Aside from C(sp^2)-X (X = Cl, Br, I, OTf), benzylic halides can be used in Pd^0/X-Phos couplings (e.g., PhCH$_2$Br is methylated in 72% yield). One limitation of the methylation procedure is its present inability to tolerate the methylation of electron deficient heterocycles (e.g. halopyridines and isoquinolines; entries 34-35, Table 1). The low yields of methylated products isolated in these cases are believed to be due to competing Chichibabin-type processes. Additionally, enolizable substrates capable of self-condensation are not always tolerated (entry 26, Table 1). The use of 0.8 equiv of DABAL-Me$_3$ per coupled C-X, together with Pd^0/X-Phos makes for a highly robust procedure. In small-scale reactions (0.25 mmol) the procedure can be carried out in undried THF under standard reflux conditions in air – the DABAL-Me$_3$ (1) acting simultaneously as a drying agent. However, the slower rates of heating in large-scale reactions (such as that presented here) slows the solvent drying/deoxygenation process too much, leading to non-reproducibility. Couplings with DABAL-Et$_3$ allow ethylation procedures in >85% yield without any issues associated with β-elimination. The ethyl analogue is not stable in air and is best prepared in situ from DABCO and AlEt$_3$.

In addition to the cross-coupling procedure illustrated here, DABAL-Me$_3$ (1) has found application in a wide range of transformations including: 1,2-additions to aldehydes[9] and enones,[10] 1,4-additions to enones,[11] methylation of allylic electrophiles[12] and direct conversion of esters to amides.[13]

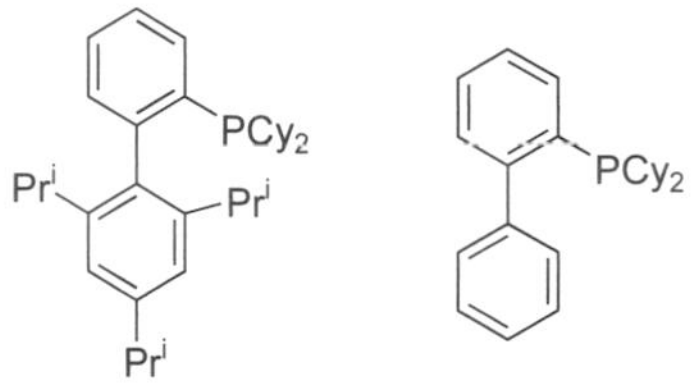

X-Phos   Cy-JohnPhos

Table 1. Methylation of aryl and vinyl halides and pseudohalides with DABAL-Me$_3$ **1**.[a]

Entry	Substrate	Product	Yield [%][b]
1	C$_6$H$_5$Br	C$_6$H$_5$Me	>99
2	C$_6$H$_5$OTf	C$_6$H$_5$Me	>99
3	4-MeC$_6$H$_4$Br	1,4-Me$_2$C$_6$H$_4$	>99
4	4-MeC$_6$H$_4$Cl	1,4-Me$_2$C$_6$H$_4$	>99
5	4-FC$_6$H$_4$Br	4-FC$_6$H$_4$Me	>99
6	4-ClC$_6$H$_4$Br	4-ClC$_6$H$_4$Me	96[c]
7	4-(CF$_3$)C$_6$H$_4$Br	4-(CF$_3$)C$_6$H$_4$Me	>99
8	4-tBuC$_6$H$_4$Br	4-tBuC$_6$H$_4$Me	96
9	4-NCC$_6$H$_4$Br	4-NCC$_6$H$_4$Me	95
10	4-NCC$_6$H$_4$Cl	4-NCC$_6$H$_4$Me	95
11	3-NCC$_6$H$_4$Br	3-NCC$_6$H$_4$Me	95
12	2-NCC$_6$H$_4$Br	2-NCC$_6$H$_4$Me	96
13	4-(H$_2$C=CH)C$_6$H$_4$Br	4-(H$_2$C=CH)C$_6$H$_4$Me	>99
14	4-(H$_2$C=CH)C$_6$H$_4$Cl	4-(H$_2$C=CH)C$_6$H$_4$Me	98
15	4-(MeO)C$_6$H$_4$Br	4-(MeO)C$_6$H$_4$Me	>99
16	4-(MeO)C$_6$H$_4$Cl	4-(MeO)C$_6$H$_4$Me	>99
17	4-(MeO)C$_6$H$_4$OTf	4-(MeO)C$_6$H$_4$Me	>99
18	2-(MeO)C$_6$H$_4$Br	2-(MeO)C$_6$H$_4$Me	>99
19	4-(EtO$_2$C)C$_6$H$_4$Br	4-(EtO$_2$C)C$_6$H$_4$Me	99
20	4-(EtO$_2$C)C$_6$H$_4$Cl	4-(EtO$_2$C)C$_6$H$_4$Me	98
21	4-HOC$_6$H$_4$Br	4-HOC$_6$H$_4$Me	94[d]
22	4-(O$_2$N)C$_6$H$_4$Br	4-(O$_2$N)C$_6$N$_4$Me	76
23	4-(O$_2$N)C$_6$H$_4$Cl	4-(O$_2$N)C$_6$N$_4$Me	81
24	4-(O$_2$N)C$_6$H$_4$OTf	4-(O$_2$N)C$_6$N$_4$Me	59
25	4-(HOCH$_2$)C$_6$H$_4$Br	4-(HOCH$_2$)C$_6$H$_4$Me	79
26	4-(MeOC)C$_6$H$_4$Br	4-(MeOC)C$_6$H$_4$Me	0[e]
27	4-(CHO)C$_6$H$_4$Br	4-(CHO)C$_6$H$_4$Me	88[f]
28	C$_6$H$_5$CH$_2$Br	C$_6$H$_5$CH$_2$Me	72
29	1-C$_{10}$H$_7$OTf	1-C$_{10}$H$_7$Me	>99
30	2-C$_{10}$H$_7$OTf	2-C$_{10}$H$_7$Me	98
31	1-C$_{10}$H$_7$Cl	1-C$_{10}$H$_7$Me	90
32	2-C$_{10}$H$_7$Cl	2-C$_{10}$H$_7$Me	98

Table 1. (continued)

Entry	Substrate	Product	Yield [%][b]
33	Br, thiophene	Me, thiophene	90
34	Br, pyridine	Me, pyridine	16
35	Br, isoquinoline	Me, isoquinoline	59
36	OTf, cyclohexenyl	Me, cyclohexenyl	>99
37	OTf, CO$_2$Et, cyclopentenyl	Me, CO$_2$Et, cyclopentenyl	98

[a] Reactions performed on a 0.25 mmol scale using 1.5mol % Pd$_2$(dba)$_3$, 3 mol % X-Phos, 0.8 equiv. DABAL-Me$_3$ (1), THF, N$_2$, 80 °C, 4 h. In all cases quantitative conversions were attained. [b] Yields determined by GC *vs.* internal standard. In reactions run at larger scale the isolated yields were directly comparable and typically within 5% of the GC yields. [c] 1.0 equivalents of DABAL-Me$_3$ 1 used; using Cy-JohnPhos, <2% dimethylation observed. [d] 1.6 equivalents of DABAL-Me$_3$ 1 used. [e] Major products are self-aldol derived. [f] 0.5 equivalents of DABAL-Me$_3$ 1 used.

1. School of Chemistry, The University of Nottingham, NG7 2RD United Kingdom; e-mail simon.woodward@nottingham.ac.uk
2. Bennett, B. K.; Richmond, T. G. *J. Chem. Ed.* **1998**, *75*, 1034.
3. Milani, B.; Anzilutti, A.; Vicentini, L.; o Santi, A. S.; Zangrando, E.; Geremia, S.; Mestroni, G. *Organometallics* **1997**, *16*, 5064-5075.
4. Minami, T., Nishimura, K.; Hirao, I.; Suganuma, H.; Agawa, T. *J. Org. Chem.* **1982**, *47*, 2360-2363.
5. Bradford, A. M.; Bradley, D. C.; Hursthouse, M. B.; Motevalli, M. *Organometallics* **1992**, *11*, 111-115.
6. Huang, X.; Anderson, K. W.; Zim, D.; Jiang, L.; Klapars, A.; Buchwald, S. L. *J. Am. Chem. Soc.* **2003**, *125*, 6653-6655.

7. Cooper, T.; Novak, A.; Humphreys, L. D.; Walker, M. D.; Woodward, S. *Adv. Synth. Catal.* **2006**, *348*, 686-690.

8. Wolfe, J. P.; Singer, R. A.; Yang, B. H.; Buchwald, S. L. *J. Am. Chem. Soc.* **1999**, *121*, 9550-9561. Cy-JohnPhos (247940-06-3) is also widely commercially available in research quantities.

9. (a) Biswas, K.; Prieto, O.; Goldsmith, P. J.; Woodward S. *Angew. Chem., Int. Ed.* **2005**, *44*, 2232-2234. (b) Mata, Y.; Diéguez, M.; Pàmies, O.; Woodward, S. *J. Org. Chem.* **2006**, *71*, 8159-8165. (c) Biswas, K.; Chapron, A.; Cooper, T.; Fraser, P. K.; Novak, A.; Prieto, O.; Woodward, S. *Pure Appl. Chem.* **2006**, *78*, 511-518.

10. Siewert, J.; Sandmann, R.; von Zezschqitz, P. *Angew. Chem., Int. Ed.* **2007**, *46*, 7122-7124.

11. Alexakis, A.; Albrow, V.; Biswas, K.; d'Augustin, M.; Prieto, O.; Woodward, S. *Chem. Commun.* **2005**, 2843-2845.

12. (a) Novak, A.; Fryatt, R.; Woodward. S. *C. R. Chimie* **2007**, *10*, 206-212. (b) Novak, A.; Calhorda, M. J.; Costa, P. J.; Woodward, S. *Eur. J. Org. Chem.* **2009**, 898-903.

13. (a) Novak, A.; Humphreys, L. D.; Walker, M. D.; Woodward, S. *Tetrahedron Lett.* **2006**, *47*, 5767-5769. (b) Glynn, D.; Bernier, D.; Woodward, S. *Tetrahedron Lett.* **2008**, *49*, 5687-5688.

Appendix
Chemical Abstracts Nomenclature; (Registry Number)

Trimethylaluminium: (75-24-1)

DABCO: 1,4-Diazabicyclo[2.2.2]octane; (280-57-9)

Pd_2(dba)$_3$: tris(dibenzylideneacetone)dipalladium(0); (51364-51-3)

X-Phos: 2-di-cyclo-hexylphosphino-2',4',6'-triisopropylbiphenyl; (564483-18-7)

Ethyl 4-bromobenzoate: (5798-75-4)

Simon Woodward undertook both his undergraduate and Ph.D. at the University of Sheffield, the latter in the group of Dr. Mark J. Winter. After a period as a Fulbright Scholar in the group of Prof. M. David Curtis at the University of Michigan he was a postdoctoral researcher with Dr. John M. Brown, FRS at the University of Oxford.   Initially appointed to a Lectureship in Organometallic and Catalytic Chemistry at The University of Hull, he moved to the University of Nottingham in 1999. His research group is interested in chemo, regio and stereoselective catalysis of organic reactions, under convenient/simple conditions.

Andrej Vinogradov studied Chemistry at the University of Münster. He did his Ph.D. at the Inorganic Chemistry Department of the University of Münster under supervision of Prof. Dr. Werner Uhl. Since 2009 he is a postdoctoral research fellow at the University of Nottingham with Prof. Simon Woodward. Dr. Vinogradov's research interests centre round the preparation and use of organoalanes.

Denise Rageot was born in 1985 in Basel, Switzerland. She studied Chemistry at the University of Basel, where she obtained her M.S. in 2008 under the supervision of Prof. Andreas Pfaltz. She began her Ph. D. work in summer 2008 in the group of Prof. Andreas Pfaltz, where she is currently working on the synthesis of new chiral ligands for asymmetric metal catalysis.

SYNTHESIS OF DIMETHYL 2-PHENYLCYCLOPROPANE-1,1-DICARBOXYLATE USING AN IODONIUM YLIDE DERIVED FROM DIMETHYL MALONATE

A.

B.

Submitted by Sébastien R. Goudreau, David Marcoux and André B. Charette.[1]

Checked by David Hughes.

1. Procedure

A. Bis(methoxycarbonyl)(phenyliodinio)methanide (1). A 1-L, 3-necked round-bottomed flask (Note 1) equipped with a 3-cm oval Teflon-coated magnetic stir bar (Note 2) is fitted with a gas inlet adapter connected to a nitrogen line and a gas bubbler, and a septum through which is inserted a thermocouple probe (Note 3). The third neck is fitted with a ground glass stopper, which is removed for solvent and solids additions. To the flask is added potassium hydroxide (30.2 g, 85%, 0.45 mol, 5.0 equiv) (Note 4) and acetonitrile (300 mL) (Note 5). The solution is cooled to 5 °C with an ice/water bath and then dimethyl malonate (11.9 g, 90.0 mmol, 1.00 equiv) is added via a weighed 20-mL syringe over 5 min, forming a viscous slurry (Note 2). PhI(OAc)$_2$ (32.0 g, 99.3 mmol, 1.10 equiv) is added in one portion. The viscous heterogeneous mixture is stirred vigorously for 2 h at 0 to 5 °C and gradually becomes a viscous creamy suspension (Note 6). Water (150 mL) is added in one portion, resulting in a temperature rise to 10 °C. The mixture is stirred for 2 min at this temperature (Note 7), then filtered through a 350-mL medium porosity sintered glass funnel (Note 8). The flask is rinsed with room temperature water (100 mL), and this rinse is used to slurry wash the filter cake (Note 9). The cake is washed a second time with water (100 mL) (Note 10). The white solid is next slurry-washed with Et$_2$O (150 mL). The solids are dried to constant weight under vacuum (20 mmHg) at 30 °C for 25 h to afford bis(methoxycarbonyl)(phenyliodinio)methanide **1** (23.9 g, 79 %) (Note 11) as an off-white crystalline solid (Notes 12, 13, and 14).

   B. Dimethyl 2-phenylcyclopropane-1,1-dicarboxylate (2). A 250-mL 3-necked round-bottomed flask equipped with a 2-cm oval Teflon-coated magnetic stir bar is fitted with a gas inlet adapter connected to a nitrogen line and a gas bubbler, and a septum through which is inserted a thermocouple probe (Note 3). The third neck is fitted with a ground glass stopper, which is removed for solvent and solids additions. To the flask is added bis[rhodium(α,α,α′,α′-tetramethyl-1,3-benzenedipropionic acid)] (5 mg, 0.007 mmol, 0.02 mol %) (Note 15), dichloromethane (80 mL) and styrene (3.90 g, 37.4 mmol, 1.00 equiv), and the stirred solution is cooled to 2 °C using an ice/water bath. Iodonium ylide **1** (14.7 g, 43.9 mmol, 1.2 equiv) is added in four portions (3.5 – 4.0 g each) at 5 min intervals (Notes 16 and 17). After the last addition, the ice/water bath is removed. The mixture is allowed to warm to room temperature and is stirred for 1 h at 18–20 °C (Note 18). Aqueous 10% thiourea (50 mL) is added in one portion and the biphasic solution is stirred for 15 min at 20 °C. The mixture is transferred to a 250-mL separatory funnel along with a dichloromethane rinse (15 mL) of the flask. The layers are separated and the aqueous layer is extracted with dichloromethane (2 × 20 mL) (Note 19). The organic layers are combined and vacuum filtered through a bed of Na$_2$SO$_4$ (20 g) on top of a Celite$^©$ cake (20 g) in a 150-mL sintered-glass funnel. The cake is washed with dichloromethane (2 x 50 mL). The filtrate is concentrated by rotary evaporation (40 °C bath, 100 mmHg initial, lowered to 20 mmHg) to afford 18 g of a yellow, oily mixture of cyclopropane **2** and iodobenzene. This oil is purified by chromatography on SiO$_2$ (Note 20) to afford dimethyl 2-phenylcyclopropane-1,1-dicarboxylate (**2**) (8.07–8.34 g, 92–95% yield) as a yellowish oil (Notes 21 and 22).

2. Notes

   1.   No special exclusion of water is needed.
   2.   The procedure was checked using both magnetic stirring and mechanical stirring. The reaction mixture is viscous and poorly mixed using magnetic stirring at 700 rpm, but no difference in yield was obtained with

116

the two methods of agitation. Adding more solvent for better stirring proved to be detrimental for the yield.

3. The internal temperature is monitored using a J-Kem Gemini digital thermometer with a Teflon-coated T-Type thermocouple probe (12-inch length, 1/8 inch outer diameter, temperature range –200 to +250 °C).

4. The submitters used potassium hydroxide pellets (88% purity) purchased from American Chemicals Ltd. and used as received. The checkers used powdered KOH (85%) obtained from Sigma Aldrich. An excess of potassium hydroxide was optimal for both yield and reproducibility. The checker obtained a 69% yield at this scale using 2.5 equiv KOH instead of 5 equiv.

5. The following reagents and solvents in Step A were obtained from Sigma-Aldrich and used without further purification: acetonitrile (Chromatosolv, 99.9%), dimethyl malonate (98%), and diethyl ether (ACS reagent, anhydrous, BHT-inhibited). Iodosobenzene diacetate (98%) was purchased from Acros and used as received. Deionized tap water was used throughout.

6. Reaction time of 2 h provided the optimal yield. The reaction was not monitored.

7. Vigorous stirring for 2 min is sufficient to dissolve all remaining potassium hydroxide.

8. The filtration took about 10 min.

9. The cake is slurry-washed as follows: 1) the vacuum line is disconnected from the filter flask, 2) water (100 mL) is added to the cake, 3) the cake is mixed for approx. 20 s with a spatula to ensure full contact of all solids with water, 4) the vacuum line is reconnected to the filter flask to remove the cake wash. It is important that the solvent wash be completely removed between each wash.

10. The second cake wash is carried out as a displacement wash, with the vacuum remaining on and no mixing of the cake.

11. After drying to constant weight, the solids contained approx 2 wt% water based on ^{1}H NMR analysis in dry DMSO-d_6. Corrected yield is 77%.

12. Bis(methoxycarbonyl)(phenyliodinio)methanide **1** has the following physical and spectroscopic data: mp 100–104 °C (softens to gum; full melt at 120 °C); IR (neat) 3081, 2982, 2948, 2899, 1668, 1585, 1575, 1562, 1435, 1425, 1320, 1060, 990 cm^{-1}; ^{1}H NMR (400 MHz, DMSO-d_6) δ: 3.75 (s, 6 H), 7.40–7.44 (m, 2 H), 7.52–7.56 (m, 1 H), 7.74–7.76 (m, 2 H);

^{13}C NMR (100 MHz, DMSO-d$_6$) δ: 50.7, 58.0 (broad), 116.0, 130.5, 131.1, 131.2, 166.0.

13. The ylide **1** is poorly soluble and unstable in organic solvents. DMSO-d$_6$ was the optimum solvent for NMR analysis. Samples analyzed within 10 min of dissolution in this solvent indicated product purity >98%. The resonance of the C(2) carbon is broad (δ: 58 ppm) and requires acquisition for several hours to obtain an observable signal. The resonance of this carbon is similar to that reported for the phenyliodonium ylide derived from methyl acetoacetate (68 ppm).[2] In solution, the product decomposes to iodobenzene and dimer **3** (Note 14).

14. The product showed no degradation based on NMR analysis after drying for 25 h at 30 °C. As a dry solid, the product degrades slowly at room temperature (over 2 weeks) to form mainly dimer **3** and iodobenzene, but it can be stored at −20 °C for months without significant degradation. The submitters observed rapid degradation of the ylide when not properly dried. However, its degradation never led to an explosion. Indeed, the exothermic degradation can be described as a rapid "melting" of the ylide to produce a hot iodobenzene solution, fumes, and black solids. Care should be taken when an oily solid is obtained, meaning the workup/drying process was not correctly done. The checkers ran the procedure 3 times as written or with slight variations with no issues with decomposition.

3

Dimer **3** was not isolated and characterized, but ^{1}H and ^{13}C NMR spectral data are consistent with reported literature NMR data.[3]

15. The following reagents and solvents used in Step B were obtained from Sigma-Aldrich and used without further purification: bis[rhodium(α,α,α′,α′-tetramethyl-1,3-benzenedipropionic acid)] (abbr. as Rh$_2$(esp)$_2$), dichloromethane (ACS reagent, 99.5%), thiourea, Celite 545, silica gel (230-400 mesh, 60 Å), ethyl acetate (ACS reagent, >99.5%), and hexanes (ACS reagent, >98.5%). Styrene (99.5%) was purchased from Acros and used as received. Sodium sulfate was obtained from Fisher Scientific.

118

16. After each addition the mixture warms to 9–11 °C within one minute, then is cooled to <5 °C prior to the next addition of the ylide.

17. The mixture is heterogeneous after addition of the ylide and is stirred at 500 rpm.

18. The submitters followed the progress of the reaction by GC analysis (Agilent 6890 Series GC, Column: HP-5HS30m x 0.25 mm, Agilent 5973EI mass detector, temperature of the injector: 250 °C, temperature of the detector: 250 °C) using a 5 °C/min gradient from 40 to 270 °C and 63 psi H_2. Styrene has a t_r = 5.48 min, iodobenzene has a t_r = 7.73 min, cyclopropane 2 has t_r = 15.9 min. The checker followed the reaction by ^{1}H NMR as follows. A drop of the reaction mixture was added to 1 mL of $CDCl_3$ and analyzed by ^{1}H NMR. The styrene resonances at 5.8 and 6.8 ppm were integrated relative to the cyclopropane resonances at 1.8 and 2.2 ppm of the product to assess conversion. If more than 2% styrene remained, an additional amount of the ylide was added at room temperature to drive the reaction to completion.

19. The solids present in these extractions are kept with the aq. phase.

20. A 5 cm glass column is wet-packed (hexanes) with SiO_2 (170 g) topped with 0.5 cm sand. The crude reaction product 2 is loaded neat on the column and eluted as follows: hexanes (200 mL), 10 % EtOAc in hexanes (500 mL), 20% EtOAc in hexanes (800 mL). The first 5 fractions are collected as 100 mL fractions, then the remaining as 50 mL fractions. TLC (UV visualization) is used to follow the chromatography. The Rf value of the title compound is 0.5 (20% EtOAc/hexanes). Fractions 13-19 are concentrated by rotary evaporation (40 °C bath, 20 mmHg), then vacuum dried (20 mmHg) at 22 °C for 20 h to constant weight (8.07–8.34 g). The product contains <0.2 wt % EtOAc by ^{1}H NMR analysis.

21. An analytically pure sample is prepared by crystallization of 2 as follows. Compound 2 (600 mg, isolated by chromatography as described in Note 20) is dissolved in MeOH (5 mL) at 22 °C, transferred to a 10-mL syringe, then filtered through a 0.45 micron PTFE syringe filter into a 25-mL Erlenmeyer flask. Water (1.5 mL) is added and the solution is swirled to mix. The flask is capped and held in a –15 °C freezer for 2 days, affording crystals. The crystalline solid is vacuum filtered using a 15-mL medium porosity sintered glass funnel, then dried for 15 h at 22 °C in a vacuum oven (20 mmHg) to constant weight (420 mg, 70% yield).

22. Dimethyl 2-phenylcyclopropane-1,1-dicarboxylate (2) has the following physical and spectroscopic properties: mp 40–41°C; IR (neat)

2953, 1726, 1436, 1332, 1277, 1217, 1130 cm^{-1}; ^{1}H NMR (400 MHz, CDCl$_3$) δ: 1.76 (dd, J = 9.2, 5.3 Hz, 1 H), 2.22 (dd, J = 8.1, 5.3 Hz, 1 H), 3.25 (t, J = 8.7 Hz, 1 H), 3.37 (s, 3 H), 3.81 (s, 3 H), 7.20–7.30 (m, 5 H); ^{13}C NMR (100 MHz, CDCl$_3$) δ: 19.3, 32.7, 37.4, 52.4, 53.0, 127.6, 128.3, 128.6, 134.8, 167.2, 170.4; HRMS (EI) m/z calcd. for C$_{13}$H$_{14}$O$_4$ ([M]$^+$) 234.09; found 234.20; Anal. calcd for C$_{13}$H$_{14}$O$_4$: C, 66.66; H, 6.02; found: C, 66.67; H, 5.88.

Safety and Waste Disposal Information

All hazardous materials should be handled and disposed of in accordance with "Prudent Practices in the Laboratory"; National Academy Press; Washington, DC, 1995.

3. Discussion

Activated cyclopropanes represent an important class of synthons, often acting as electrophiles or involved in cycloaddition reactions. The most straightforward and convergent approach to these cyclopropanes remains the reaction of olefins with metal carbenes, which are derived from the decomposition of diazo compounds or iodonium ylides (Scheme 1).[4,5] In the case of the synthesis of 1,1-cyclopropane diesters, the decomposition of diazomalonate derivatives is well known and has been proven to be efficient. González-Bobes and co-workers have developed a very efficient method to synthesize these cyclopropanes[6] using 0.1 mol% of Rh$_2$(esp)$_2$, 1.3 equivalent of the diazomalonate and 1.0 equivalent of the alkene to achieve good to excellent yields. The reaction proceeds smoothly with mono-substituted aliphatic or aromatic alkenes and good reactivity was observed with disubstituted alkenes. Lower yields were obtained with styrene derivatives bearing strong electron-withdrawing groups.

Although dimethyl diazomalonate is a relatively stable compound, it is both explosive and shock sensitive. Moreover, its synthesis requires at least one equivalent of azide, which is also explosive.[4a] This presents a challenge when reactions are performed on large scale. Nevertheless, these problems can be overcome through the use of iodonium ylides.[7] Indeed, differential scanning calorimetry (DSC) analyses indicate that the iodonium ylide derived from dimethyl malonate is safer than the corresponding diazo

120

compound. Moreover, it is readily accessible in one step from iodosobenzene diacetate, which itself is a stable white solid.[8]

Previous methods to synthesize this compound reported by Müller[9] and Hadjiarapoglou[2] described the treatment dimethyl malonate with iodosobenzene diacetate in presence of an excess of potassium hydroxide in methanol. The product was extracted with a large amount of dichloromethane and concentrated under reduced pressure to yield an off-white solid that was purified by dissolving it in a minimum amount of chloroform followed by precipitation with hexane. However, when the reaction is performed on larger scale, reduced yields are observed. We determined that the lower yield was due to the workup and the removal of the solvent. Once the product crystallizes, it becomes very insoluble and dissolving the solid leads to degradation. Furthermore, a large amount of dichloromethane is needed during the workup and this leads to a longer concentration time. Since the product is not very stable in solution, the excess time and the temperature needed on larger scale to remove the methanol and the dichloromethane results in a larger percentage of degradation, hence lower yields.

The synthesis of the iodonium ylide of dimethyl malonate herein described has many advantages. First, the isolated yield is higher (77–79%) compared to the previous methods (56% and 61% for Müller and Hadjiarapoglou respectively), and no extraction/concentration steps are needed. Second, the product is obtained from a simple filtration and without need for further purification. Finally, it is suitable for large scale synthesis.

As discussed above, the cyclopropanation reaction using 1.1 to 1.2 equivalent of iodonium ylide derived from dimethyl malonate and 1.0 equivalent of styrene gives an excellent yield (92–95%). The low catalyst loading required (0.02 mol% of $Rh_2(esp)_2$) makes this process a very efficient and inexpensive way to prepare dimethyl 2-phenylcyclopropane-1,1-dicarboxylate. With other alkenes, similar efficiency (0.1 mol% of $Rh_2(esp)_2$) was obtained using the González-Bobes conditions while substituting the diazomalonate for the iodonium ylide (Table 1).[6]

A drawback of using iodonium ylides compare to diazo compounds is the stoichiometric formation of iodobenzene byproduct instead of molecular nitrogen. However, the volatility and low polarity of iodobenzene compared to 1,1-cyclopropane diesters permits facile separation by evaporation or by chromatography on silica gel. Consequently, we have not encountered problems in its removal from the desired cyclopropanes.

Table 1. Scope of the Cyclopropanation Using Rh$_2$(esp)$_2$ and Iodonium Ylide 1

entry	product	yield (%)[a]
1	**3a**	78
2	**3b**	95
3	**3c**	95
4	**3d**	94
5	**3e**	93
6	**3f**	69
7[b]	**3g**	84

[a] Isolated yields. [b] Reaction time = 16 h.

1. Department of Chemistry, Université de Montréal, P.O. Box 6128, Station Downtown, Montréal, Québec, Canada H3C 3J7, andre.charette@umontreal.ca

2. Batsila, C.; Kostakis, G.; Hadjiarapoglou, L. P. *Tetrahedron Lett.* **2002**, *43*, 5997–6000.

3. (a) Barfield, M.; Gotoh, T.; Hall, H. K. *Mag. Res. Chem.* **1985**, *23*, 705–709; (b) Hall, H. K.; Daly, R. C. *Macromol.* **1975**, *8*, 22–31.

4. (a) Doyle, M. P.; McKervey, M. A.; Ye, T. *Modern Catalytic Methods for Organic Synthesis with Diazo Compounds: From Cyclopropanes to Ylides*; Wiley: New York, 1998. (b) Pellissier, H. *Tetrahedron* **2008**, *64*, 7041–7095. (c) Lebel, H.; Marcoux, J.-F.; Molinaro, C.; Charette, A. B. *Chem. Rev.* **2003**, *103*, 977–1050.

5. For examples of cyclopropanations using iodonium ylides, see: (a) Bonge, H. T.; Hansen, T. *Synlett* **2007**, 55–58. (b) Müller, P; Allenbach, Y. F.; Chappelet, S.; Ghanem, A. *Synthesis* **2006**, 1689–1696. (c) Moreau, B.; Charette, A. B. *J. Am. Chem. Soc.* **2005**, *127*, 18014–18015. (d) Wurz, R. P.; Charette, A. B. *Org. Lett.* **2005**, *7*, 2313–2316. (e) Ghanem, A.; Lacrampe, F.; Schuring, V. *Helv. Chim. Acta* **2005**, *88*, 216–239. (f) Ghanem, A.; Müller, P.; *Chirality* **2005**, *17*, 44–50. (g) Müller, P.; Ghanem, A. *Org. Lett.* **2004**, *6*, 4347–4350. (h) Müller, P. *Acc. Chem. Res.* **2004**, *37*, 243–251. (i) Müller, P.; Allenbach, Y.; Robert, E. *Tetrahedron: Asymmetry* **2003**, *14*, 779–785. (j) Müller, P.; Ghanem, A. *Synlett* **2003**, *12*, 1830–1833. (k) Wurz, R. P.; Charette, A. B. *Org. Lett.* **2003**, *5*, 2327–2329. (l) Müller, P.; Bolea, C. *Helv. Chim. Acta* **2001**, *84*, 1093–1111. (m) Georgakopoulou, G.; Kalogiros, C.; Hadjiarapoglou, L. P. *Synlett* **2001**, 1843–1846. (n) Dauban, P.; Saniere, L.; Tarrade, A.; Dodd, R. H. *J. Am. Chem. Soc.* **2001**, *123*, 7707–7708. (o) Müller, P.; Fernández, D. *Helv. Chim. Acta* **1995**, *78*, 947–958. (p) Moriarty, R. M.; Vaid, R. K. *Synthesis* **1990**, 431–447. and references cited therein.

6. González-Bobes, F.; Fenster, M. D. B.; Kiau, S.; Kolla, L.; Kolotuchin, S.; Soumeillant, M. *Adv. Synth. Catal.* **2008**, *350*, 813–816.

7. Zhdankin, V. V.; Stang, P. J. *Chem. Rev.* **2008**, *108*, 5299–5358.

8. Goudreau, S. R.; Marcoux, D.; Charette A. B. *J. Org. Chem.* **2009**, *74*, 470–473. Iodonium ylide **1** exhibits 4 exothermic peaks by DSC: (1) 76.5–103.3 °C, –41.4 J/g; (2) 106.3–151.6 °C, –329 J/g, (3) 276.8–339.5 °C, –151.5 J/g; (4) 369.5–400 °C, –169.4 J/g. The corresponding

dimethyl 2-diazomalonate has 2 large DSC exotherms: (1) 103.3–
211.7 °C, –719.7 J/g; (2) 215–300 °C, –220.1 J/g.
9. Müller, P.; Fernández, D. *Helv. Chim. Acta* **1995**, *78*, 947–958.

Appendix
Chemical Abstracts Nomenclature; (Registry Number)

Potassium hydroxide; (1310-58-3)
Dimethyl malonate; (108-59-8)
Iodosobenzene diacetate; (3240-34-4)
Bis[rhodium($\alpha,\alpha,\alpha',\alpha'$-tetramethyl-1,3-benzenedipropionic acid)]; (819050-89-0)
Styrene; (100-42-5)
Thiourea; (62-56-6)

André B. Charette was born in 1961 in Montréal, Quebec. Upon completion of his B.Sc. from Université de Montréal in 1983, he pursued his graduate studies at the University of Rochester, earning his M.Sc. (1985) and Ph.D. (1987) with Robert Boeckman Jr. Following NSERC postdoctoral fellowship at Harvard University with D. A. Evans, he began his academic career at Université Laval in 1989. In 1992, he returned to his alma mater, where he is today Full Professor and holder of an NSERC/Merck Frosst/Boehringer Ingelheim Industrial Chair and a Canada Research Chair. His research focuses on the development of new methods for the stereoselective synthesis of organic compounds. Recent honors include a Cope Scholar Award (2007), the Prix Marie-Victorin (2008) and the Alfred Bader Award (2009).

124

Sébastien R. Goudreau was born in 1981 in Sherbrooke (Canada). He completed his undergraduate degree in chemistry at the Université de Sherbrooke in 2005. Following these studies he then joined the Charette group at the Université de Montréal where he is currently completing his Ph.D. studies as a NSERC postgraduate fellow. His Ph.D. research focuses on the development of new enantioselective methods to access activated cyclopropanes, and the application of these cyclopropanes in synthesis.

David Marcoux received his B.Sc. degree in chemistry at the Université du Québec à Trois-Rivières (Canada) in 2005. He then moved to the Université de Montréal as a NSERC predoctoral fellow where he received his Ph.D. degree in chemistry in 2009 under the mentorship of Professor André B. Charette. His graduate work is oriented towards the synthesis and used of tetraarylphophonium salts as a solubility control group in organic chemistry as well as towards the stereoselective Rh(II)-catalyzed cyclopropanation of olefins using diazo reagents bearing two acceptor groups. Since 2010, he has joined Professor David A. Evans at Harvard University as a NSERC postdoctoral fellow.

NEW, CONVENIENT ROUTE FOR TRIFLUOROMETHYLATION OF STEROIDAL MOLECULES

A.

B.

Submitted by Xiang-Shu Fei, Wei-Sheng Tian, Kai Ding, Yun Wang and Qing-Yun Chen.[1]
Checked by Takayuki Yamakawa and Tohru Fukuyama.

1. Procedure

Caution! Br₂ develops harmful vapors and should be handled only in an efficient fumehood.

A. *4-Bromoandrost-4-ene-3,17-dione.* A modification of Kirk's procedure was employed (Note 1). A 250-mL, single-necked, round-bottomed flask is well wrapped with aluminum foil and fitted with a 3.0-cm octagon-shaped stir bar and a 100-mL pressure-equalizing dropping funnel connected to an oil bubbler (Note 2). The flask is charged with androst-4-en-3,17-dione **1** (11.4 g, 40 mmol) (Note 3) and propylene oxide (120 mL) and cooled to 0 °C (Note 3). A solution of bromine (12 g, 75 mmol) (Notes 3 and 4) in acetic acid (60 mL) (Note 3) is added over 30 min via the dropping funnel. After completion of the addition, the mixture is allowed to stir for 2 h in the dark at 0 °C. After complete consumption of the starting material (as

Org. Synth. **2010**, *87*, 126-136
Published on the web 2/5/2010

monitored by TLC, Note 4), the reaction is quenched with the addition of saturated aq. Na_2SO_3 solution until the red color disappears. This mixture is diluted with EtOAc (300 mL) and washed with H_2O (200 mL × 2). The aqueous layer is re-extracted with EtOAc (150 mL) and the combined organic phases are washed successively with saturated aq. sodium hydrogen carbonate (200 mL) and brine (200 mL). The organic phase is dried over Na_2SO_4, filtered through a glass funnel with a cotton wool plug into a 2-L round-bottomed flask and concentrated under reduced pressure with a rotary evaporator (Note 5). A 300-mL single-necked round-bottomed flask fitted with a 3.0-cm octagon-shaped stir bar is charged with the residue, CH_2Cl_2 (100 mL) and Et_3N (5 mL). (Notes 3 and 6). After stirring for 5 h, the mixture is diluted with 200 mL of EtOAc and washed successively with saturated aq. sodium hydrogen carbonate (100 mL), 2 M hydrochloric acid (100 mL) and brine (100 mL), and dried with anhydrous Na_2SO_4. The organic phase is concentrated by rotary evaporation to a thick slurry, which was added to a 100-mL single-necked round-bottomed flask fitted with a 2.0-cm octagon shaped stir bar. The slurry is diluted through the addition of ethyl acetate (15 mL) and hexanes (30 mL). After a few minutes of stirring, the resulting solid is collected by filtration, then washed with hexanes and dried under high-vacuum (0.2 mmHg) to afford compound **2** (8.0 g) as yellow solid. The mother liquor is concentrated under reduced pressure, and the yellow residue is purified by chromatography on SiO_2 to afford an additional 1.7 g of compound **2** (Note 7). The combined crystals are dissolved in boiling methanol (50 mL) and allowed to cool to 0 °C. Colorless needles are collected by suction filtration to afford the pure compound **2** as colorless needles (7.7 g, 52%). The mother liquor is concentrated to 10 mL and cooled to 0 °C to give additional product (1.6 g, 11%) (Note 8).

B. *4-(Trifluoromethyl)androst-4-ene-3,17-dione.* A flame-dried, 50-mL single-necked round-bottomed flask, containing a 2.0-cm octagon-shaped stir bar and equipped with a rubber septum and nitrogen inlet, is charged with compound **2** (3.65 g, 10 mmol), CuI (2.3 g, 12 mmol) (Note 3) and methyl fluorosulfonyldifluoroacetate (8.0 mL, 63 mmol) (MFSDA, Note 3) in dry NMP (10 mL) (Notes 3 and 9). After stirring for 5 h at 80 °C (Note 10), the dark red solution is allowed to cool to room temperature. Additional MFSDA (2 mL, 16 mmol) is added *via* syringe (Note 11). The solution is stirred at 80 °C for an additional 10 h. After complete consumption of the starting material (as monitored by ^{1}H NMR, Note 12), the mixture is filtered through a Celite pad and washed with ethyl acetate (300 mL). The solution is washed with H_2O (100 mL × 3) and the aqueous layer is re-extracted with EtOAc (50 mL). The combined organic layer is washed with brine (50 mL) and dried over Na_2SO_4. After the Na_2SO_4 is removed by filtration through a glass funnel with a cotton wool plug, the solution is concentrated. The residue is eluted through a short silica gel column to remove the colored impurities (Note 13). The eluent is concentrated to afford the crude product as yellow solid. Recrystallization with ethyl acetate and hexanes provides compound **3** as a white powder (2.8 g, 79%) (Note 14). The mother liquor is concentrated under reduced pressure, and the pink residue is eluted through a short silica gel column to afford additional **3** (0.12 g, 3.4%) (Notes 15 and 16).

2. Notes

1. Kirk provided two methods for the 4-bromination of androst-4-en-3,17-dione.[2] The submitters performed the procedure with an excess of Br_2/collidine for 48 h to obtain compound **2**.[3] Because of the unpleasant odor of collidine and the laborious purification by chromatography, an alternative method[2,4] with propylene oxide as an acid scavenger was used to prepare compound **2** on large scale.

2. The use of an oil bubbler prevents the leakage of harmful vapors.

3. (Submitter) Androst-4-en-3,17-dione **1** was obtained from Xinchang Pharmaceutical Company and was used after recrystallization from ethyl acetate (>95% purity of **1**). Compound **1** can also be purchased from Acros Organics or Aldrich Chemical Company. Acetic acid (99.5+%), propylene oxide (99.5+%), Et_3N (99+%), CH_2Cl_2 (99.5+%), Br_2 (99.5+%, freshly opened bottle), CuI (99.5+%, freshly opened bottle) were purchased from Sinopharm Chemical Reagent Co., Ltd and were used as received. DMF, HMPA, NMP (99.5+%) were purchased from Shanghai Chemical Reagent Co., Ltd and were distilled from CaH_2 prior to use. Methyl fluorosulfonyldifluoroacetate (MFSDA), purchased from Acros Organics or Aldrich Chemical Company, was distilled prior to use (66 °C/85 mmHg).

(Checker) Androst-4-en-3,17-dione **1** was purchased from Tokyo Chemical Industry Co., Ltd and was used after recrystallization from ethyl acetate. Acetic acid (>99.7%), propylene oxide (>99.5%), and Et_3N (>98.0%) were purchased from Kanto Chemical Co. Inc., and CH_2Cl_2 (>99.0%), Br_2 (>99.0%, freshly opened bottle), CuI (>99.5%) were purchased from Wako Pure Chemical Industried, Ltd and were used as received. NMP (>99.0%) was purchased from Kanto Chemical Co., Inc. and was distilled prior to use. Methylfluorosulfonyldifluoroacetate (MFSDA) was purchased from Tokyo Chemical Industry Co. Ltd and used as received. Instead of hexanes, *n*-hexane was used during the checking process.

4. TLC: starting material **1**, R_f= 0.40; compound **2**, R_f= 0.45 (SiO$_2$, GF254, hexane/ethyl acetate, 3:1). Spots are visualized with UV light or by staining with aqueous ceric ammonium molybdate solution followed by heating

5. Concentration at high temperature (60 °C) results in a significant formation of a more polar by-product.

6. The additional step promotes the elimination of bromide, which improves the yield and purity of product.

7. The concentrated mother liquid, a mixture of compound **2** and a polymer of propylene oxide, is mixed with SiO$_2$ (30 g, 100-200 mesh) and washed with hexane through a plug of SiO$_2$ (10 g, 100-200 mesh). After the non-polar polymer is eluted, the column is eluted with CH$_2$Cl$_2$/ethyl acetate/hexane (1:1:5). The fractions containing the product are combined and evaporated under reduced pressure to give an additional 1.6 g of compound **2**. Checkers used the procedure as follows; the residue is purified on SiO$_2$ (120 g) and eluted with EtOAc/*n*-hexane (10% to 15%). After concentration on a rotary evaporator, a magnetic stir bar is added to the flask and the slurry is diluted with ethyl acetate (15 mL) and *n*-hexane (30 mL). After a few minutes of stirring, the resulting solid is collected by filtration, washed with hexanes, and dried under high-vacuum (0.2 mmHg) to afford compound **2** (1.7 g) as a yellow solid.

8. An alternative workup procedure was less laborious but gave a lower yield on a 40 mmol-scale preparation: The crude product was dissolved in ethyl acetate (10 mL). Petroleum ether was added until the resulting solid became sticky. The semi-solid (14 g) was crystallized from ethyl acetate and petroleum ether twice to afford a yellow solid (7.3 g, 50%), which was pure enough for further use. Compound **2** has the following physical and spectroscopic properties: mp 164.0–165.0 °C (The submitter reported a mp of 149.0–149.3 °C); $[\alpha]_D^{25}$ = +203.3 (c 1.16, CHCl$_3$); IR (KBr): 2944, 2857, 1738, 1685, 1574, 1053, 1011, 915, 802 cm^{-1}; ^{1}H NMR (400 MHz, CDCl$_3$) δ: 0.91 (s, 3 H), 1.04 (ddd, J = 11.9, 11.0, 4.1 Hz, 1H), 1.15 (dddd, J = 14.0, 12.5, 12.5, 4.1 Hz, 1H), 1.25 (s, 3 H), 1.23-1.33 (m, 2H),

1.46 (dddd, J = 13.7, 13.7, 11.9, 3.7 Hz, 1H), 1.58 (dddd, J = 12.5, 12.5, 9.2, 8.9 Hz, 1H), 1.65-1.82 (m, 3H), 1.86 (ddd, J = 12.8, 3.7, 2.7 Hz, 1H), 1.93-2.06 (m, 3H), 2.10 (ddd, J = 19.5, 9.2, 9.2 Hz, 1H), 2.29 (ddd, J = 15.1, 14.0, 5.3 Hz, 1H), 2.47 (dd, J = 19.5, 8.9 Hz, 1H), 2.54-2.68 (m, 2H), 3.32 (ddd, J = 15.1, 4.1, 2.8 Hz, 1H); ^{13}C NMR (101 MHz, CDCl$_3$) δ: 13.7, 17.8, 20.4, 21.6, 30.0, 31.2, 32.5, 33.9, 34.5, 34.8, 35.7, 42.4, 47.4, 50.7, 53.9, 122.2, 167.1, 190.4, 220.0; LRMS (EI) m/z 364 (M$^+$, 11), 285 (M-Br, 100); Anal. calcd. for C$_{19}$H$_{25}$BrO$_2$: C, 62.47; H, 6.90; found: C, 62.52; H, 6.96.

9.   Submitters used DMF as the reaction solvent. MFSDA reacts with DMF to produce a byproduct CF$_3$SO$_2$CH=CHNMe$_2$,[5] which was removed by sublimation. The checkers changed the solvent from DMF to NMP in order to avoid the need for the sublimation. Removal of the byproduct by chromatography would necessitate a laborious purification for a large-scale preparation due to the similar R$_f$ value of compound 3 and the byproduct. When HMPA is used as the solvent, the formation of byproduct CF$_3$SO$_2$CH=CHNMe$_2$ is avoided; however, the use of toxic HMPA is undesirable. The reaction could not be carried out in DMSO or DMA.

10.   The submitters used a Schlenk tube for the transformation. The stopcock of the Schlenk tube must be closed to prevent traces of solvent and reagents from entering the nitrogen system while heating the reaction. The reaction is carried out under sealed conditions.

11.   Addition of MFSDA in one batch results in incomplete conversion (about 10–20% starting material after 20 h).

12.   The reaction process cannot be monitored by TLC because compounds 2 and 3 have identical R$_f$ values.

13.   The residue is eluted with CH$_2$Cl$_2$/petroleum ether/ethyl acetate (1:4:2) through a plug of 20 g of SiO$_2$ (200-300 mesh). The Checkers used a different eluent with the same amount of SiO$_2$ (EtOAc/n-hexane, 10% to 20%)

14.   The crude material is dissolved in as little ethyl acetate as possible (~3 mL, 80 °C). 20 mL of hexane is added with stirring. The suspension is stirred at 80 °C for 30 min, and then cooled to room temperature. The resulting solid is collected by filtration and dried in vacuo to afford 2.8 g of compound 3 as a white powder.

15.   Washing with petroleum ether through a plug of 5 g of SiO$_2$ (200-300 mesh) removes low polar impurities to prevent aggregation. Then finely powdered product is obtained after eluting with CH$_2$Cl$_2$/petroleum

ether/ethyl acetate (1:6:2). The Checkers used *n*-hexane instead of petroleum ether.

16. Purification by chromatography provides a higher yield (85%) on a 5 mmol scale preparation with HMPA as solvent. Compound **3** has the following characteristics: mp 148.1–148.9 °C; $[\alpha]_D^{25}$ 201.1 (*c* 1.06, CHCl$_3$); IR (KBr): 2944, 2857, 1738, 1685, 1574, 1053, 1011, 915, 802 cm^{-1}; ^{1}H NMR (400 MHz, CDCl$_3$) δ: 0.92 (s, 3 H), 1.12 (ddd, *J* = 11.9, 11.0, 4.1 Hz, 1 H), 1.14-1.26 (m, 1 H), 1.28 (s, 3 H), 1.25-1.34 (m, 2 H), 1.47 (dddd, *J* = 14.3, 14.3, 11.0, 3.6 Hz, 1 H), 1.58 (dddd, *J* = 12.5, 12.5, 9.2, 8.9 Hz, 1 H), 1.67-1.84 (m, 3 H), 1.87 (ddd, *J* = 14.2, 3.6, 3.6 Hz, 1 H), 1.92-2.06 (m, 3 H), 2.10 (ddd, *J* = 19.5, 9.2, 9.2 Hz, 1 H), 2.19-2.30 (m, 1 H), 2.41-2.52 (m, 3 H), 3.09 (ddd, *J* = 14.7, 3.2, 3.2 Hz, 1 H); ^{13}C NMR (101 MHz, CDCl$_3$) δ: 13.7, 18.0, 20.5, 21.6, 28.2, 30.9, 31.2, 33.9, 33.9, 34.9, 35.7, 41.1, 47.4, 50.6, 54.3, 123.0 (q, *J* = 279.3 Hz), 125.3 (q, *J* = 25.9 Hz), 174.7, 193.0, 219.9; ^{19}F NMR (282 MHz, CDCl$_3$) δ: −56.4 (s); LRMS (EI) *m/z* 354 (M$^+$, 70), 192(100); Anal. calcd. for C$_{20}$H$_{25}$F$_3$O$_2$: C, 67.78; H, 7.11; found: C, 67.53; H, 7.13.

Safety and Waste Disposal Information

All hazardous materials should be handled and disposed of in accordance with "Prudent Practices in the Laboratory"; National Academy Press; Washington, DC, 1995.

3. Discussion

The unique properties of the trifluoromethyl group, e.g. its high electronegativity, stability and lipophilicity, have ensured that trifluoromethylated compounds play an increasingly important role in bioorganic and medicinal chemistry.[6] With respect to the synthesis of structurally simple molecules, the selection of a trifluoromethylated building block followed by several necessary synthetic steps seems to be straightforward and convenient.[7] However, for the synthesis of complicated trifluoromethylated natural products, such as steroids, terpenes and alkaloids, the building block strategy is very difficult, if not impossible, to apply. Trifluoromethylation of halides offers a direct approach to introduce the trifluoromethyl group into complex molecules. Although numerous methods have been employed for the trifluoromethylation of aryl halides,[8] these methods often suffer from (1) low reactivity and yield, (2) expensive and unstable reagents, (3) harsh conditions, and (4) poor generality.

We previously found that $FO_2SCF_2CO_2Me$ (MFSDA), an inexpensive, stable and low-toxicity trifluoromethylating agent, can successfully replace a halogen in allyl, aryl and alkenyl halides under mild conditions.[9] The method is also efficient for the trifluoromethylation of some complex steroidal molecules and can be used to provide a series of trifluoromethylated natural products (Table 1).[3] The preparation on a 15 mmol-scale, as described in this procedure, provides the desired trifluoromethylated product with no drop in yield.

The mechanism of this trifluoromethylation is suggested to involve a difluorocarbene intermediate and proceed as shown in Scheme 1. Methyl difluoro(fluorosulfonyl)acetate reacts with copper(I) iodide. After elimination of SO_2 and CO_2, difluorocarbene and fluoride ion furnish $[CF_3Cu]$ in the presence of CuI. Nucleophilic trifluoromethylation of aryl or alkenyl halide leads to RCF_3.

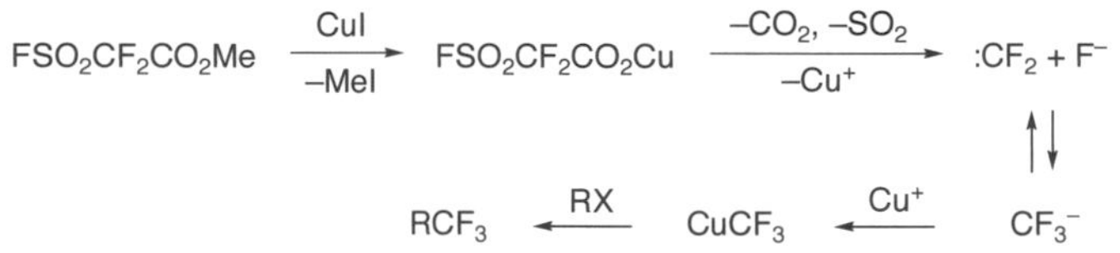

$$FSO_2CF_2CO_2Me \xrightarrow[-MeI]{CuI} FSO_2CF_2CO_2Cu \xrightarrow[-Cu^+]{-CO_2,\ -SO_2} :CF_2 + F^-$$

$$RCF_3 \xleftarrow{RX} CuCF_3 \xleftarrow{Cu^+} CF_3^-$$

Scheme 1

Table 1 Trifluoromethylation of 4-Bromo-3-oxo-Δ^4-steroids.[a]

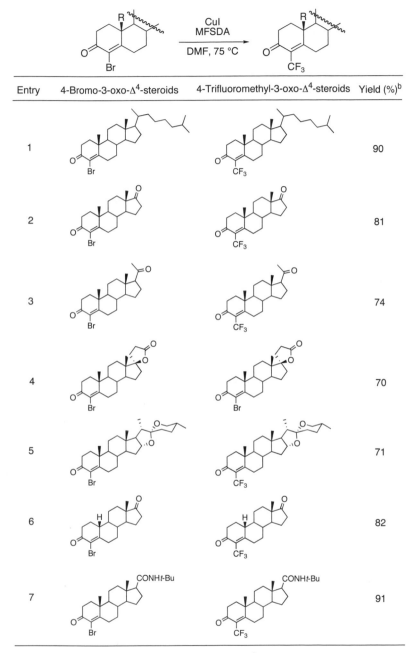

Entry	4-Bromo-3-oxo-Δ^4-steroids	4-Trifluoromethyl-3-oxo-Δ^4-steroids	Yield (%)[b]
1			90
2			81
3			74
4			70
5			71
6			82
7			91

[a] Reactions performed on 100 mg scale. [b] Isolated yield.

1. Key Laboratory of Organofluorine Chemistry, Shanghai Institute of Organic Chemistry, Chinese Academy of Sciences, 345 Lingling Road, Shanghai, China. Email: chenqy@mail.sioc.ac.cn.
2. Kirk, D. N.; Patel, D. K.; Petrow, V. *J. Chem. Soc.* **1956**, 627-629.
3. (a) Fei, X. S.; Tian, W. S.; Chen, Q. Y. *J. Chem. Soc., Perkin Trans. 1* **1998**, 1139-1142. (b) Fei, X. S.; Tian, W. S.; Chen, Q. Y. *Bioorg. Med. Chem. Lett.* **1997**, *7*, 3113 -3118.
4. (a) Lukashev, N. V.; Latyshev, G. V.; Donez, P. A.; Skryabin, G. A.; Beletskaya, I. P. *Synthesis* **2005**, 1578-1580. (b) Hodson, H. F.; Madge, D. J.; Widdowson, D. A. *J. Chem. Soc., Perkin Trans. 1* **1995**, 2965-2968.
5. Chen, Q. Y.; Yang, G. Y.; Wu, S. W. *J. Fluorine Chem.* **1991**, *55*, 291-298.
6. (a) Begue, J. P.; Bonnet-Delpon, D. *Bioorganic and Medicinal Chemistry of Fluorine* John Wiley & Sons, Inc., Hoboken, NJ . **2008**. (b) Welch, J. T. and Eswarakrishnan, S. *Fluorine in Bioorganic Chemistry* Wiley, New York, **1991**;
7. (a) Usachev, B. I.; Obydennov, D. L.; Röschenthaler, G. V.; Sosnovskikh, V. Y. *Org. Lett.* **2008**, 10, 2857-2859. (b) Massicot, F.; Monnier-Benoit, N.; Deka, N.; Plantier-Royon, R.; Portella, C. *J. Org. Chem.* **2007**, *72*, 1174-1180.
8. (a) McClinton, M. A.; McClinton, D. A. *Tetrahedron* **1992**, *48*, 6555-6666. (b) Man, E. H.; Coffman, D. D.; Muetterties, E. L. *J. Am. Chem. Soc.* **1959**, *81*, 3575-3577. (c) Emeleus, H. J.; Macduffi, D. E. *J. Chem. Soc.* **1961**, 2597-2599. (d) Kondratenko, N. V.; Kolomeytsev, A. A.; Popov, V. I.; Yagupolskii, L. M. *Synthesis* **1985**, 667-669. (e) Harris, J. F. *J. Org. Chem.* **1967**, *32*, 2063-2074.
9. Chen, Q. Y.; Wu, S. W. *J. Chem. Soc., Chem. Commun.* **1989**, 705-706.

Appendix
Chemical Abstracts Nomenclature; (Registry Number)

Androst-4-ene-3,17-dione; (63-05-8)
Oxirane, 2-methyl-; (75-56-9)
Bromine; (7726-95-6)
Androst-4-ene-3,17-dione, 4-bromo-; (19793-14-7)
Acetic acid, 2,2-difluoro-2-(fluorosulfonyl)-, methyl ester; (680-15-9)

Copper iodide; (7681-65-4)

Phosphoric triamide, N,N,N',N',N'',N''-hexamethyl- ; (680-31-9)

Androst-4-ene-3,17-dione, 4-(trifluoromethyl)-; (201664-30-4)

2-Pyrrolidinone, 1-methyl-; (872-50-4)

Qing-Yun Chen graduated from Peking University in 1952. He was a graduate student in 1956-1960 at Institute of Elementoorganic Compounds USSR and received candidate Ph.D. in 1960. He joined the faculty team of Shanghai Institute of Organic Chemistry (SIOC), Chinese Academy of Sciences in 1963 and was elected the academician of Chinese Academy of Sciences in 1993. His research interests focus on perfluoroalkanesulfonic acids and their derivatives, difluorocarbene, trifluoromethylation, single electron transfer reactions and fluorinated porphyrins.

Xiang-Shu Fei received his B.S. degree and M.S. degree from Xiamen University in 1991 and 1994, respectively. During 1994-1997 he pursued his Ph.D. at the Shanghai Institute of Organic Chemistry (SIOC), Chinese Academy of Sciences with Professor Qing-Yun Chen and Wei-Sheng Tian. He did postdoctoral research at Texas A&M University with Professor Sir Derek H. R. Barton from 1997-1998. He also served as a post-doctoral researcher at Wayne State University with Professor Aloke K. Dutta from 1999-2001.

Wei-Sheng Tian received his Ph.D. degree from Shanghai Institute of Organic Chemistry (SIOC), Chinese Academy of Sciences in 1985 under the supervision of Professor Weishan Zhou. He worked as a postdoctoral fellow for two years (1986-1988) with Professor T. Livinghouse at University of Minnesota and Montana State University. He was appointed as an Associate Professor at SIOC In 1990 and promoted to Full Professor in 1995. His research interests focus on the synthetic of natural and/or artificial resource molecules (Resource Chemistry).

Kai Ding was born in 1976 in Hubei, China. He obtained his B.S. degree in 1996 and M.S. degree in 1999 from East China University of Science and Technology. He joined the Shanghai Institute of Organic Chemistry to obtain his doctorate in 2003 studying natural products total synthesis. After being a research assistant for two years, he joined the group of Professor Albert, S.C. Chan at the Hong Kong Polytechnic University as a postdoctoral research associate. In 2008, he moved back to Shanghai and became associate professor at Shanghai Institute of Organic Chemistry. His research focused on natural product total synthesis.

Yun Wang was born in 1978 in Shanghai, China. She received her M.S. degree in 2003 from the Shanghai University. Then she moved to Shanghai Institute of Organic Chemistry, Chinese Academy of Sciences, where she is now a research assistant of organic chemistry. Her current interest is in the synthesis and application of steroid drugs.

Takayuki Yamakawa was born in 1983 in Hiroshima, Japan. He graduated in 2007 and received his M.S. degree in 2009 from University of Tokyo under the direction of Professor Tohru Fukuyama. The same year he started his Ph.D. study under the supervision of Professor Fukuyama. His research interest is total synthesis of natural products.

PREPARATION OF
ETHYL 1-BENZYL-4-FLUOROPIPERIDINE-4-CARBOXYLATE

Submitted by Jianshe Kong,[1*] Tao Meng,[1] Pauline Ting,[1] and Jesse Wong.[1]
Checked by Chaofeng Huang and Kay M. Brummond.

1. Procedure

A flame-dried, 1-L, three-necked, round-bottomed flask is equipped with a mechanical-stirrer, a thermometer holder fitted with an internal thermometer, a rubber septum and a nitrogen inlet. The flask is charged with diisopropylamine (13.7 mL, 9.9 g, 97.8 mmol, 1.2 equiv) and THF (200 mL) via syringe. The rubber septum is replaced with a pressure-equalizing funnel equipped with a nitrogen inlet. The solution is then cooled to –78 °C with a cryocool and 2.5 M n-BuLi in hexanes (39 mL, 97.5 mmol, 1.2 equiv (Note 1) is added dropwise via the addition funnel over 20 min with stirring. The solution is then warmed to 0 °C by placing the flask in an ice-water bath. The addition funnel is removed and replaced by another pressure-equalizing addition funnel. The solution is continually stirred at 0 °C for 30 min and then cooled to –40 °C by placement of the flask in a cryocool.

To the above freshly prepared solution of LDA is added a solution of ethyl 1-benzylpiperidine-4-carboxylate (20.0 g, 80.9 mmol, 1.0 equiv) (Notes 2 and 3) in THF (150 mL) dropwise via the addition funnel over 30 min such that the solution is kept between –30 and –40 °C. The addition funnel is then removed and replaced with a different pressure-equalizing addition funnel. The resulting reaction mixture is continually stirred for 1 h between –10 °C and –15 °C before being cooled to –78 °C. A solution of N-fluorobenzenesulfonimide (NFSI) (Note 4) (26.7 g, 84.7 mmol, 1.05 equiv) in THF (150 mL) is added to the reaction via the addition funnel over 40 min while keeping the solution between –65 °C and –78 °C. After the addition, the reaction mixture is stirred for 1 h at –78 °C, then the temperature is allowed to warm to –50 °C over a 2 h period.

The reaction mixture is poured into a 2-L separatory funnel containing ice-H_2O (600 mL) and diluted with ethyl acetate (600 mL). Layers are separated and the aqueous layer is extracted with ethyl acetate (2 x 200 mL). The combined organic layers are washed with brine (200 mL), dried over $MgSO_4$ (20 g), filtered (Note 5), and concentrated by rotary evaporation (45 °C, 35 mmHg), and dried under vacuum (25 °C, 15 mmHg) to afford a deep brown oil. The crude product is purified by flash column chromatography on SiO_2 (Note 6), eluting with ethyl acetate and hexanes (0% to 15%). The fractions containing the product (TLC: $R_f = 0.39$) (Note 7) are combined and concentrated by rotary evaporation (45 °C, 20 mmHg) to yield 15.7 g (73%) of pure ethyl 1-benzyl-4-fluoropiperidine-4-carboxylate as a brown oil (Note 8).

2. Notes

1. A 2.5 M solution of *n*-butyllithium in hexanes was purchased from Aldrich Chemical Co., Inc. Diisopropylamine (99.5%) was purchased from Aldrich Chemical Co., Inc. THF (99+%) was purchased from Aldrich Chemical Co., Inc. Diethyl ether (99.9%) was purchased from Fisher Scientific. Ethyl acetate (99.5%) was purchased from EMD Chemicals. Hexanes (98.5%) was purchased from Mallinckrodt Chemicals. All of these reagents and solvent were used as received.

2. Ethyl 1-benzylpiperidine-4-carboxylate is commercially available. Both commercial material (purchased from Oakwood Inc) and freshly prepared of ethyl 1-benzylpiperidine-4-carboxylate provided similar results in the fluorination reaction. Ethyl 1-benzylpiperidine-4-carboxylate is prepared from ethyl isonipecotate and benzyl bromide (Note 3). Both ethyl isonipecotate and benzyl bromide are purchased from Aldrich Chemical Co., Inc. and used directly without further purification.

3. A 500-mL, two-necked, round-bottomed flask equipped with a magnetic-stirring bar and a pressure-equalizing addition funnel is charged with ethyl isonipecotate (21.6 g, 137 mmol, 1 equiv), K_2CO_3 (38.0 g, 275 mmol, 2 equiv) and DMF (100 mL). The solution is cooled to 0 °C (external temperature) in an ice-bath. A solution of benzyl bromide (23.5 g, 137 mmol) in DMF (100 mL) is added dropwise via the addition funnel over 20 min. The cooling bath is removed after addition, and the reaction mixture is stirred at room temperature for 14 h. The solid is filtered and washed with diethyl ether (3 x 250 mL). The combined organic layers are washed with

138

H_2O (4 x 150 mL), dried over $MgSO_4$ (20 g), filtered, and concentrated by rotary evaporation to afford a light-yellow oil (31.0 g, 96%), which is used directly for the fluorination reaction without further purification. The product exhibits the following physicochemical properties: ^1H NMR (500 MHz, $CDCl_3$) δ: 1.24 (t, 3 H, J = 7.0 Hz), 1.75-1.80 (m, 2 H), 1.85-1.88 (m, 2 H), 2.01 (dt, 2 H, J = 2.0, 11.5 Hz), 2.23-2.29 (m, 1 H), 2.84 (d, 2 H, J = 11.5 Hz), 3.48 (s, 2 H), 4.11 (q, 2 H, J = 7.0 Hz), 7.24-7.31 (m, 5 H); ^{13}C NMR (126 MHz, $CDCl_3$) δ: 14.3, 28.3, 41.3, 53.0, 60.3, 63.3, 127.0, 128.2, 129.1, 138.5, 175.3.

4.   N-Fluorobenzenesulfonimide (NFSI) was purchased from Aldrich Chemical Co., Inc.

5.   A 150 mL filter funnel (M) charged with 20 g Celite was used.

6.   A disposable column filled with SiO_2 (330 g) provided by Teledyne Isco, Inc. was used for the purification.

7.   TLC plates (UV_{254} active) were purchased from EMD Chemicals, Inc. and 15% ethyl acetate/hexanes was used as an eluent.

8.   The title compound shows the following analytical and spectroscopic data: FTIR (film): 2929, 2819, 2778, 1756, 1736, 1286, 1139, 1070, 1017, 735, 698 cm^{-1}; ^1H NMR (500 MHz, $CDCl_3$) δ: 1.30 (t, 3 H, J = 7.5 Hz), 1.90–1.96 (m, 2 H), 2.14 (dtd, 2 H, J = 37.5 ($^3J_{HF}$), 13.0, 4.5 Hz), 2.33 (dt, 2 H, J = 1.5, 12.0 Hz), 2.74–2.77 (m, 2 H), 3.53 (s, 2 H), 4.23 (q, 2 H, J = 7.5 Hz), 7.25–7.35 (m, 5 H); ^{13}C NMR (125.76 MHz, $CDCl_3$) δ: 14.1, 32.6 (d, J = 21.4 Hz), 48.3, 61.6, 63.0, 92.0 (d, J = 186.0 Hz), 127.1, 128.3, 129.1, 138.3, 171.6 (d, J = 25.2 Hz); MS m/z (relative intensity): 265 (M^+, 75), 244 (15), 236 (30), 192 (40), 188 (70), 174 (20), 160 (10), 154 (82), 91 (100); HRMS (EI) m/z calcd. for $C_{15}H_{20}FNO_2$ 265.1471; found: 265.1467. Anal. calcd. for $C_{15}H_{20}FNO_2$: C, 67.90; H, 7.60; N, 5.28; F, 7.16; found: C, 68.02; H, 7.65; N, 5.27; F, 7.14.

Safety and Waste Disposal Information

All hazardous materials should be handled and disposed of in accordance with "Prudent Practices in the Laboratory"; National Academy Press; Washington, DC, 1995.

3. Discussion

The title compound and derivatives are important pharmaceutical building blocks.[2a-k] N-Boc-4-fluoropiperidine-4-carboxylate was reported recently. However, the product using the patent procedure[3] suffered from very poor purity (50%). Unreacted starting material (~50%) could not be separated by either distillation or column chromatography. The benzyl-protected 4-fluoropiperidine-4-carboxylate ester was developed as an improved version. Our synthetic procedure provides a practical, efficient method for the preparation of the title compound on larger scale, in high yield and purity.

Although other fluorinating reagents[4] are available for the synthesis of α-fluorocarbonyl-type compounds, N-fluorobenzenesulfonimide[5] was found to be the best with respect to reactivity, safety, and availability.

1. Merck Research Laboratories, 2015 Galloping Hill Road, K-15-2 (2800), Kenilworth, NJ 07033-1300, jianshe.kong@merck.com
2. (a) Kurose, N.; Hayashi, M.; Ogawa, T.; Masuda, K.; Kojima, E., WO 2007/058346. (b) De Lera Ruiz, M.; Aslanian, R.; Berlin, M.; Mccormick, K.; Celly, C., US 2007/066644 A1 20070322. (c) Kuang, R.; Blythin, D.; Shih, N.; Shue, H.; Chen, X.; Cao, J.; Gu, D.; Huang, Y.; Schwerdt, J.; Ting, P.; Wong, S.; Xiao, L., WO 2005/116009. (d) Zoller, G.; Petry, S.; Mueller, G.; Heuer, H.; Baringhaus, K., WO 2005/073199. (e) Inami, H.; Kawaguchi, K.; Kubota, H.; Yamasaki, S.; Matsuzawa, T.; Kaga, D.; Seki, N.; Morio, H., WO 2005/073183. (f) Alvaro, G.; Cardullo, F.; D'adamo, L.; Piga, E.; Seri, C., WO 2004/005255. (g) Ozaki, F.; Ono, M.; Kawano, K.; Norimine, Y.; Onogi, T.; Yoshinaga, T.; Kobayashi, K.; Suzuki, H.; Minami, H.; Sawada, K., WO 2003/084948. (h) Barrow, J.; Lindsley, C.; Shipe, W.; Yang, Z.; Wisnoski, D., WO 2007/002884. (i) Zeng, Q.; Aslanian, R.; Berlin, M.; Boyce, C.; Cao, J.; Kozlowski, J.; Mangiaracina, P.; McCormick, K.; Mutahi, M.; Rosenblum, S.; Shih, N.; Solomon, D.; Tom, W., WO 2003/088967. (j) Friary, R.; Kozlowski, J.; Shankar, B.; Wong, M.; Zhou, G.; Lavey, B.; Shih, N.; Tong, L.; Chen, L.; Shu, Y., WO 2003/042174. (k) Aslanian, R.; Shih, N.; Ting, P.; Berlin, M.; Rosenblum, S.; McCormick, K.; Tom, W.; Boyce, C.; Mangiaracina, P.; Mutahi, M.; Piwinski, J., WO 2002/032893.

3. Aslanian, R.; Berlin, M.; Mangiaracina, P.; McCormick, K.; Mutahi, M.; Rosenblum, S., WO 2004/000831.
4. (a) Sankar Lal, G.; Pez, G.; Syvret, R. *Chem. Rev.* **1996**, *5*, 1737-1756. (b) Taylor, S.; Kotoris, C.; Hum, G. *Tetrahedron* **1999**, *55*, 12431-12477.
5. Differding, E.; Ofner, H. *Synlett* **1991**, 187-189.

Appendix
Chemical Abstracts Nomenclature (Registry Number)

Diisopropylamine (108-18-9)
n-Butyllithium (109-72-8)
Ethyl isonipecotate (1126-09-6)
Benzyl bromide (100-39-0)
Ethyl 1-benzylpiperidine-4-carboxylate (24228-40-8)
N-Fluorobenzenesulfonimide (133745-75-2)

Jianshe Kong received his B.S. from Nankai University and Ph.D. from Oviedo University of Spain in 1993 under the direction of the Professor José Barluenga. After doing postdoctoral studies with Professors Jeff Aubé and Robert Hanzlik at University of Kansas and Professor Robert Coleman at The Ohio State University, he joined Schering-Plough Research Institute in 1998. He is currently working in the Discovery Synthetic Group as a Senior Principal Scientist of Merck Research Laboratories at Kenilworth.

Tao Meng went to Middle Tennessee State University in 1998 and studied organic chemistry under the guidance of Dr. Norma K. Dunlap. He received the Master of Science degree in Organic Chemistry in 2000. After graduation, he joined the Discovery Synthetic Group in Schering-Plough Research Institute at Kenilworth, which now is Merck Research Laboratories.

Pauline Ting received her bachelor's degree in chemistry from the University of Illinois at Champaign-Urbana and her Ph.D. degree in organic chemistry from the University of California at Berkeley with Professor Paul Bartlett. She is currently a Director in the Merck Research Laboratories at Kenilworth, New Jsersey.

Jesse K. Wong received his B.S. and M.S. at Rutgers University. He joined Schering Plough Research Institute in 1980, which now is Merck Research Laboratories, as a medicinal chemist. Four years later he received his Ph.D. with Professor Hugh W. Thompson at Rutgers while working at Schering. In 1993 he was appointed to lead the Discovery Synthetic Group where he developed many new techniques in large scale synthesis. He has 21 patents and 27 publications.

Chaofeng Huang was born in 1978 in China. After completing his Ph.D. under the supervision of Prof. Michael Harmata at the University of Missouri-Columbia, he works as a post-doctoral fellow in the group of Prof. Kay Brummond at the University of Pittsburgh. Currently, his research efforts are focused on extending the synthetic utility of the rhodium(I)-catalyzed cyclocarbonylation reaction to a class of highly oxygenated sesquiterpene lactones.

LITHIUM AMIDES AS HOMOCHIRAL AMMONIA EQUIVALENTS FOR CONJUGATE ADDITIONS TO α,β-UNSATURATED ESTERS: ASYMMETRIC SYNTHESIS OF (S)-β-LEUCINE

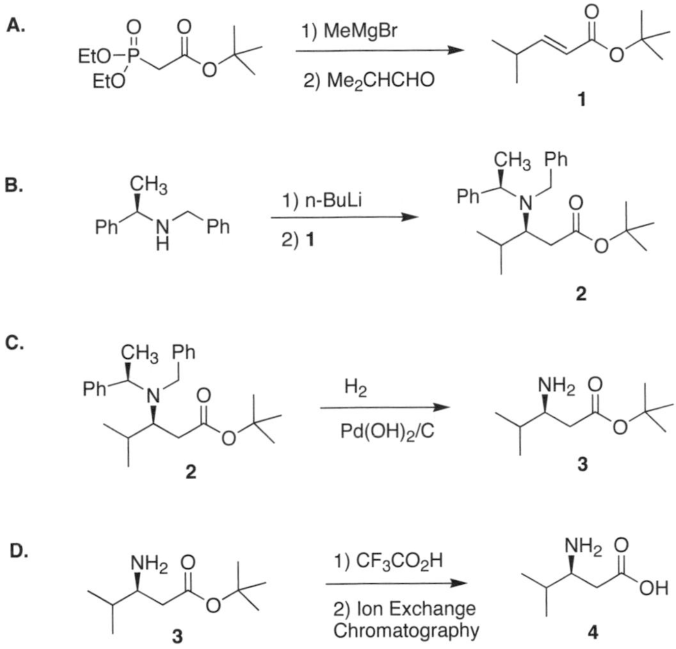

A.

B.

C.

D.

Submitted by Stephen G. Davies,[1] Ai M. Fletcher, and Paul M. Roberts.
Checked by David Hughes.[2]

1. Procedure

A. t-Butyl (E)-4-methylpent-2-enoate (**1**). An oven-dried 1-L, three-necked, round-bottomed flask equipped with a 3-cm oval Teflon-coated stir bar is fitted with two septa on the outer necks and a reflux condenser on the center neck. The reflux condenser is fitted with a nitrogen gas inlet adapter connected to a nitrogen line and a gas bubbler. A thermocouple thermometer probe is inserted through one of the septa. The flask is allowed to cool under a flow of nitrogen. It is charged with *t*-butyl diethylphosphonoacetate (17.8

g, 70.4 mmol, uncorrected for purity, 1.0 equiv) (Note 1) followed by anhydrous tetrahydrofuran (150 mL) (Note 2) and cooled to 2 °C using an ice bath. A 2.93 M solution of methylmagnesium bromide in diethyl ether (25.3 g, 24.5 mL, 71.7 mmol, 1.02 equiv) (Note 3) is added via syringe to the stirred solution over a period of 15 min, resulting in an exotherm to 8 °C and the evolution of gas (gentle bubbling of the reaction mixture). After the addition is complete, stirring is continued at 0–5 °C for 30 min. Isobutyraldehyde (5.73 g, 79.5 mmol, 1.1 equiv) is added dropwise to the reaction mixture via syringe over 5 min, resulting in a slight exotherm from 2 °C to 6 °C. The ice-bath is replaced with a heating mantle and the mixture is heated to reflux (62 °C) for 2 h. The heating mantle is removed and the reaction flask is allowed to cool to 23 °C over 30 min (Note 4). The reflux condenser is replaced with a 250-mL pressure-equalizing dropping funnel, which is charged with saturated aqueous ammonium chloride solution (200 mL). The reaction mixture is quenched by the dropwise addition of the saturated aqueous ammonium chloride solution over a period of 10 min, with vigorous stirring being continued throughout (Note 5). After the addition is complete, diethyl ether (100 mL) is added and the mixture is transferred to a 1-L separatory funnel. The layers are separated and the aqueous layer is extracted with diethyl ether (2 x 100 mL). The combined organic layers are washed with saturated aqueous sodium chloride (100 mL), filtered through a bed of anhydrous sodium sulfate (50 g, rinsed with 2 x 75 mL diethyl ether) and concentrated by rotary evaporation (20 °C water bath, 100 mmHg, decreased to 20 mmHg) (Note 6) to give the crude reaction product 1 (20 g) as a colorless oil. The crude product is purified by chromatography on silica gel (Note 7) to afford *t*-butyl (*E*)-4-methylpent-2-enoate 1 (11.4 g, 95% yield) as a colorless oil having an *E/Z* ratio of 142:1 (Notes 8 and 9).

  B. t-Butyl · (3S,αR)-3-[N-benzyl-N-(α-methylbenzyl)amino]-4-methylpentanoate (**2**). An oven-dried 500-mL, three-necked, round-bottomed flask with a 3-cm oval Teflon-coated magnetic stir bar is fitted with a nitrogen gas inlet adapter connected to a nitrogen line and gas bubbler and is allowed to cool under a stream of nitrogen. The remaining two necks are fitted with septa. A thermocouple thermometer probe is inserted through one of the septa. The flask is charged with (*R*)-*N*-benzyl-*N*-(α-methylbenzyl)amine (20.1 g, 95 mmol, 1.6 equiv) (Note 10) and anhydrous tetrahydrofuran (200 mL) (Note 11) and cooled to −72 °C (internal temperature) using a dry-ice/acetone bath in a dish-shaped Dewar flask. A

Org. Synth. **2010**, *87*, 143-160

2.37 M solution of *n*-butyllithium in hexanes (27.1 g, 39.1 mL, 93 mmol, 1.55 equiv) (Note 12) is added dropwise via syringe (Note 13) over 10 min. The temperature increases to –64 °C at the end of the addition (Note 14). After the addition is complete, stirring is continued at –75 to –70 °C for 30 min. *t*-Butyl (*E*)-4-methylpent-2-enoate **1** (10.1 g, 59.4 mmol, 1.0 equiv) is added dropwise by syringe (Note 13) over 15 min, keeping the internal temperature between –76 to –74 °C (Note 15). When the addition is complete, stirring is continued at –76 to –74 °C for 3 h (Note 16). The dry-ice bath is removed and the reaction mixture is allowed to warm to –50 °C over 20 min. The nitrogen gas adapter is replaced with a 100-mL pressure-equalizing dropping funnel and the gas adapter is moved to the top of the dropping funnel. The dropping funnel is charged with saturated aqueous ammonium chloride solution (60 mL). The reaction mixture is quenched by the dropwise addition of the saturated aqueous ammonium chloride solution over a period of 10 min, with vigorous stirring being continued throughout. The mixture warms to –25 °C during the quench (Note 17). The dark orange solution turns pale yellow and a white precipitate is formed. After the addition is complete, a water bath is used to warm the mixture to room temperature over 15 min. The mixture is diluted with saturated aqueous sodium chloride (60 mL) and stirred for 15 min. The white precipitate dissolves. The mixture is transferred to a 1-L separatory funnel. The layers are separated and the aqueous layer is extracted with diethyl ether (2 x 100 mL). The combined organic layers are concentrated by rotary evaporation (40 °C water bath, 150 mmHg, decreased to 20 mmHg) to provide 34 g of crude product (Note 18). The resultant pale yellow oil is dissolved in dichloromethane (300 mL) and washed sequentially with 0.5 M aqueous citric acid solution (2 x 200 mL) (Note 19), saturated aqueous sodium bicarbonate (200 mL), and saturated aqueous sodium chloride (200 mL). The organic layer is filtered through a bed of sodium sulfate (50 g, rinsed with 2 x 75 mL of dichloromethane) and the filtrate is concentrated by rotary evaporation (40 °C water bath, 150 mmHg, decreased to 20 mmHg) to give the crude reaction product **2** (22.2 g) as a viscous, pale yellow oil. The crude product is purified by chromatography on silica gel (Note 20) to afford *t*-butyl (3*S*,α*R*)-3-[*N*-benzyl-*N*-(α-methylbenzyl)amino]-4-methylpentanoate **2** (20.6 g, containing 2 wt% hexane by ^{1}H NMR analysis, 89% corrected yield) as a viscous, colorless oil (Note 21).

C. t-Butyl (S)-3-amino-4-methylpentanoate (**3**). *t*-Butyl (3*S*,α*R*)-3-[*N*-benzyl-*N*-(α-methylbenzyl)amino]-4-methylpentanoate **2** (19.1 g, 50.0 mmol) and methanol (200 mL) (Note 22) are added to a 500-mL round-bottomed flask and the contents are swirled to completely dissolve the oil. The solution is transferred to a 500-mL borosilicate Parr bottle and 20% wet palladium hydroxide on carbon (5.0 g) is added. The bottle is connected to a Parr shaker hydrogenation apparatus and flushed by 3 vacuum purge/hydrogen fill cycles. The hydrogenation reaction is then carried out under 40 psi hydrogen pressure with shaking at 21–24 °C for 6 h (Note 23). The vessel is vented and 3 nitrogen fill/vacuum purge cycles are carried out to remove hydrogen from the solution. The solution is vacuum filtered through a bed of Solka-Floc (50 g) in a 350-mL medium porosity sintered glass funnel and washed with methanol (3 x 100 mL). The catalyst cake is kept wet with methanol throughout the filtration to minimize the potential for the palladium to ignite (Note 24). The hazy filtrate is transferred to a 1-L round bottomed flask and is concentrated by rotary evaporation (40 °C water bath, 20 mmHg) to give the crude reaction product **3** (9.6 g) as a pale yellow oil. The crude product is purified by chromatography on silica gel (Note 25) to afford *t*-butyl (*S*)-3-amino-4-methylpentanoate **3** (7.8 g, 83% yield) as pale yellow oil (Note 26).

D. (S)-3-Amino-4-methylpentanoic acid [(S)-β-leucine] (**4**). *t*-Butyl (*S*)-3-amino-4-methylpentanoate **3** (6.60 g, 35.2 mmol) and dichloromethane (15 mL) (Note 27) are added to a 250-mL round-bottomed flask equipped with a 2-cm Teflon-coated magnetic stir bar. The flask is fitted with a septum, through which a thermocouple temperature probe and an 18-gauge needle connected to a nitrogen line and gas bubbler are inserted. The solution is cooled to 5 °C (internal temperature) using an ice bath. Trifluoroacetic acid (15 mL) is added dropwise via syringe over a period of 10 min, resulting in a temperature rise to 9 °C. The cooling bath is removed and the reaction mixture is stirred at 21–23 °C for 13 h. The solution is concentrated by rotary evaporation (40 °C water bath, 20 mmHg) to give a pale orange residue. Toluene (50 mL) is added to the residue and the mixture is concentrated by rotary evaporation (60 °C water bath, 20 mmHg). A further 50 mL of toluene is added and the evaporation process is repeated to afford crude product as a gum (12 g) (Note 28). Diethyl ether (30 mL) and a 2-cm oval Teflon-coated magnetic stir bar are added to the flask. The flask is fitted with a septum through which a thermocouple temperature probe and an 18-gauge needle with an outlet to a nitrogen line and gas bubbler are

146

inserted. A 2 M solution of hydrogen chloride in diethyl ether (30 mL) is added dropwise via syringe to the stirred solution over a 5 min period. The temperature rises to 35 °C and the mixture becomes a thick white slurry, which is stirred for 1 h at ambient temperature (35 °C initial, cooling in air to 23 °C). The stir bar is removed and the mixture is concentrated by rotary evaporation (40 °C water bath, 100 mmHg, decreased to 20 mmHg) to afford the crude hydrochloride salt of **4** (10.7 g) as a white solid (Note 29). The crude product is purified by ion-exchange resin chromatography (Note 30) to afford (*S*)-3-amino-4-methylpentanoic acid [(*S*)-β-leucine] **4** (4.06–4.30 g, 88–93% yield) as a tan crystalline solid (Notes 31-34).

2. Notes

1.   *t*-Butyl diethylphosphonoacetate (95%) was obtained from Alfa Aesar. The submitters prepared it by heating *t*-butyl bromoacetate (Sigma-Aldrich, 98%) and triethylphosphite (Sigma-Aldrich, 98%) for 2 h without solvent, followed by distillation under vacuum.[3]

2.   The following materials used in step A were obtained from Sigma-Aldrich: tetrahydrofuran (anhydrous, >99.9%, inhibitor-free), methylmagnesium bromide (3.0 M in diethyl ether), isobutyraldehyde (≥99%), pentane (Chromasolv, >99%), diethyl ether (ACS reagent, anhydrous, BHT-inhibited), and silica gel (230-400 mesh, 60 Å). De-ionized tap water was used throughout.

3.   Methylmagnesium bromide was titrated before use as follows: *n*-Butanol (350 mg, 4.72 mmol) and 1,10-phenanthroline (2 mg) are dissolved in tetrahydrofuran (7 mL) in a 25-mL round-bottomed flask equipped with a 0.5-cm oval Teflon-coated magnetic stir bar and a septum pierced with an 18-gauge nitrogen-inlet needle connected to a nitrogen line and gas bubbler. The stirred solution is cooled in an ice-bath, then 3 M methylmagnesium bromide in diethyl ether is added dropwise via a weighed syringe to a persistent pink endpoint. The syringe is weighed before and after addition (difference 1.67 g, 1.61 mL, corresponding to a solution molarity of 2.93). The submitters titrated against salicylaldehyde phenylhydrazone following the procedure described by Love and Jones.[4]

4.   The reaction is >98% complete as determined by ^{1}H NMR analysis as follows: A 0.1 mL aliquot of the reaction mixture is added to 1 mL of CDCl$_3$ and 0.5 mL of saturated aqueous ammonium chloride, shaken, then the bottom organic phase is filtered through a cotton plug into an NMR tube

for analysis. The doublet of the CH_2 group at 2.9 ppm and the quartet of the CH_2 group of the ethyl group at 4.2 ppm of the starting material compared to the olefinic resonances of the product at 5.7 and 6.8 ppm are diagnostic for reaction completion.

5. When ammonium chloride solution is first added to the reaction mixture a white precipitate is formed; this subsequently dissolves when addition of all the ammonium chloride solution is complete, giving a clear solution. The mixture warms to 30 °C during the quench with no external cooling.

6. The product 1 is volatile and the temperature of the water bath should not be increased above 20 °C while the solution is being concentrated on the rotary evaporator.

7. A 3-cm glass column is wet-packed (pentane) with 80 g of silica gel topped with 0.5 cm of sand. The crude reaction product 1 is loaded neat on the column and eluted with 500 mL of a 5:1 mixture of pentane:diethyl ether, taking 50 mL fractions. The chromatography is monitored by TLC (R_f = 0.3 in pentane:diethyl ether, 50:1). The product elutes in fractions 3-5. The eluent is concentrated by rotary evaporation (20 °C water bath, 150 mm Hg, decreased to 20 mm Hg) (Note 6) to constant weight. The pure product 1 contains 0.7 wt% residual tetrahydrofuran by ^{1}H NMR analysis.

8. The geometric isomer purity of 1 is assessed by peak integration of the ^{13}C-^{1}H satellite peaks corresponding to C(2)H of the (E)-isomer (δ_H 5.68 ppm) against the ^{12}C-^{1}H peaks corresponding to C(2)H and C(3)H of the (Z)-isomer [δ_H 5.56 ppm (dd, J 11.5, 1.0 Hz) and 5.90 ppm (dd, J 11.5, 9.9 Hz), respectively] in the quantitative ^{1}H NMR spectra of the crude reaction mixture and the pure product.[5]

9. t-Butyl (E)-4-methylpent-2-enoate 1 has the following physical and spectroscopic data: ν_{max} (thin film) 2968, 2873, 1715, 1652, 1460, 1367, 1301, 1155 cm^{-1}; ^{1}H NMR (400 MHz, CDCl$_3$) δ: 1.04 (6 H, d, J = 6.8 Hz, C(4)Me_2), 1.48 (9 H, s, CMe_3), 2.40–2.45 (1 H, m, C(4)H), 5.68 (1 H, dd, J = 15.7, 1.5 Hz, C(2)H), 6.83 (1 H, dd, J = 15.7, 6.6 Hz, C(3)H); ^{13}C NMR (100 MHz, CDCl$_3$) δ: 21.5 (C(4)Me_2), 28.4 (CMe_3), 31.0 (C(4)), 80.1 (CMe_3), 120.5 (C(2)), 154.3 (C(3)), 166.6 (C(1)); MS (ESI$^+$) m/z 193 ([M+Na]$^+$, 100%); HRMS (ESI$^+$) m/z calcd. for C$_{10}$H$_{18}$NaO$_2^+$ ([M+Na]$^+$) 193.1199; found 193.1205.

10. (R)-N-Benzyl-N-(α-methylbenzyl)amine was obtained from Sigma-Aldrich and used without purification. The enantiomeric excess

148

quoted by the vendor was 97.1%. The submitters prepared it from reductive alkylation of (R)-α-methylbenzylamine (Acros Organics, >99%, 99% ee).

11. The following materials used in step B were obtained from Sigma-Aldrich: tetrahydrofuran (anhydrous, >99.9%, inhibitor-free), 2.5 M butyllithium in hexanes, citric acid (ACS reagent grade, >99.5%), dichloromethane (ACS reagent grade, >99.5%), ethyl acetate (ACS reagent grade, >99.5%), and hexanes (ACS reagent grade, >98.5%). Ammonium chloride, sodium bicarbonate, and sodium sulfate were sourced from Fisher.

12. n-Butyllithium was titrated against diphenylacetic acid before use as follows: Diphenylacetic acid (0.766 g, 3.61 mmol) is dissolved in tetrahydrofuran (10 mL) in a 25-mL round-bottomed flask equipped with a 0.5 cm oval Teflon-coated magnetic stir bar and a septum pierced with an 18-gauge nitrogen-inlet needle connected to a nitrogen line and gas bubbler. The stirred solution is cooled in an ice-bath. n-Butyllithium is added dropwise via a weighed syringe until a yellow color persists. The syringe is weighed before and after addition (difference 1.05 g, 1.52 mL, corresponding to a solution molarity of 2.37).

13. The syringe is weighed before and after addition to determine the amount of n-butyllithium and t-butyl (E)-4-methylpent-2-enoate 1 added.

14. With the addition of the first few drops of n-butyllithium the reaction mixture turns from clear and colorless to clear and light pink, and darkens as further n-butyllithium solution is added.

15. With the addition of the first few drops of t-butyl (E)-4-methylpent-2-enoate 1 solution, the reaction mixture turns from dark pink to bright orange, which persists as further t-butyl (E)-4-methylpent-2-enoate 1 solution is added. The solution remains orange for the rest of the reaction.

16. The reaction is followed by ^{1}H NMR as follows. A 0.1 mL reaction aliquot is quenched into a mixture of 1 mL of CDCl$_3$ and 1 mL of saturated aqueous ammonium chloride. The organic layer is separated and filtered through a plug of sodium sulfate and cotton into an NMR tube. The olefin peaks at 5.7 and 6.8 ppm of the starting material are compared to the product peak at 3.2 ppm to assess reaction completion. After a 1.5 h reaction time, 13% starting material remained.

17. The submitters quenched the reaction at −78 °C after 2 h. In cases where the intermediate lithium β-amino enolate is unstable with respect to retro-conjugate addition at elevated temperatures, the quench should be performed at −78 °C.

18. ^{1}H NMR analysis of the crude product indicated about 1% unreacted starting material.

19. The citric acid wash removes >95% of the unreacted (*R*)-*N*-benzyl-*N*-(α-methylbenzyl)amine.

20. A 3-cm glass column is wet-packed (hexanes) with silica gel (100 g) topped with 0.5 cm sand. The crude reaction product **2** is dissolved in dichloromethane (10 mL), loaded on the column, and eluted with 1 L of a 97:3 mixture of hexanes:ethyl acetate, taking 75 mL fractions. The chromatography is monitored by TLC (R_f = 0.5 in ethyl acetate:hexanes, 5:95). The product elutes in fractions 3-10. The eluent is concentrated by rotary evaporation (40 °C water bath, 20 mmHg) to constant weight.

21. *t*-Butyl (3*S*,α*R*)-3-[*N*-benzyl-*N*-(α-methylbenzyl)amino]-4-methyl-pentanoate **2** has the following physical and spectroscopic data: $[\alpha]^{22}_D$ −1.7 (*c* 2.0, chloroform); IR (thin film) v_{max} 3102, 3080, 2973, 1947, 1875, 1807, 1731, 1601, 1454, 1369, 950 cm^{-1}; ^{1}H NMR (400 MHz, CDCl$_3$) δ: 0.89 (3 H, d, *J* = 6.8 Hz, C(4)*Me*$_A$), 1.12 (3 H, d, *J* = 6.8 Hz, C(4)*Me*$_B$), 1.41 (3 H, d, *J* = 7.1 Hz, C(α)*Me*), 1.42 (9 H, s, C*Me*$_3$), 1.67–1.73 (1 H, m, C(4)*H*), 1.81 (1 H, dd, *J* = 16.1, 2.0 Hz, C(2)*H*$_A$), 1.97 (1 H, dd *J* = 16.1, 9.5 Hz, C(2)*H*$_B$), 3.24–3.28 (1 H, m, C(3)*H*), 3.49 (1 H, d, *J* = 15.0 Hz, NC*H*$_A$), 3.77 (1 H, d, *J* = 15.0 Hz, NC*H*$_B$), 3.74–3.81 (1 H, m, C(α)*H*), 7.22–7.48 (10 H, m, *Ph*); ^{13}C NMR (100 MHz, CDCl$_3$) δ: 19.8 (C(4)*Me*$_A$), 20.4 (C(α)*Me*), 21.3 (C(4)*Me*$_B$), 28.2 (C*Me*$_3$), 33.0 (C(4)), 36.5 (C(2)), 51.4 (NCH$_2$), 58.0 (C(3)), 58.2 (C(α)), 80.1 (*C*Me$_3$), 126.8, 127.1 (*p-Ph*), 128.2, 128.4, 128.5 (2 degenerate peaks) (*o-*, *m-Ph*), 141.8, 142.1 (*i-Ph*), 172.6 (C(1)); MS (ESI$^+$) *m/z* 382 ([M+H]$^+$, 100%); HRMS (ESI$^+$) calcd. for C$_{25}$H$_{36}$NO$_2^+$ ([M+H]$^+$) 382.2741; found 382.2742. An analytically pure sample was prepared by dissolving 200 mg of the product oil in 5 mL of 50:50 pentane:diethyl ether, filtering through a 0.45 micron PTFE syringe filter, concentrating under vacuum, and vacuum drying at 50 °C for 20 h: Anal. calcd. for C$_{25}$H$_{35}$NO$_2$: C, 78.70; H, 9.25; N, 3.67; found: C, 78.34; H, 9.15; N, 3.70.

22. The following materials used in step C were obtained from Sigma-Aldrich: methanol (ACS reagent grade, 99.8%), triethylamine (>99.5%, SureSeal bottle), ethyl acetate (ACS reagent grade, >99.5%), and hexanes (ACS reagent grade, >98.5%). 20% Palladium hydroxide (wet) on carbon was sourced from BASF.

23. The reaction was monitored by hydrogen pressure drop. Hydrogen uptake was complete in 1 h, but the hydrogenation was continued for 6 h to ensure complete conversion.

24. Once the cake is thoroughly rinsed with methanol to wash off all product, it is wetted with water and transferred as a water slurry to a PTFE bottle to hold the palladium waste for recycling.

25. A 3-cm glass column is wet-packed (1.5% triethylamine in hexanes) with silica gel (150 g) topped with 0.5 cm sea sand. The crude product is loaded neat to the column and the flask rinsed with dichloromethane (2 x 5 mL) to ensure complete transfer. The column is eluted as follows: (1) 400 mL of 1:3 ethyl acetate:hexanes containing 1.5% triethylamine, (2) 400 mL of 1:1 ethyl acetate:hexanes containing 1.5% triethylamine, (3) 400 mL ethyl acetate containing 2 % triethylamine, taking 50 mL fractions. The product **3** (R_f = 0.25 in 1:2 ethyl acetate:hexanes containing 1.5% triethylamine, visualized using both potassium permanganate (yellow spot on purple background) and iodine) is obtained in fractions 9-20, which are concentrated by rotary evaporation (40 °C water bath, 20 mmHg) in a 500 mL round-bottomed flask to constant weight (7.82 g, 83% yield). ^{1}H NMR analysis indicated <1 % ethyl acetate in the product. Fractions 8 and 21-25 were combined and concentrated to constant weight (0.55 g). The purity of these fractions was assessed by ^{1}H NMR as approx. 80% (0.44 g corrected for purity, 5% yield).

26. t-Butyl (S)-3-amino-4-methylpentanoate **3** has the following physical and spectroscopic data: $[\alpha]^{22}_D$ −24 (c 2.0, chloroform); IR (thin film) v_{max} 3387, 3321, 2964, 2933, 2874, 1727, 1597, 1466, 1392, 1367, 1152 cm^{-1}; ^{1}H NMR (400 MHz, CDCl$_3$) δ: 0.90 (3 H, d, J = 6.7 Hz, C(4)Me_A), 0.91 (3 H, d, J = 6.7 Hz, C(4)Me_B), 1.30 (2 H, br s, NH_2), 1.45 (9 H, s, CMe_3), 1.58-1.63 (1 H, m, C(4)H), 2.14 (1 H, dd, J = 15.3, 10.0 Hz, C(2)H_A), 2.37 (1 H, dd J = 15.3, 3.5 Hz, C(2)H_B), 2.98 (1 H, ddd, J = 10.0, 15.3, 3.5 Hz, C(3)H); ^{13}C NMR (100 MHz, CDCl$_3$) δ: 17.9, 19.0 (C(4)Me_2), 28.3 (CMe_3), 33.5 (C(4)), 41.2 (C(2)), 53.8 (C(3)), 80.6 (CMe_3), 172.6 (C(1)); MS (ESI$^+$) m/z 210 ([M+Na]$^+$, 100%); HRMS (ESI$^+$) m/z calcd. for C$_{10}$H$_{21}$NNaO$_2^+$ ([M+Na]$^+$) 210.1465; found 210.1469. An analytical sample is prepared by re-chromatographing a portion of the rich cut of the original chromatography, as follows: silica gel (15 g) is wet-packed (3:1 hexanes:ethyl acetate containing 1.5% triethylamine) in a 1-cm column topped with 0.5 cm sand. Compound **3** (0.53 g) is added neat to the column and is eluted sequentially with 30 mL 3:1 hexanes: ethyl acetate containing 1.5% triethylamine, 60 mL 1:1 hexanes: ethyl acetate containing 1.5% triethylamine, and 90 mL 1:2 hexanes: ethyl acetate, taking 10 mL fractions. Fractions 6 and 7 are combined, filtered through a 0.45 micron syringe filter

and concentrated by rotary evaporation to constant weight (290 mg). Anal. calcd. for $C_{10}H_{21}NO_2$: C, 64.13; H, 11.30; N, 7.48; found: C, 63.87; H, 11.20; N, 7.48.

27. The following materials used in step D were sourced from Sigma-Aldrich and used as received: trifluoroacetic acid (ReagentPlus, >99%), dichloromethane (ACS reagent grade, >99.5%), 2 M hydrogen chloride in diethyl ether, Dowex resin 50WX4-200, and racemic β-leucine. Ammonium hydroxide was sourced from Fisher.

28. ^{1}H NMR analysis (CD$_3$OD) indicated approx 8% unreacted starting material.

29. ^{1}H NMR analysis (CD$_3$OD) indicated approx 4% unreacted starting material and an equimolar quantity of diethyl ether.

30. A glass column (3 cm internal diameter) is wet-packed (water) with Dowex® 50WX4-200 ion-exchange resin (100 g). The column is equilibrated by sequential elution with water (200 mL), methanol (200 mL), water (200 mL), 1 M aqueous hydrogen chloride solution (200 mL) and water (500 mL). The crude reaction product **4** is diluted with distilled water (20 mL) and the resultant solution is loaded on to the column. The column is eluted with 100 mL of distilled water followed by 800 mL of 1 M aqueous ammonium hydroxide solution, taking 150 mL fractions. The fractions containing product are identified by spotting on a silica TLC plate and staining with potassium permanganate solution (yellow spot on purple background upon heating). Fractions 3-5 are concentrated by rotary evaporation (60 °C water bath, 20 mmHg) to constant weight in a 500-mL round-bottomed flask to afford 4.19–4.30 g (88-93% yield) of product **4**. This material contained 0.5 wt% water by Karl Fisher titration. Fractions 2, 6, and 7 are combined and concentrated to afford 4.75 g of a white solid that contained no product by ^{1}H NMR (peak at –75.5 ppm by ^{19}F analysis indicates the presence of a trifluoromethyl group).

31. Due to lack of a chromophore and lack of volatility, the purity of **4** could not be assessed by HPLC or GC. Therefore, the weight percent purity of **4** recovered from the ion-exchange chromatography was determined by ^{1}H NMR using both ethylene glycol and 1,2-dimethoxyethane as internal standards, as follows. β-Leucine (41.1 mg, 31.3 μmol) and 1,2-dimethoxyethane (67.1 mg, 74.5 μmol) are accurately weighed in a 5-mm NMR tube, then D$_2$O (0.7 mL) is added. The ^{1}H NMR analysis is carried out with a 5 second delay to ensure complete relaxation (10 second delay gave the same results, while a 0.1 second delay gave a 6% lower response for 1,2-

dimethoxyethane). The dimethyl resonances (apparent triplet) of the product at 0.9 ppm are integrated vs. the 4 CH_2 protons at 3.6 ppm for 1,2-dimethoxyethane. The average of several integrations gave a molar ratio of product:standard of 0.414 vs. the 0.420 ratio of the weighed samples, indicating a wt % of 98.6%. An additional weighing was carried out with 1,2-dimethoxyethane and 2 weighings and NMR analyses were carried out using ethylene glycol as standard, providing an average wt% assay of 98.4 ± 0.7% for material isolated directly from the ion-exchange chromatography.

32. A chiral HPLC assay was developed on the β-leucine N-p-toluenesulfonamide derivative **5**. The derivatization procedure is as follows:

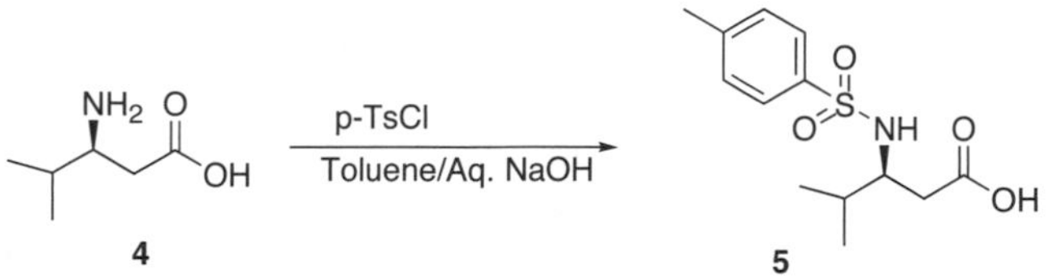

β-Leucine (108 mg, 0.82 mmol), p-toluenesulfonyl chloride (1.48 g, 7.8 mmol, 10 equiv), 2 N sodium hydroxide (7 mL, 14 mmol), toluene (7 mL), and a 1-cm oval Teflon-coated magnetic stir bar are added to a 50-mL round-bottomed flask sealed with a septum through which is inserted a thermocouple thermometer probe and an 18-gauge syringe needle connected to a nitrogen line and gas bubbler. The mixture is warmed to 55–60 °C using a heating mantle and vigorously stirred for 22 h at this temperature. The mixture is cooled, and transferred to a 50-mL separatory funnel along with water (5 mL) and toluene (5 mL). The layers are separated. 6 M aqueous hydrogen chloride (3 mL) is added to the aqueous layer. The mixture is transferred to a 50-mL separatory funnel and extracted with diethyl ether (2 x 15 mL). The organic layer is washed with water (10 mL), then filtered through a bed of magnesium sulfate and concentrated by rotary evaporation (40 °C water bath, 20 mmHg) to constant weight to afford (S)-3-[N-(p-toluenesulfonyl)amino]-4-methylpentanoic acid **5** (132 mg, 56% yield). This non-purified material is used to determine the optical purity of β-leucine using the chiral SFC method described below. A portion of **5** is recrystallized (90 mg dissolved in 3 mL toluene at 80 °C, cooled to ambient temperature, held for 15 h, filtered to afford 70 mg) to provide **5** with the following physical and spectroscopic data: mp 131–133 °C; ^{1}H NMR (400 MHz, CDCl$_3$) δ: 0.87 (6 H, app t, J = 6.7 Hz, C(4)Me_2), 1.88 (1 H, app octet, J = 6.8 Hz, C(4)H), 2.45 (1 H, dd, J = 16.2, 5.7 Hz, C(2)H_A), 2.46 (3 H, s,

CH_3), 2.54 (1 H, dd, J = 16.2, 5.0 Hz, C(2)H_B), 3.32–3.41 (1 H, m, C(3)H), 5.17 (1 H, d, J = 9.2 Hz, NH), 7.30–7.34 (2 H, m, Ar), 7.78–7.81 (2 H, m, Ar); ^{13}C NMR (100 MHz, CDCl$_3$) δ: 18.6, 19.0 (C(4)Me_2), 21.6 (ArMe), 31.6 (C(4)), 36.1 (C(2)), 56.0 (C(3)), 127.2, 129.7, 137.8, 143.5 (Ar), 175.8 (C(1)). The racemic derivative was similarly prepared and recrystallized from toluene, providing racemic **5** with mp 117–119 °C. A chiral supercritical fluid chromatography (SFC) method was developed for chiral analysis of the 4-toluenesulfonamide derivative (**5**): AD-H (250 x 4.6 mm, 5 µm) column, gradient method using methanol with 25mM *i*-butylamine, 4% methanol /CO$_2$ for 4 min then ramp at 6%/min to 40% methanol /CO$_2$, hold at 40% for 5 minutes, 3.0 mL/min, 200 bar, 35 °C, 230 nm, 15 minutes run time; minor enantiomer elutes at 7.3 min, major at 8.0 min. The enantiomeric excess of the derivatized, non-recrystallized **5** derived from β-leucine isolated directly from the resin column was 96% ee. Since the enantiomeric excess of (*R*)-*N*-benzyl-*N*-(α-methylbenzyl)amine used in the conjugate addition was 97%, the reaction diastereoselectivity of this step was 99%.

33. An analytically pure sample is prepared by recrystallization. β-Leucine **4** (1.04 g) is added to methanol (20 mL) in a 100-mL round bottomed flask with a 1-cm oval Teflon-coated stir bar and warmed in a 50 °C oil bath with stirring to dissolve the solids. The solution is quickly poured into a 50-mL syringe and hot filtered through a 0.45 micron PTFE syringe filter into a 100-mL round-bottomed flask. The solution is concentrated by rotary evaporation (40 °C water bath, 20 mmHg) to 5 mL. Partial crystallization occurs during this concentration step. After concentration, a 1-cm oval Teflon-coated stir bar is added to the flask, then *t*-butyl methyl ether (15 mL) is added at 22 °C over 20 min to the stirred mixture, resulting in further crystallization. After the addition is complete, the mixture is stirred for 1 h at 21–22 °C, then vacuum filtered through a 30-mL medium porosity sintered glass funnel, washed with *t*-butyl methyl ether (5 mL) and dried in a vacuum oven for 3 h at 60 °C to afford (*S*)-3-amino-4-methylpentanoic acid **4** (0.82 g, 79% yield) as a white crystalline powder. The ee of the recrystallized β-leucine is only marginally improved over the crude material (97 vs 96%).

34. (*S*)-3-Amino-4-methylpentanoic acid **4** purified by recrystallization has the following physical and spectroscopic data: ee 97%; mp 197–198 °C, lit.[6a] 182 °C, lit.[6b] 201-202.5 °C, lit.[6c] 202–210 °C, lit.[6d] 206 °C, lit.[6e] 212 °C; $[\alpha]_D^{20}$ –53 (*c* 2.0, water), –52 (c 0.5, water), –40 (c 0.5, 1.N hydrochloric

acid), lit.[6a] +40.3 [(R)-isomer, c 1, water], lit.[6b] +55.2 [(R)-isomer, c 1, water], lit.[6c] −39.2 [(S)-isomer, c 0.5, water], lit.[6d] +47 [(R)-isomer, c 1, water], lit.[6e] +52.3 [reported as (S)-isomer, but drawn as (R)-isomer, c 0.6 water], lit.[6f] +51.5 [reported as (S)-isomer, c 0.6, water]; IR (KBr) v_{max} 3397, 2965, 2936, 1649, 1621, 1467, 1326, 1262 cm^{-1}; ^{1}H NMR (400 MHz, D$_2$O) δ: 0.92 (6 H, app t, J = 7.0 Hz, C(4)Me_2), 1.88 (1 H, app octet, J = 6.8 Hz, C(4)H), 2.33 (1 H, dd, J = 16.7, 9.3 Hz, C(2)H_A), 2.50 (1 H, dd J = 16.7, 4.2 Hz, C(2)H_B), 3.24–3.28 (1 H, m, C(3)H); ^{13}C NMR (100 MHz, D$_2$O) δ: 17.3, 17.4 (C(4)Me_2), 30.0 (C(4)), 36.0 (C(2)), 54.9 (C(3)), 178.5 (C(1)); m/z (ESI^{+}) 154 ([M+Na]$^{+}$, 100%); HRMS (ESI^{+}) m/z calcd. for C$_6$H$_{13}$NNaO$_2$$^{+}$ ([M+Na]$^{+}$) 154.0838; found 154.0841; Anal. calcd. for C$_6$H$_{13}$NO$_2$: C, 54.94; H, 9.99; N, 10.68; found: C, 54.84; H, 9.95; N, 10.64.

Safety and Waste Disposal Information

All hazardous materials should be handled and disposed of in accordance with "Prudent Practices in the Laboratory"; National Academy Press; Washington, DC; 1995.

3. Discussion

The conjugate addition reaction was first reported by Komnenos in 1883, who demonstrated the 1,4-addition of diethyl sodiomalonate to diethyl ethylidenemalonate.[7] A range of carbon and heteroatom based nucleophiles have since been shown to participate in this reaction manifold. In 1991, Davies and Ichihara described the highly diastereoselective conjugate addition of lithium (R)-N-benzyl-N-(α-methylbenzyl)amide 6 to benzyl crotonate 12, which gave β-amino ester (3R,αR)-18 in 95% de. Global hydrogenolytic N-deprotection of β-amino ester mediated by Pearlman's catalyst [Pd(OH)$_2$/C] proceeded under 5 atm of hydrogen to give the corresponding free β-amino acid, (R)-3-aminobutanoic acid, in quantitative yield and >95% ee.[8] This methodology has since been developed into a generally applicable synthesis of either enantiomer of homochiral β-amino acids, via conjugate addition of homochiral lithium N-benzyl-N-(α-methylbenzyl)amide to an α,β-unsaturated t-butyl ester,[9] followed by hydrogenolytic N-debenzylation and ester hydrolysis.[10] A range of homochiral lithium amides that are readily derived from commercially available, homochiral α-methylbenzylamine derivatives has been developed, which allow for either differential *N*-deprotection or further elaboration in synthesis. All members of this family of lithium amides undergo highly diastereoselective conjugate addition to a wide range of α,β-unsaturated esters and amides to give the corresponding diastereo- and enantiomerically pure, homochiral β-amino ester or amide product (Table 1). This lithium amide conjugate addition methodology has been expanded to allow the stereoselective, in situ elaboration of the intermediate lithium (*Z*)-β-amino enolate to give access to homochiral α-substituted-β-amino esters; it has been employed for the synthesis of hundreds of β-amino esters, amides and acids in enantiomerically pure form, and has found utility in a plethora of synthetic applications, including total syntheses, initiation of tandem asymmetric processes, and molecular recognition phenomena. Its scope and utility was comprehensively reviewed in 2005.[11]

Table 1 Representative members of the lithium amide family **6-11** and α,β-unsaturated esters **12-17** for conjugate addition.

Lithium amide	α,β-Unsaturated carbonyl	Conjugate addition product	Yield (de)	Reference
(R)-6	12	18	88 (95)	8
(R)-7	13	19	72 (90)	12
(R)-8	14	20	73(>96)	13
(R)-9	15	21	71 (91)	14
(R)-10	16	22	86 (90)	15
(S)-11	17	23	82 (>97)	16

1. Department of Chemistry, Chemistry Research Laboratory, University of Oxford, Mansfield Road, Oxford, OX1 3TA, UK. E-mail: steve.davies@chem.ox.ac.uk.

2. The checker would like to thank Zainab Pirzada for development of the chiral HPLC assay, Scott Hoerrner and Anthony Houck for assistance with the hydrogenation experiments, and Mirlinda Biba for determination of the specific rotations.

3. For a representative experimental procedure see: Harwood, L. M.; Moody, C. J.; Percy, J. M. *Experimental Organic Chemistry – Standard and Microscale*, **1999**, 2nd ed, Blackwell Publishing, p. 569-570.

4. Love, B. E.; Jones, E. G. *J. Org. Chem.* **1999**, *64*, 3755-3756.

5. Claridge, T. D. W.; Davies, S. G.; Polywka, M. E. C.; Roberts, P. M.; Russell, A. J.; Savory, E. D.; Smith, A. D. *Org. Lett.* **2008**, *10*, 5433-5436.

6. (a) Enders, D.; Wahl, H.; Bettray, W. *Angew. Chem. Int. Ed.* **1995**, *34*, 455-457. Reported ee 98%. (b) Yamada, T.; Kuwata, S.; Watanabe, H. *Tetrahedron Lett.* **1978**, *21*, 1813-1816. (c) Balenovic, K.; Dvornik, D. *J. Chem. Soc.* **1954**, 2976. (d) Callens, R.; Larcheveque, M.; Pousset, C.; Patent #FR2853315 (A1) August 10, 2004. (e) Okamoto, S.; Harada, T.; Tai, A. *Bull. Chem. Soc. Jpn.* **1979**, *52*, 2670-2673. (f) Evans, D. A.; Wu, L. D.; Wiener, J. J. M.; Johnson, J. S.; Ripin, D. H. B.; Tedrow, J. S. *J. Org. Chem.* **1999**, *64*, 6411-6417. Reported ee 98%. Designation of (*S*)-isomer for β-leucine having a positive rotation is inconsistent with other literature reports.

7. Komnenos, T. *Liebigs Ann. Chem.* **1883**, *218*, 145-169.

8. Davies, S. G.; Ichihara, O. *Tetrahedron: Asymmetry* **1991**, *2*, 183-186.

9. Claridge, T. D. W.; Davies, S. G.; Lee, J. A.; Nicholson, R. L.; Roberts, P. M.; Russell, A. J.; Smith, A. D.; Toms, S. M. *Org. Lett.* **2008**, *10*, 5437-5440.

10. (a) Davies, S. G.; Garrido, N. M.; Kruchinin, D.; Ichihara, O.; Kotchie, L. J.; Price, P. D.; Price Mortimer, A. J.; Russell, A. J.; Smith, A. D. *Tetrahedron: Asymmetry* **2006**, *17*, 1793-1811; (b) Davies, S. G.; Mulvaney, A. W.; Russell, A. J.; Smith, A. D. *Tetrahedron: Asymmetry* **2007**, *18*, 1554-1566.

11. Davies, S. G.; Smith, A. D.; Price, P. D. *Tetrahedron: Asymmetry* **2005**, *16*, 2833-2891.

12. Okamoto, S.; Iwakubo, M.; Kobayashi, K.; Sato, F. *J. Am. Chem. Soc.* **1997**, *119*, 6984-6990.

13. Davies, S. G.; Haggitt, J. R.; Ichihara, O.; Kelly, R. J.; Leech, M. A.; Price Mortimer, A. J.; Roberts, P. M.; Smith, A. D. *Org. Biomol. Chem.* **2004**, *2*, 2630-2649.
14. Davies, S. G.; Smyth, G. D. *Tetrahedron: Asymmetry* **1996**, *7*, 1005-1006.
15. Bull, S. D.; Davies, S. G.; Delgado-Ballester, S.; Kelly, P. M.; Kotchie, L. J.; Gianotti, M.; Laderas, M.; Smith, A. D. *J. Chem. Soc., Perkin Trans. 1* **2001**, 3112-3121.
16. Ma, D.; Sun, H. *Tetrahedron Lett.* **2000**, *41*, 1947-1950.

Appendix
Chemical Abstracts Nomenclature (Registry Number)

t-Butyl diethylphosphonoacetate (27784-76-5)

Methylmagnesium bromide (75-16-1)

Isobutyraldehyde; 2-methylpropionaldehyde; 2-methylpropanal (78-84-2)

t-Butyl (*E*)-4-methylbut-2-enoate (87776-18-9)

(*R*)-α-Methylbenzylamine; (*R*)-α-phenylethylamine; (*R*)-1-phenylethylamine (3886-69-9)

(*R*)-*N*-Benzyl-*N*-(α-methylbenzyl)amine; (*R*)-*N*-benzyl-α-phenylethylamine; (*R*)-*N*-benzyl-1-phenylethylamine (38235-77-7)

Butyllithium; lithium-1-butanide (109-72-8)

t-Butyl 3-[*N*-benzyl-*N*-(α-methylbenzyl)amine]-4-methylpentanoate (38235-77-7)

Palladium hydroxide on carbon; Pearlman's catalyst (12135-22-7)

t-Butyl 3-amino-4-methylpentanoate (202072-47-7)

Trifluoroacetic acid (76-05-1)

(*S*)-3-Amino-4-methylpentanoic acid; (*S*)-β-homovaline; (*S*)-β-leucine (40469-85-0)

(*S*)-3-[*N*-(*p*-toluenesulfonyl)amino]-4-methylpentanoic acid (936012-07-6)

Steve Davies obtained a B.A. in Chemistry (1973) and D.Phil. in Organic Synthesis (1975) at the University of Oxford, following which he undertook postdoctoral fellowship positions with Professor M. L. H. Green at Oxford and with Professor Sir Derek Barton in France. He began his independent research career first as a member of the C.N.R.S. (1978-80) in France, followed by appointment as a Lecturer in Organic Chemistry (1980-96), Professor of Chemistry (1996-2006), and Chairman of Chemistry and Waynflete Professor of Chemistry (2006 to date) at Oxford. The development of novel and efficient methods for the production of enantiomerically pure compounds has formed the main focus of his research; highlights include stereoselective organometallic chemistry, the development of chiral relay networks, and homochiral ammonia equivalents.

Paul Roberts graduated with an M.Chem. from Jesus College, Oxford, in 2000, which was followed by a D.Phil. with Professor Steve Davies in the area of the asymmetric synthesis of piperidine alkaloids employing a ring closing metathesis approach. In 2005, he took up a post-doctoral position with Professor Davies at Oxford, where his research interests centre upon natural product synthesis and the development of new stereoselective methodologies, for example to effect the chemo- and stereoselective functionalisation of allylic amines with a range of electrophilic reagents.

After graduating from Keio University, Japan, in 2000, Ai (Matsuno) Fletcher studied for a Ph.D. at Imperial College London under supervision of Dr Chris Braddock with an O.R.S. award. During her Ph.D. she developed two novel methodologies: "one-pot" cascade catalysis of allylic isomerisation and olefin metathesis, and the cyclopropyl methyl silane terminated Prins reaction. Since completing her Ph.D. in 2004, she explored a range of chemistry, such as enantioselective synthesis of DNA analogues and palladium catalysis as a post-doctoral researcher at the University of Regensburg, and at the University of Bath. In 2007 she joined the group of Professor Steve Davies in Oxford, where she has been involved with the development of asymmetric synthetic methodology and novel ammonia-based energy technology.

160

CYCLOHEXENE IMINE
(7-AZA-BICYCLO[4.1.0]HEPTANE)

A.

B.

Submitted by Iain D. G. Watson, Nicholas Afagh and Andrei K. Yudin.[1]
Checked by Lars Troendlin and Andreas Pfaltz.

1. Procedure

WARNING: Organic azides can be unstable and/or explosive.

A. trans-2-Azido-cyclohexanol. In a 500-mL, one-necked, round-bottomed flask equipped with a magnetic stir bar (4 cm, cylindrical-shaped) is placed sodium azide (25.2 g, 388 mmol, 2.5 equiv) (Note 1). Water (85 mL) is added and the solution is stirred at room temperature until all solid has dissolved. Acetone (60 mL) is added followed by a slow and continuous dropwise addition of cyclohexene oxide (15.5 mL, 153 mmol, 1.0 equiv) in acetone (25 mL) using a pressure equalizing dropping funnel over 30 min (Note 2). The flask is fitted with a reflux condenser and the solution is heated to reflux for 18 h (85 °C), after which acetone is evaporated *in vacuo* and the residual solution is extracted with diethyl ether (3 x 200 mL). The combined organic phases are washed with brine (100 mL) and dried over Na_2SO_4. The solvent is removed *in vacuo* (Note 3) and *trans*-2-azido-cyclohexanol is obtained in 99% yield as a yellow oil (21.6 g, 153 mmol), which is reacted immediately and without further purification (Note 4).

B. Cyclohexene imine. Into a flame-dried, 500-mL, three-necked, round bottomed flask equipped with a magnetic stir bar (4 cm, cylindrical-shaped), reflux condenser, a pressure equalizing dropping funnel, internal thermometer and high-vacuum-argon connection is added the azido alcohol (21.6 g, 153 mmol, 1.0 equiv) followed by freshly distilled methyl *tert*-butyl ether (MTBE) (40 mL) (Note 5). Triphenylphosphine (40.2 g, 153 mmol, 1.0 equiv) is dissolved in MTBE (100 mL) and the solution is transferred into the dropping funnel. The solution is added slowly and continuously over 45 min (Note 6). The addition is accompanied by the vigorous evolution of nitrogen gas and a temperature increase to 40 °C. After the addition of the triphenylphosphine solution, the dropping funnel is washed with TBME (20 mL). After nitrogen evolution has ceased, the solution is heated to reflux (70 °C) for 16 h under a nitrogen atmosphere. The solution is then transferred in a well ventilated hood to a single-necked, 1-L evaporating flask (Note 7). MTBE is removed at 40 °C under reduced pressure (360 mmHg) on a rotary evaporator (Note 8) leaving an off-white solid (Ph_3PO), within which is trapped the cyclohexene imine product. The cyclohexene imine is distilled *in vacuo* (0.01 mmHg) at 160 °C from the resulting residue using a short path distillation apparatus to afford a clear, colorless liquid (Note 9). Residual MTBE, which will inevitably co-distill with the product, can be removed by freeze-drying the resulting liquid. In this procedure, the flask containing the cyclohexene imine is immersed in a -12 °C sodium chloride/ice cold bath and the flask is rotated such that as cyclohexene imine begins to solidify, it forms a thin coat along the interior of the flask effectively maximizing the surface area and facilitating removal of MTBE. While still immersed in the cold-bath, a spatula is used to scratch the sides of the flask to break up the thin layer of cyclohexene imine to afford fine flakes. The residual MTBE is then removed *in vacuo* (0.13 mbar) while the flask remains immersed in the cold-bath to afford pure cyclohexene imine (13.6 g, 140 mmol) in 91% yield as fine white crystals below 5 °C and a clear colorless liquid above this temperature (Notes 10 and 11; for GC data, see Note 12).

2. Notes

1.   All reagents were purchased from Sigma-Aldrich. MTBE was purchased from Sigma-Aldrich; all other solvents were purchased from EMD Chemicals. Sodium azide must be measured with a plastic or glass spoon. Do not allow sodium azide to come in contact with metal or chlorinated solvents.

2.   If cyclohexene oxide is added neat, all at once, a white insoluble polymeric material is formed which is deposited on the side of the flask, negatively affecting the isolated yield.

3.   It is important that the product is dried as water will negatively affect subsequent synthetic manipulations. After isolation using the rotary evaporator, residual water is removed from the product using a vacuum pump (0.1 mmHg).

4.   *WARNING: organic azides can be unstable and/or explosive.* ^{1}H NMR (CDCl$_3$, 400 MHz) δ: 1.23–1.39 (m, 4 H), 1.71–1.78 (m, 2 H), 1.99–2.08 (m, 2 H), 2.31 (bs, 1 H), 3.15–3.21 (m, 1 H), 3.35–3.40 (m, 1 H); ^{13}C NMR (CDCl$_3$, 100 MHz) δ: 23.8, 24.2, 29.7, 33.0, 67.1, 73.6. If desired, the intermediate *trans*-2-azido-cyclohexanol can be further purified by flash chromatography on SiO$_2$ (R$_f$= 0.15, 95:5 hexanes/EtOAc).

5.   Anhydrous MTBE was obtained by distillation over 4Å molecular sieves (30 g for 300 mL of MTBE).

6.   The slow, continuous addition of triphenylphosphine under an atmosphere of N$_2$ was found to be essential in obtaining high yields. The submitters added solid triphenylphosphine using a pressure-equalizing powder addition funnel.

7.   Transferring the solution to the larger 1 L evaporating flask avoids complications arising from bumping of the residual solid during the two distillation steps.

8.   The checkers used a rotary evaporator for removing the solvent. The submitters performed the distillation in a short path apparatus. The connector between the flasks should have a sharp angle to prevent bumped triphenylphosphine oxide from contaminating the collected product. A piece of glass wool may be optionally inserted into the connector to avoid contamination of the collected product with bumped impurities. The collecting flask must be *completely* immersed in dry ice/acetone to prevent loss of volatile cyclohexene imine.

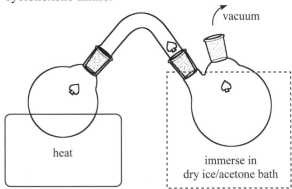

9. Due to the low melting point of cyclohexene imine, it will occasionally freeze in the condenser (indicated by the arrow) at which point the distillation will arrest. If this occurs, then the vacuum can be temporarily removed and the water flowing through the condenser can be shut off and the condenser area gently heated with a heat-gun to melt the frozen product into the collecting flask. This opportunity can also be used to break up the solid in the evaporating flask using a metal spatula.

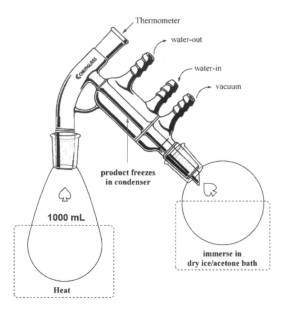

10. Cyclohexene imine is stored over KOH pellets (4 pellets for 13.6 g of product) at –20 °C. Aziridines should always be considered as highly toxic compounds and treated with care in ventilated spaces.

11. ^{1}H NMR (CDCl$_3$, 400 MHz) δ: 0.41 (br, 1 H), 1.16–1.36 (m, 4 H), 1.77 (s, 4 H), 2.14 (s, 2 H), 3.21 (d, J = 2 Hz); ^{13}C NMR (CDCl$_3$, 100 MHz) δ: 20.4, 24.5, 29.2. MS (ESI) m/z (rel. intensity) 196 ([2M]$^+$, 75), 98 ([M+1]$^+$, 100), 65 (30).

12. GC analysis was conducted immediately after distillation of cyclohexene imine. A sample of cyclohexene imine was dissolved in acetonitrile (HPLC grade) and subjected to GC analysis, which revealed the presence of 4 peaks (relative to MeCN blank). The major peak (t_R = 9.750 min) accounted for 98.9% of the total area of all peaks. Gas-phase chromatography was performed on an HP-6890 series instrument using an HP-5 column (crosslinked 5% phenyl methyl siloxane, 30 m x 0.32 mm x

164

0.25 µm film thickness). The oven was heated at 30 °C for 5 min followed by a temperature gradient of 5 °C/min to 120 °C. Inlet temperature and pressure were 200 °C and 4.88 psi respectively, with a split ratio of 50:1. Hydrogen was the carrier gas.

The checkers used a Carlo Erba Instrument GC 8000Top, an Restek Rtx-1701 column (14% cyanpropylphenyl - 86% dimethylpolysiloxan, 30 m x 0.25 mm x 0.25 µm film thickness). A sample of cyclohexene imine was dissolved in TBME. The oven temperature was kept at 80 °C for 5 min followed by a temperature gradient of 5 °C/min to 130 °C and then 10 °C/min to 250 °C and, finally, at 250 °C for 5 min. Inlet temperature was 220 °C and the pressure was 60 kPa, helium was the carrier gas. The major peak (t_R = 10.64 min) accounted for 96.6% of the total area of all peaks.

Safety and Waste Disposal Information

Sodium azide should not be allowed to come into contact with any metal or chlorinated solvents. The organic azide intermediate should be treated as potentially explosive and reacted immediately. Aziridines should always be considered as highly toxic compounds and to only be used in ventilated fume hoods.

All hazardous materials should be handled and disposed of in accordance with "Prudent Practices in the Laboratory;" National Academy Press; Washington, DC, 1995.

3. Discussion

N-Unsubstituted, or NH-aziridines have been used as key substrates in a number of different reactions over the course of our work in this area.[2] These substrates may be prepared in a number of different ways: by Staudinger, Wenker, Gabriel or conjugate addition strategies.[3] Starting materials for NH-aziridines may include epoxides, aminoalcohols, or α,β-unsubstituted carbonyl compounds.

Over the course of the work on the functionalization of aziridines, cyclohexene imine was used extensively as a test substrate (Figure 1). This molecule contains several favorable properties, including relative ease of preparation. It can be distilled below 60 °C under reduced pressure allowing for easy purification. The material is not volatile enough to evaporate while heating over the course of long reaction times. This aziridine is also less likely to open than monosubstituted aziridines, allowing the use of harsher reaction conditions without loss of selectivity from background ring-opening of starting materials or products.

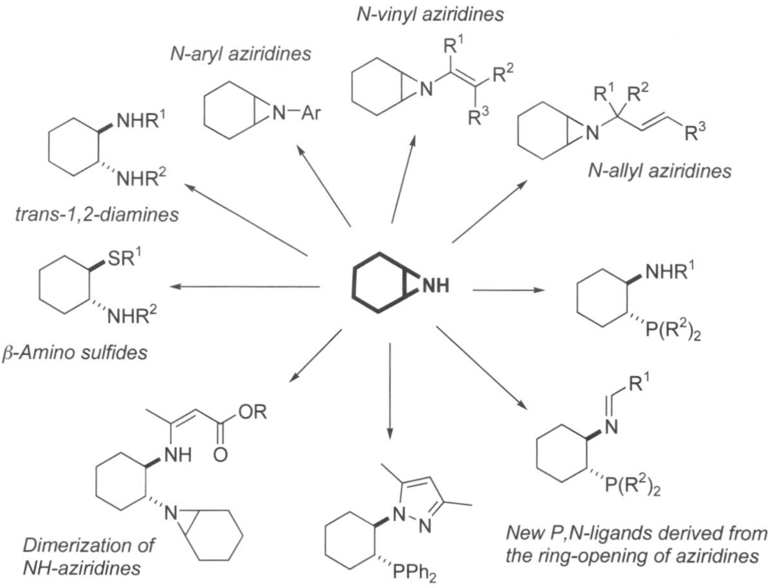

Figure 1. Cyclohexene imine as a precursor for a wide variety of nitrogen containing functionalities.

The synthesis of cyclohexene imine was accomplished by a two-step protocol.[4] First of all, the ring opening of cyclohexene oxide with sodium azide provided the azidoalcohol intermediate in quantitative yield. This intermediate, isolated by extraction, was then reacted without further purification by the addition of one equivalent of triphenylphosphine.

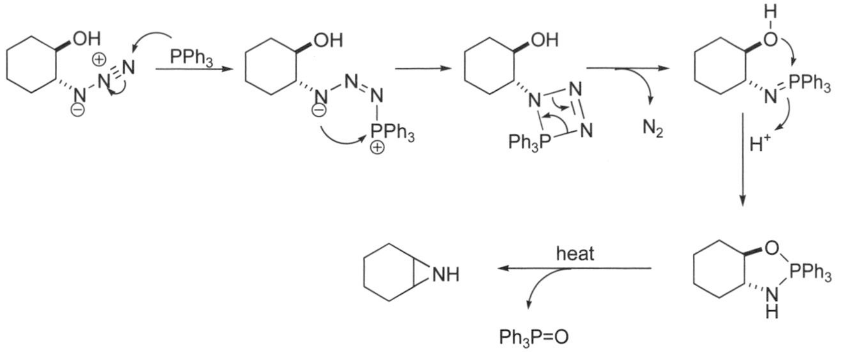

Figure 2. Mechanism of the Staudinger reaction, formation of cyclohexene imine.

The second step of the protocol, the Staudinger reaction, proceeds via initial reduction of the azide group to generate a five-membered intermediate (Figure 2). Decomposition of the intermediate then generates the aziridine by exclusion of triphenylphosphine oxide. Reduction of the azide group occurs immediately upon addition of triphenylphosphine, evident by bubbling caused by the evolution of nitrogen gas. However, heating the solution is necessary for the production of the aziridine. In fact, when the reaction was performed in diethyl ether, yields were much lower than when THF or methyl *tert*-butyl ether were used, owing to the higher boiling points of the latter solvents.

1. Davenport Research Laboratories, Department of Chemistry, The University of Toronto, 80 St. George Street, Toronto, Ontario, Canada, M5S 3H6.
2. (a) Watson, I. D. G.; Yu, L.; Yudin, A. K. *Acc. Chem. Res.* **2006**, 39, 194-206. (b) Watson, I. D. G.; Yudin, A. K. *J. Am. Chem. Soc.* **2005**, 127, 17516 - 17529. (c) Watson, I. D. G.; Styler, S. A.; Yudin, A. K. *J. Am. Chem. Soc.* **2004**, 126, 5086-5087. (d) Alphonse, F.-A.; Yudin, A. K. *J. Am. Chem. Soc.* **2006**, 128, 11754-11755. (e) Watson, I. D. G.; Yudin, A. K. *J. Org. Chem.* **2003**, 68, 5160-5167. (f) Dalili, S.; Yudin, A. K. *Org. Lett.* **2005**, 7, 1161-1164. (g) Dalili, S.; Caiazzo, A.; Yudin, A. K. *J. Organomet. Chem.* **2004**, 689, 3604-3611. (h) Sasaki, M.; Dalili, S.; Yudin, A. K. *J. Org. Chem.* **2003**, 68, 2045-2047. (i) Caiazzo, A.; Dalili,

S.; Yudin, A. K. *Synlett* **2003**, 2198-2202. (j) Caiazzo, A.; Dalili, S.; Yudin, A. K. *Org. Lett.* **2002**, *4*, 2597-2600.

3. *Aziridines and Epoxides in Organic Synthesis*; Yudin, A. K., Ed.; Wiley-VCH: Weinheim, Germany, 2006.

4. (a) Staudinger, H.; Meyer, J. *Helv. Chim. Acta.* **1919**, *2*, 635-646. (b) Ittah, Y.; Sasson, Y.; Shahak, I.; Tsaroom, S.; Blum, J. *J. Org. Chem.* **1978**, *43*, 4271-4273. (c) Christoffers, J.; Schulze, Y.; Pickardt, J. *Tetrahedron* **2001**, *57*, 1765-1769.

Appendix
Chemical Abstracts Nominclature; (Registry Number)

Cyclohexene oxide, 98%; (286-20-4)
Triphenylphosphine, 99%; (603-35-0)
Sodium azide, ≥99.5%; (26628-22-8)

Professor Andrei K. Yudin obtained his B.Sc. degree at Moscow State University and his Ph.D. degree at the University of Southern California under the direction of Professors G. K. Surya Prakash and George A. Olah. He subsequently took up a postdoctoral position in the laboratory of Professor K. Barry Sharpless at the Scripps Research Institute. In 1998, he started his independent career at the University of Toronto. He received early tenure in 2002 and became Full Professor in 2007. He is a recipient of a number of awards and his research interests are in development and application of novel synthetic methods that enable discovery of functionally significant molecules.

Iain Watson was born in Toronto, Canada. He is a medicinal chemist at the Ontario Institute for Cancer Research (OICR) in Toronto. Iain worked on the development of asymmetric gold(I)-catalyzed cycloisomerization reactions during his postdoctoral studies at the University of California, Berkeley with Professor F. Dean Toste. He received his Ph.D. from the University of Toronto in 2006, working with Professor Andrei Yudin in the fields of amination and palladium catalysis.

168

Nicholas A. Afagh received a B.Sc. in biochemistry from the University of Ottawa in 2007, working under the supervision of Prof. Robert N. Ben. He is currently a graduate student in the group of Professor Andrei K. Yudin at the University of Toronto, where he is investigating the synthetic applications of *N*-alkenyl aziridines.

Lars Tröndlin was born in 1977 in Lörrach, Germany. He studied Chemistry at the University of Basel where he obtained his M.Sc. in 2006 under the supervision of Prof. Andreas Pfaltz. He began his Ph.D. work in summer 2006 in the group of Prof. Andreas Pfaltz, where he is currently working on synthesis of new chiral ligands for metal-catalyzed reactions.

PREPARATION OF (*E*)-(2-IODOVINYL)BENZENE FROM BENZYL BROMIDE AND DIIODOMETHANE [(*E*)-β-STYRYL IODIDE]

a) NaHMDS, CH$_2$I$_2$
   THF/ether, −78 °C
b) addition of PhCH$_2$Br
   THF, −78 °C to rt
c) DBU, rt

Submitted by James A. Bull, James J. Mousseau and André B. Charette.[1]
Checked by Dominik J. Frank and Andreas Pfaltz.

1. Procedure

(E)-(2-Iodovinyl)benzene. A flame-dried, 500-mL, single-necked, round-bottomed flask is charged with sodium bis(trimethylsilyl)amide (19.8 g, 108 mmol, 3.60 equiv) (Note 1), and equipped with a 2.5-cm egg-shaped stir bar and with a rubber septum pierced by an argon inlet needle. Anhydrous tetrahydrofuran (72 mL) (Note 2) is added to the flask via syringe and stirring commenced. Once the base is completely dissolved, giving a yellow solution, anhydrous diethyl ether (72 mL) (Note 3) is added to the flask in two portions using a 50-mL syringe. The solution is cooled to −78 °C (bath temperature) in a dry ice/acetone bath (Note 4).

Concurrently, solutions of diiodomethane and benzyl bromide in tetrahydrofuran are prepared. *Diiodomethane*: A flame-dried, 50-mL, single-necked, round-bottomed flask equipped with a rubber septum pierced by an argon inlet needle is charged with diiodomethane (3.62 mL, 45.0 mmol, 1.50 equiv) via syringe through the septum (Note 5). Anhydrous tetrahydrofuran (15 mL) is added to the flask via syringe and the flask swirled to obtain a uniform solution. *Benzyl bromide:* A flame-dried, 50-mL, single-necked, round-bottomed flask equipped with a rubber septum and fitted with an argon inlet needle is charged with benzyl bromide (3.56 mL, 30.0 mmol, 1.00 equiv) via syringe through the septum (Note 6). Anhydrous tetrahydrofuran (15 mL) is added to the flask via syringe and the flask swirled to obtain a uniform solution.

The solution of diiodomethane in THF is cooled to −78 °C (bath temperature) in a dry ice/acetone bath, then added via cannula to the stirred solution of NaHMDS in THF/ether over 5 to 8 min in the dark (Notes 7 and

Org. Synth. **2010**, *87*, 170-177
Published on the Web 3/11/2010

8). The dark red/brown mixture is stirred at −78 °C. The solution of benzyl bromide in THF is cooled to −78 °C (bath temperature) in a dry ice/acetone bath.

Twenty minutes after completion of the addition of diiodomethane, the cold benzyl bromide solution is added via cannula to the reaction flask over 5 to 8 min in the dark. After complete addition, the stirred mixture is allowed to slowly warm to room temperature over 14 h to 18 h (Note 9). 1,8-Diazabicyclo[5.4.0]undec-7-ene (DBU) (4.48 mL, 30.0 mmol, 1.00 equiv) is then added to the dark suspension and the mixture is stirred for a further 1 h at room temperature (Note 10).

Water (200 mL) and diethyl ether (100 mL) are added to the reaction flask and the biphasic mixture is filtered through a plug of cotton wool in a filtration funnel, under vacuum, rinsing with diethyl ether to aid transfer (50 mL) (Note 11). The mixture is transferred to a 1-L separatory funnel, and the layers separated. The aqueous layer is extracted with ether (2 × 200 mL). The combined organic layers are then washed with brine (100 mL), dried over anhydrous $MgSO_4$ (20.0 g), filtered, washing with diethyl ether (100 mL), and concentrated at 30 °C by rotary evaporation (500 mbar, then 20 mbar for 15 min) (Note 12). The resulting brown oil is purified by chromatography on SiO_2 (150.0 g) (Note 13) packed in a 6-cm diameter column. The sample is loaded in 20 mL of n-pentane and eluted with n-pentane. The solvent is collected in 20 mL fractions, and the product elutes within the first 30 fractions (Note 14). (E)-(2-Iodovinyl)benzene is obtained as a yellow oil (5.97 g, 87%, ≥98/2 E/Z (^{1}H NMR)) (Notes 15 and 16).

2. Notes

1.   Solid sodium bis(trimethylsilyl)amide (95%) was purchased from Aldrich and used as received. Submitters used 3.00 equiv (16.5 g, 90.0 mmol) of sodium bis(trimethylsilyl)amide. With 3.00 equiv, checkers did not obtain full conversion.

2.   Anhydrous tetrahydrofuran was obtained by filtration through a drying column on an Innovative Technology system (Newburyport, MA). Submitters used 60 mL anhydrous tetrahydrofuran which was obtained by filtration through a drying column on a Glass Contour system (Irvine, CA).

3.   Anhydrous diethyl ether was obtained by filtration through a drying column on an Innovative Technology system (Newburyport, MA).

Submitters used 60 mL of anhydrous diethyl ether obtained by filtration through a drying column on a Glass Contour system (Irvine, CA).

4. An appropriately sized dewar was used to ensure the solvent level of the reaction mixture was well below the acetone dry-ice. The dewar was well loaded with dry ice and the solution was cooled for at least 15 min prior to the first addition.

5. Diiodomethane (99% stab. with copper wire) was purchased from Aldrich (submitters from Alfa Aesar) and used as received.

6. Benzyl bromide was purchased from Aldrich (submitters from Alfa Aesar) and purified by distillation prior to use.

7. The reaction flask (and dewar) are covered with a sheet of aluminum foil during the reaction time due to the potential sensitivity to light of the title compound and the 1,1-diiodoalkane intermediate.

8. Throughout the reaction, the mixture should be well stirred to ensure good mixing and cooling. However, over-vigorous stirring, causing splashing of the reaction mixture, should be avoided.

9. The reaction flask was left in the dry ice/acetone bath overnight, without subsequent addition of dry ice and had in general reached room temperature by morning.

10. DBU was purchased from Fluka (submitters from Alfa Aesar) and purified by distillation prior to use. The addition of DBU at this stage ensures the complete elimination of HI from the diiodide intermediate, leading to a higher yield and easier purification.

11. Filtration through a cotton plug removes insoluble by-products that cause problems in the separation of phases in the work up procedure.

12. The submitters used the following pressure profile: 250 mmHg then to 20 mmHg.

13. Silica purchased from Merck, 230-400 mesh (submitters purchased from Silicycle, 230-400 mesh).

14. Rf 0.44 (n-pentane) (submitters loaded the brown oil in 30 mL of hexanes and eluted with hexanes collecting 40-50 mL fractions, the product is eluted within the first 25 fractions. Rf 0.65 (hexanes).

15. The (E)-(2-iodovinyl)benzene displays the following spectral properties: IR (film) 3059, 3037, 1946, 1873, 1802, 1740, 1702, 1677, 1596, 1569, 1524, 1494, 1469, 1444, 1299, 1278, 1170, 1070, 1029, 1001, 983, 946, 688 cm^{-1}; ^{1}H NMR (400 MHz; CDCl$_3$) δ: 6.83 (d, J = 14.9 Hz, 1 H), 7.27–7.36 (m, 5 H), 7.44 (d, J = 14.9 Hz, 1 H); ^{13}C NMR (101 MHz; CDCl$_3$) δ: 76.8, 126.1, 128.5, 128.9, 137.8, 145.2; MS m/z (relative intensity) 230

(M$^+$, 100), 103 (91), 77 (48), 51 (19); Anal. calcd. for C$_8$H$_7$I: C, 41.77; H, 3.07; found: C, 41.57; H, 3.19.

16. The title compound is light sensitive and will discolor from bright yellow, becoming increasingly orange to red. However, it can be stored in the freezer for several months without noticeable decomposition by ^{1}H NMR.

Safety and Waste Disposal Information

All hazardous materials should be handled and disposed of in accordance with "Prudent Practices in the Laboratory"; National Academy Press; Washington, DC, 1995.

3. Discussion

Vinyl halides have become increasingly important reactants in recent years due to the advances in metal-catalyzed cross-coupling reactions, allowing the introduction of carbon-carbon double bonds with controlled regio- and stereochemistry. Vinyl halides are much less commercially available than aryl halides and there has been much interest in efficient means for their stereoselective preparation. The development of facile methods to access vinyl halides with high levels of stereocontrol is still required.

(E)-β-Aryl vinyl halides are often attractive substrates in synthetic studies and for the synthesis of compounds of biological relevance.[2] (E)-β-Styryl iodide has been prepared by several methods, including the powerful Takai-Utimoto reaction, which uses benzaldehyde and iodoform with stoichiometric (87% yield, 94/6 E/Z) or catalytic (78% yield, 93/7 E/Z) quantities of chromium complexes.[3] Other methods require the formation of a suitable alkene precursor followed by the installation of the halide, such as the Hunsdiecker reaction,[4] or hydrometallation of an alkyne and trapping with an electrophilic halide source.[5]

The procedure above provides a high yielding one-pot synthesis of (E)-β-styryl iodide from benzyl bromide and diiodomethane.[6] This facile homologative method forms the C-C bond and installs the E-stereochemistry. Alkylation of NaCHI$_2$ with benzyl bromide,[7] followed by an *in situ* elimination of HI from the *gem*-diiodide intermediate in the presence of excess base provides the (E)-β-styryl iodide with high stereoselective. This method has been demonstrated to provide excellent *E*-selectivities using benzyl bromides with a range of steric and electronic requirements (Table 1). Related methods allow the use of sensitive substrates and the synthesis of *E*-styryl chlorides and bromides.[6]

This one-pot procedure for the synthesis of (*E*)-β-styryl iodide provides improved selectivity and minimizes waste and the need for toxic reagents.

174

Table 1. Synthesis of styryl iodides from benzyl bromides and diiodomethane

Entry[a]	Substrate	Product	Yield (%)[b]	E/Z[c]
1	(4-Me benzyl bromide)	(4-Me styryl iodide)	93	99:1
2	(2-Me benzyl bromide)	(2-Me styryl iodide)	90	99:1
3	(naphthylmethyl bromide)	(naphthyl styryl iodide)	70	98:2
4[d,e]	(4-MeO benzyl bromide)	(4-MeO styryl iodide)	92	97:3
5[e]	(3-MeO benzyl bromide)	(3-MeO styryl iodide)	95	99:1
6[e]	(methylenedioxy benzyl bromide)	(methylenedioxy styryl iodide)	93	99:1
7	(4-F benzyl bromide)	(4-F styryl iodide)	85	98:2
8	(2-Cl benzyl bromide)	(2-Cl styryl iodide)	78	98:2
9	(3-Br benzyl bromide)	(3-Br styryl iodide)	62	99:1
10[f]	(1,3-bis-benzyl bromide)	(bis-styryl iodide)	73%	97:3 EE:EZ

[a] Reactions performed on a 4.0 mmol scale; CH$_2$I$_2$ (1.5 equiv), NaHMDS (3 equiv), [0.2 M], −78 °C (1 h 30 min) to rt (30 min) then DBU (1 equiv) for 1 h; [b] Yield of isolated product. [c] E:Z ratio determined by ^{1}H NMR. [d] Performed on a 1.0 mmol scale. [e] Used 1.5 equiv DBU. [f] Used 0.5 equiv of bis-benzyl bromide.

1. Département de Chimie, Université de Montréal, P.O. Box 6128, Station Downtown, Montréal (Québec) Canada, H3C 3J7. Email: andre.charette@umontreal.ca.

2. For recent relevant examples see: (a) Kabir, M. S.; Van Linn, M. L.; Monte, A.; Cook, J. M. *Org. Lett.* **2008**, *10*, 3363-3366. (b) Besselièvre, F.; Piguel, S.; Mahuteau-Betzer, F.; Grierson, D. S. *Org. Lett.* **2008**, *10*, 4029-4032.

3. (a) Takai, K.; Nitta, K.; Utimoto, K. *J. Am. Chem. Soc.* **1986**, *108*, 7408-7410. (b) Takai, K.; Ichiguchi, T.; Hikasa, S. *Synlett* **1999**, 1268-1270.

4 (a) Kuang, C.; Yang, Q.; Senboku, H.; Tokuda, M. *Synthesis* **2005**, 1319-1325. (b) You, H.-W.; Lee, K.-J. *Synlett* **2001**, 105-107. (c) Das, J. P.; Roy. S. *J. Org. Chem.* **2002**, *67*, 7861-7864. (d) Kuang, C.; Senboku, H.; Tokuda, M. *Tetrahedron* **2005**, *61*, 637-642.

5. Brown, H. C.; Hamaoka, T.; Ravindran, N.; Subrahmanyam, C.; Somayaji, V.; Bhat, N. G. *J. Org. Chem.* **1989**, *54*, 6075-6079.

6. Bull, J. A.; Mousseau, J. J.; Charette, A. B. *Org. Lett.* **2008**, *10*, 5485-5488.

7. Bull, J. A.; Charette, A. B. *J. Org. Chem.* **2008**, *73*, 8097-8100.

Appendix
Chemical Abstracts Nomenclature; (Registry Number)

Sodium bis(trimethylsilyl)amide; (1070-89-9)
Benzyl bromide: Benzene, (bromomethyl)-; (100-39-0)
1,8-Diazabicyclo[5.4.0]undec-7-ene (DBU); (6674-22-2)
(*E*)-(2-Iodovinyl)benzene; (42599-24-6)

André B. Charette received his B.Sc. from Université de Montréal in 1983, after which he pursued his graduate studies at the University of Rochester, earning his M.Sc. (1985) and Ph.D. (1987) with Robert Boeckman, Jr. Following NSERC postdoctoral fellowship at Harvard University with David A. Evans, he began his academic career at Université Laval in 1989. In 1992, he returned to his alma mater, where he is Full Professor and holder of an NSERC/Merck Frosst/Boehringer Ingelheim Industrial Research Chair and a Canada Research Chair. His research focuses on the development of new methods for the stereoselective synthesis of organic compounds. Honors include a Cope Scholar Award (2007), the Prix Marie-Victorin (2008) and the Alfred Bader Award (2009).

James A. Bull was born in 1978 in Birmingham, England. He obtained his M.Sci. in Natural Sciences, Chemistry (first class honours) from the University of Cambridge. Following a year working in the pharmaceutical industry, he returned to Cambridge for his Ph.D. studies with Professor Steven V Ley, where he completed the synthesis of bisoxazole natural product bengazole A. In 2007 he joined the group of Professor André B. Charette as a postdoctoral fellow where he has developed improved conditions for the synthesis of *gem*-diiodides and an intramolecular Simmons-Smith cyclopropanation reaction.

James J. Mousseau was born in Montréal, Quebec in 1981. Upon completing his B.Sc. in Honours Biochemistry in 2004 at Concordia University he continued his M.Sc. studies at Concordia under the supervision of Prof. Louis A. Cuccia and was involved in the synthesis of novel crescent shaped urea-linked heterocyclic foldamers. In 2007 he joined the group of Prof. André B. Charette at Université de Montréal to pursue his Ph.D. studies. His research currently focuses on the direct functionalization of *N*-iminopyridinium ylides.

Dominik Frank was born in Bad Säckingen (Germany) in 1980. He studied Chemistry at the University of Basel where he obtained his M. Sc. in 2006 under the supervision of Prof. Andreas Pfaltz. He began his Ph. D. work in summer 2006 in the group of Prof. Andreas Pfaltz, where he is currently working on the synthesis of new chiral ligands for metal-catalyzed reactions.

SYNTHESIS OF TETRAORGANOSILANES:
(CHLOROMETHYL)DIMETHYLPHENYLSILANE

Submitted by Kei Murakami, Hideki Yorimitsu, and Koichiro Oshima.[1]
Checked by Jane Panteleev and Mark Lautens.

1. Procedure

(Chloromethyl)dimethylphenylsilane. A flame-dried 1-L, three-necked, round-bottomed flask is equipped with a 500-mL pressure equalizing dropping funnel fitted with a septum, a two-way stopcock with an argon inlet, an internal temperature probe, and a 5-cm egg-shaped stirring bar. Dichloro(N,N,N',N'-tetramethylethylenediamine)zinc (0.64 g, 2.5 mmol, 1 mol%) (Note 1) is placed in the flask, and the apparatus is purged with argon. 1,4-Dioxane (240 mL) (Note 2) is added to the flask at 23 °C. Chloro(chloromethyl)dimethylsilane (33.8 mL, 250 mmol) (Note 1) is added to the flask through the dropping funnel at 23 °C. The dropping funnel is rinsed with 1,4-dioxane (10 mL). The mixture is cooled in an ice/water bath over 10 min. Phenylmagnesium bromide (Note 3) (1.0 M in THF, 300 mL, 300 mmol, 1.2 equiv) is then transferred to the dropping funnel using a 14 guage metal cannula and is added dropwise to the mixture over 30 min with cooling in an ice/water bath. The addition immediately leads to the formation of white salts. A gentle exothermic reaction takes place. After the completion of the addition, the resulting mixture is allowed to warm to ambient temperature (23 °C) and stirred for an additional 2 h. The reaction mixture is poured over 5 min into a rapidly stirred ice-cold saturated aqueous ammonium chloride solution (150 mL) (Note 2) in a 1-L Erlenmeyer flask equipped with a 5-cm octagonal magnetic stirring bar. The mixture is transferred to a 1-L separatory funnel, and the Erlenmeyer and round-bottomed flasks are rinsed with ethyl acetate (25 mL each) (Note 2). The organic phase is separated, and the aqueous layer is extracted with ethyl acetate (50 mL × 3). The combined organic layers are washed with brine (50 mL), dried once over anhydrous Na_2SO_4 (25 g) (Note 2), filtered through filter paper, and concentrated with a rotary evaporator (35 °C, 34–38

178

mmHg). Evaporation is stopped at the time when the volume of the mixture is reduced to approximately 75 mL (Note 4). The mixture is transferred to a 100-mL round-bottomed flask equipped with a magnetic stirring bar. The flask is then equipped with a Vigreux column (20 cm) topped with a distillation head and receiver. Vacuum (23 mmHg) is applied, and remaining 1,4-dioxane is removed until bubbling ceases. The flask is gradually heated in an oil bath to a bath temperature of 155 °C. After the temperature of the fraction reaches 115 °C, a forerun (ca. 1 mL) is collected and discarded. The desired product is then obtained, distilling at 115 °C (23 mmHg). The product weighs 37–38 g (200–203 mmol, 80–81%) and is obtained as a stable, clear, colorless liquid (Notes 5 and 6).

2. Notes

1. Dichloro(*N,N,N',N'*-tetramethylethylenediamine)zinc (98%) and chloro(chloromethyl)dimethylsilane (98%) were purchased from Aldrich Chemical Co., Inc. and used as is.

2. 1,4-Dioxane (99%, anhydrous, water <50ppm), ammonium chloride (99.5%), ethyl acetate (99%), and anhydrous sodium sulfate (99%) were obtained from Wako Pure Chemical Industries Ltd. and were used as received by the submitters. 1,4-Dioxane (99.8%, anhydrous, <0.003% water) and ethyl acetate (>99.5%) were purchased from Aldrich Chemical Co., Inc. and used as received by the checkers. Ammonium chloride (99.5%) and anhydrous sodium sulfate (99%) were purchased from ACP Chemicals Inc., and used as they were by the checkers.

3. Phenylmagnesium bromide solution (1.0M in tetrahydrofuran) was purchased from Aldrich Chemical Co., Inc, and was used as received by the checkers. The submitters synthesized phenylmagnesium bromide (1.0M in tetrahydrofuran).

4. The submitters concentrated the solution using a rotary evaporator (30 °C, 10 mmHg) and noted that the yield of (chloromethyl)-dimethylphenylsilane can be decreased when evaporation is performed under lower pressure or for a prolonged time.

5. The product exhibits the following physicochemical properties: IR (film, NaCl) 3070, 2963, 1427, 1250, 1119, 841, 698 cm^{-1}; ^{1}H NMR (CDCl$_3$, 400 MHz) δ: 0.42 (s, 6 H), 2.96 (s, 2 H), 7.35–7.44 (m, 3 H), 7.52–7.57 (m, 2 H); ^{13}C NMR (CDCl$_3$, 101 MHz) δ: –4.5, 30.4, 128.0, 129.7, 133.7, 136.1; MS (EI) *m/z* (relative intensity): 186 (2), 184 (13), 171 (9), 155

(10), 135 (100); HRMS (EI): *m/z* calcd. for $C_9H_{13}ClSi$ 184.0475; found 184.0473.

6. The purity (99%) was determined by GC using a Phenomenex ZB-5 ms column (30 m × 0.25 mm with 0.25 μm film thickness) (oven temperature: 50 °C for 5 min, ramp 50 °C per min to 300 °C; outlet flow: 1 mL/min; carrier gas: helium; retention time: 6.3 min). The submitters report: Anal. calcd. for $C_9H_{13}ClSi$: C, 58.51; H, 7.09; found: C, 58.45; H, 7.03.

7. The submitters report the reaction proceeded in an 80% yield on a 500-mmol scale.

Waste Disposal Information

All hazardous materials should be handled and disposed of in accordance with "Prudent Practices in the Laboratory"; National Academy Press; Washington, DC, 1995.

3. Discussion

Tetraorganosilanes are quite useful organometallic compounds as reagents and functional materials. The nucleophilic substitution reactions of chlorosilanes with organometallic reagents are commonly used to synthesize tetraorganosilanes. The reactions of chlorosilanes with organolithium reagents generally proceed smoothly, wherein functional group compatibility is not sufficiently wide due to the high reactivity of organolithium reagents.[2] On the other hand, the substitution reactions with the less reactive organomagnesium reagents often require prolonged reaction times and high temperatures to go to completion. Toxic cyanide or thiocyanate salts are known to catalyze the substitution reactions of chlorosilanes with organomagnesium reagents.[3] Very recently, silver nitrate proved to facilitate the substitution reaction.[4] However, the scope of the Grignard reagents in the silver-catalyzed reaction is not satisfactorily wide, i.e., limited to arylmagnesium reagents. Finally, zinc chloride, a very cheap inorganic salt, is found to catalyze the substitution reactions with a much wider variety of organomagnesium reagents (Table 1).[5] The zinc-catalyzed reaction seems to be the best method at present, comprehensively taking the efficiency, scope, operability, scalability, cost, and toxicity into account. The product, (chloromethyl)dimethylphenylsilane, is the precursor of useful

180

dimethylphenylsilylmethylmagnesium chloride[6] and serves also as an important building block in organic synthesis.[7]

Table 1. Zinc-Catalyzed Substitution Reactions of Chlorosilanes with Grignard Reagents

$$Si-Cl + RMgX \xrightarrow[\text{1,4-dioxane, 20 °C}]{\text{1 mol\% ZnCl}_2 \cdot \text{TMEDA}} Si-R$$

(0.50 mmol) (1.5 equiv)

Entry	Si	RMgX	Time [h]	Yield [%]
1	PhMe$_2$Si	2-MeC$_6$H$_4$MgBr	5	92
2	PhMe$_2$Si	4-MeOC$_6$H$_4$MgBr	1	87
3	PhMe$_2$Si	3-CF$_3$C$_6$H$_4$MgBr	3	99
4	Ph$_2$MeSi	4-MeC$_6$H$_4$MgBr	15	89
5[a]	PhMe$_2$Si	CH$_2$=CMeMgBr	3	84
6[a]	Ph$_2$MeSi	CH$_2$=CHMgBr	2	71
7[a]	i-Pr$_3$Si	CH$_2$=CHCH$_2$MgCl	12	91
8[a]	t-BuMe$_2$Si	PhCH$_2$MgCl	7	70
9[a,b]	t-BuMe$_2$Si	CH$_2$=CHCH$_2$MgCl	8	71
10	(4-NCC$_6$H$_4$)Me$_2$Si	4-MeC$_6$H$_4$MgBr	1	75

[a] THF was used as a solvent. [b] Performed on a 50-mmol scale.

1. Department of Material Chemistry, Graduate School of Engineering, Kyoto University, Katsura, Nishikyo, Kyoto 615-8510, JAPAN (oshima@orgrxn.mbox.media.kyoto-u.ac.jp)
2. (a) *Science of Synthesis (Houben-Weyl)*, Fleming, I., Ed., Georg Thieme Verlag, Stuttgart, Germany, **2002**, Vol. 4, Chapter. 4.4. (b) Brook, M. A. *Silicon in Organic, Organometallic, and Polymer Chemistry*, Wiley, New York, **2000**, Chapter 5. (c) Birkofer, L.; Stuhl, O. *The Chemistry of Organic Silicon Compounds*, Patai, S., Rappoport, Z. Eds., Wiley, New York, **1989**, Chapter 10.
3. Lennon, P. J.; Mack, D. P.; Thompson, Q. E. *Organometallics* **1989**, *8*, 1121–1122.
4. Murakami, K.; Hirano, K.; Yorimitsu, H.; Oshima, K. *Angew. Chem., Int. Ed.* **2008**, *47*, 5833–5835.
5. Murakami, K.; Yorimitsu, H.; Oshima, K. *J. Org. Chem.* **2009**, *74*, 1415–1417.
6. (a) Kobayashi, T.; Ohmiya, H.; Yorimitsu, H.; Oshima, K. *J. Am. Chem. Soc.* **2008**, *130*, 11276–11277. (b) Rodgen, S. A.; Schaus, S. E. *Angew.*

Chem., Int. Ed. **2006**, *45*, 4929–4932.

7. (a) Simov, B. P.; Wuggenig, F.; Mereiter, K.; Andres, H.; France, J.; Schnelli, P.; Hammerschmidt, F. *J. Am. Chem. Soc.* **2005**, *127*, 13934–13940. (b) Sun, H.; Martin, C.; Kesselring, D.; Keller, R.; Moeller, K. D. *J. Am. Chem. Soc.* **2006**, *128*, 13761–13771.

Appendix
Chemical Abstract Nomenclature; (Registry Number)

Bromobenzene; (108-86-1)
Magnesium; (7439-95-4)
Dichloro(*N,N,N',N'*-tetramethylethylenediamine)zinc; (28308-00-1)
Chloro(chloromethyl)dimethylsilane; (1719-57-9)
(Chloromethyl)dimethylphenylsilane; (1833-51-8)

Koichiro Oshima was born in Hyogo, Japan, in 1947. He obtained his B.S. in 1970 and Ph.D. in 1975 from Kyoto University under the guidance of Professor Hitosi Nozaki. He then worked as a postdoctoral fellow with Professor Barry Sharpless at MIT and became an Assistant Professor at Kyoto University in 1977. He was promoted to Lecturer in 1984, Associate Professor in 1986, and Professor in 1993. His research interests include the development of new reactions utilizing radical intermediates and organometallic reagents. He received the Award for Young Chemists of the Society of Synthetic Organic Chemistry, Japan in 1983, the Japan Synthetic Organic Chemistry Award in 2004, and the Chemical Society of Japan Award for 2006.

Kei Murakami was born in Osaka, Japan, in 1985. He completed his undergraduate education at Kyoto University in 2007 and is currently pursuing his Ph.D. studies under the tutelage of Professor Koichiro Oshima. He has been a JSPS research fellow since 2009, developing new metal-catalyzed reactions for carbon–carbon and carbon–silicon bond formation.

Hideki Yorimitsu was born in Kochi, Japan, in 1975. He obtained his B.S. in 1997 and Ph.D. in 2002 from Kyoto University under the supervision of Professor Koichiro Oshima. He then served as a JSPS postdoctoral fellow, working with Professor Eiichi Nakamura at the University of Tokyo. He became an Assistant Professor at Kyoto University in 2003 and has been an Associate Professor since 2008. His research program focuses on the development of new organic reactions useful for synthesizing biologically interesting compounds, novel coordinating structures, and organometallic compounds. He received the Chemical Society of Japan Award For Young Chemists for 2008.

Jane Panteleev received her Bachelor of Science degree in Biochemistry at Queen's University, Ontario, in 2007. During that time she had the opportunity to work in the research lab of Prof. Victor Snieckus. She is currently pursuing her Ph.D. degree under the supervision of Prof. Mark Lautens at the University of Toronto. Her current research is in the area of asymmetric transition metal catalysis.

A GENERAL METHOD FOR COPPER-CATALYZED ARYLATION OF ACIDIC ARENE C-H BONDS. PREPARATION OF 2-CHLORO-5-(3-METHYLPHENYL)-THIOPHENE.

Submitted by Joseph Alvarado, Hien-Quang Do, and Olafs Daugulis.[1]
Checked by Peter C. Marsden and Jonathan A. Ellman.

1. Procedure

A 50-mL oven dried Schlenk flask with a 14/20 joint equipped with a magnetic stir bar is charged with 2-chlorothiophene (7.11 g, 60.0 mmol, 2.0 equiv) and 1-iodo-3-methylbenzene (6.54 g, 30.0 mmol, 1.0 equiv). The Schlenk flask is equipped with a 14/20 thermometer adaptor, a thermometer is inserted through the adaptor into the flask so that the bulb is completely submerged in the solution, the adaptor is then tightened so that the o-ring holds the thermometer in place, and finally the joint is wrapped tightly with Teflon tape. The flask is evacuated and back filled with dry nitrogen gas three times through the side arm. After placing it inside the glove box (Note 1), the thermometer adaptor is removed from the joint, and the flask is charged with lithium *t*-butoxide (4.80 g, 60.0 mmol, 2.0 equiv), 1,10-phenanthroline (1.08 g, 6.00 mmol, 0.20 equiv), CuI (1.14 g, 6.00 mmol, 0.20 equiv), and DMPU (18 mL; tetrahydro-1,3-dimethyl-2(1H)-pyrimidinone) (Note 2). The flask is re-equipped with the thermometer and adapter, which is again wrapped tightly with Teflon tape. The Schlenk flask is taken out of the glove box. A hose attached to a Schlenk line is purged with dry nitrogen gas and connected to the side arm of the Schlenk flask. The stopcock valve is opened and the Schlenk flask is evacuated and back filled with dry nitrogen gas three times through the side arm stopcock. The flask is placed in a preheated oil bath at 125 °C and magnetically stirred for 12 h (Note 3). The brown color of the reaction mixture changes to deep red after 5-10 min and then to dark brown 30 min later.

After the reaction is complete as judged by GC analysis (Note 4), the reaction mixture is allowed to cool to room temperature. At that point,

Org. Synth. **2010**, *87*, 184-191
Published on the Web 3/17/2010

nitrogen atmosphere is not required any further. After dilution with EtOAc (30 mL) and stirring for 1 min, the reaction mixture is transferred to a 500 mL Erlenmeyer flask. The Schlenk flask is washed with EtOAc (3 x 50 mL), the EtOAc solutions are combined, and brine (100 mL) is added. The resulting mixture contains a light brown precipitate that is removed via vacuum filtration by using a 7 cm diameter filter funnel with a medium porosity fritted disk. The brown precipitate is washed with EtOAc (3 x 10 mL), and the resulting filtrate is transferred to a 500 mL separatory funnel. The top organic layer is separated from the bottom aqueous layer. The aqueous layer is extracted with additional ethyl acetate (3 x 100 mL) (Note 5). The combined organic layers are dried over $MgSO_4$ (Note 6; about 9 g) followed by vacuum filtration into a 1000 mL round bottom flask by using a 7 cm diameter filter funnel with a medium porosity fritted disk. The filtrate is concentrated by rotary evaporation (50 °C, 30 mm Hg) to give 8-10 mL of crude product. The crude product is purified by flash chromatography on SiO_2 (Note 7), using hexanes as the eluent (4.8 L) and collecting 100 mL fractions. The first 12 fractions are blank (1.2 L), followed by 6 fractions containing starting material and minor amounts of impurities (0.6 L), and then pure product for 30 fractions (3 L collected). After concentration of the fractions containing the pure product by rotary evaporation (30 °C, 30 mm Hg), 5.73 g (92%) of 2-chloro-5-(3-methylphenyl)-thiophene is obtained as an air stable, light yellow oil that crystallizes when cooled to -20 °C (Notes 8 and 9; mp 38-39 °C).

2. Notes

1.   In addition to the procedure reported above, the checkers also performed a half-scale reaction outside of the glovebox where all of the starting materials were quickly measured out in a fumehood, and then transferred to the reaction flask. The Schlenk flask was equipped with a 14/20 thermometer adaptor, a thermometer was inserted through the adaptor into the flask so that the bulb was submerged in the solution, the adaptor was then tightened so that the o-ring held the thermometer in place, and finally the joint was wrapped tightly with Teflon tape. The flask was then evacuated and backfilled with nitrogen via the sidearm stopcock. Heating, work-up and purification remained unchanged from above and 2.70 g (86%) of product was obtained. LiO*t*Bu should not be stored outside of an inert atmosphere because it is very hygroscopic.

2. All reagents were used as received. The submitters obtained 2-chlorothiophene (98%) from Matrix Scientific, 1-iodo-3-methylbenzene (99%) from Oakwood Products, Inc., lithium *t*-butoxide (98+%) from Strem Chemicals, 1,10-phenanthroline (99+%) and CuI (98+%) from Acros Organics, and DMPU (tetrahydro-1,3-dimethyl-2(1H)-pyrimidinone, 98+%) from Alfa Aesar. Lithium *t*-butoxide and CuI were stored in an argon-filled glove box. The checkers obtained 2-chlorothiophene (98%) from TCI America, 1-iodo-3-methylbenzene (99%) from Aldrich, lithium *t*-butoxide (98+%) from Strem Chemicals, 1,10-phenanthroline (99+%) and CuI (98+%) from Acros Organics, and DMPU (tetrahydro-1,3-dimethyl-2(1H)-pyrimidinone, (98+%) from Alfa Aesar.

3. Stir bar size: 2.5 cm length and 1 cm diameter. Stirring rate 1000-1100 rpm.

4. GC analyses (by the submitters) were performed on a Shimadzu CG-2010 chromatograph equipped with a Restek column (Rtx®-5, 15 m, 0.25 mm ID). The following program parameters were used: initial temperature: 80 °C (2 min), ramp at 35 °C/min to 250 °C, hold at 250 °C for 3 min. Retention times: 2-chlorothiophene 0.77 min, 1-iodo-3-methylbenzene 2.92 min, and 2-chloro-5-*m*-tolylthiophene 5.27 min. GC analyses (by the checkers) were performed on an Agilent 6890N chromatograph equipped with an Agilent column (Ultra 2, crosslinked 5% Ph Me Silicone, 25 m x 0.2 mm x 0.33 μm film thickness). The program parameters specified by the submitters gave the following retention times: 2-chlorothiophene 2.49 min, 1-iodo-3-methylbenzene 4.87 min, and 2-chloro-5-*m*-tolylthiophene 7.50 min. Aliquots from the reaction mixture were diluted with 1 mL of EtOAc and filtered for GC analysis. After 12 h, >98% of 1-iodo-3-methylbenzene was consumed.

5. Organic and inorganic layers must separate completely. There must be a sharp border between the two. Otherwise, the yield will be lower.

6. The submitters obtained magnesium sulfate (anhydrous powder) from Mallinckrodt Baker, Inc. The checkers obtained magnesium sulfate (anhydrous powder) from Fisher Chemical.

7. The submitters performed flash chromatography on 60Å silica gel (Sorbent Technologies) with 460 g of SiO_2 in a 38 cm x 8 cm column equipped with a 500 mL reservoir. The checkers performed flash chromatography on 60 Å silica gel (MP Silitech 32-63D). All solvents were HPLC grade purchased from Fisher Chemical. The product has an $R_f = 0.59$ (hexanes; visualization by UV).

186

8.	The characterization of the product is as follows: FT-IR (neat,) 1601, 1488, 1445, 1214, 1022, 796, 779, 687 cm^{-1}; ^{1}H NMR (500 MHz, CDCl$_3$) δ: 2.37 (s, 3 H), 6.86 (d, J = 3.9 Hz, 1 H), 7.04 (d, J = 3.9 Hz, 1 H), 7.10 (d, J = 7.3 Hz, 1 H), 7.23-7.27 (m, 1 H), 7.29-7.31 (m, 2 H); ^{13}C NMR (125 MHz, CDCl$_3$) δ: 21.5, 122.1, 122.7, 126.3, 127.1, 128.7, 128.92, 128.93, 133.6, 138.7, 143.1; Anal. calcd. for C$_{11}$H$_9$ClS: C, 63.30 H, 4.35; found: C, 62.96 H, 4.28.

9.	On half-scale, the checkers also obtained an 88% yield. On full scale, the submitters reported product yields of 5.68–5.81 g (90-93%).

Safety and Waste Disposal Information

All hazardous materials should be handled and disposed of in accordance with "Prudent Practices in the Laboratory"; National Academy Press; Washington, DC, 1995.

3. Discussion

The combination of copper iodide with phenanthroline ligand is a general catalyst for the arylation of acidic sp^2 carbon-hydrogen bonds.[2] It has been shown that electron-rich and electron-poor heterocycles as well as arenes possessing at least two electron-withdrawing groups can be arylated. The prerequisite for a successful arylation is an sp^2 C-H bond possessing a pKa of 35 or less (in DMSO).[3] The regioselectivity of arylation is very high, with the most acidic carbon-hydrogen bond being functionalized. A comparison between copper and palladium catalysis in C-H bond arylation is given below. The advantages of copper relative to palladium catalysis are as follows: (1) cheaper catalyst; (2) very predictable arylation regioselectivity – the most acidic C-H bond is always arylated; (3) nearly uniform reaction conditions due to simple mechanistic picture; (4) unusual regioselectivity – the most hindered C-H bond is arylated. The advantages of palladium relative to copper catalysis are: (1) possibility to use aryl chlorides as coupling partners,[4] while only ArBr, ArI, and most reactive heteroaryl chlorides can be used for copper catalysis; (2) ability to effect the arylation of less acidic C-H bonds such as nitrobenzene or indole;[5] (3) C-H bonds can be arylated in the presence of acidic N-H groups; in the case of copper catalysis, NH substituents will be arylated preferentially.[6] A mechanistically distinct procedure allows indole arylation by iodonium salts under copper catalysis.[7]

An overview of substrates that can be arylated by employing copper catalysis is given in Table 1.[2] Typically, 10 mol % of catalyst loading can be employed, although the above procedure was checked with a 20 mol % catalyst loading.[2] Electron-rich heterocycles such as thiazole, caffeine, and triazole can be arylated by aryl iodides and aryl bromides (entries 1-3). Electron-poor heterocycles such as 2-phenylpyridine oxide and pyridazine are also reactive (entries 4-5). Polyfluorobenzenes can be arylated in good yields (entries 6-7). Alkenylation is possible by employing alkenyl bromides (entry 8). Other electron-poor arenes such as polychloro-, nitro-, and cyanoarenes can be arylated at the most acidic position, which is also the most hindered one. For relatively non-acidic 1,3-dichlorobenzene, use of a hindered, strong Et_3COLi base is required to achieve reasonable product yield. For the most acidic compounds possessing pKa values of 27 or less (in DMSO), potassium phosphate base can be used. Other substrates require the use of t-BuOLi base. Arylations by aryl bromides are successful only if K_3PO_4 is employed and low conversions are observed with t-BuOLi base.

The reaction mechanism most likely involves an acid-base reaction followed by a copper-catalyzed carbon-carbon bond formation (Scheme 1). Both t-BuOCu and t-BuOLi were shown to be competent bases for the deprotonation step.[2] The deprotonation step determines overall arylation regioselectivity.

Scheme 1. Mechanistic considerations

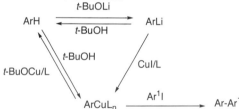

Table 1. Arylation of Acidic sp^2 C-H Bonds[a]

$$\text{Arene} \xrightarrow[\substack{\text{ArHal, solvent, base} \\ 100-125\ ^\circ\text{C}}]{\text{10 mol\% CuI/phenanthroline}} \text{Product}$$

entry	arene	aryl halide/base	product	yield(%)
1	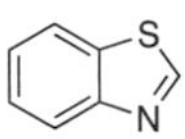	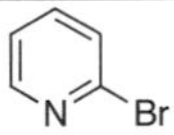 K$_3$PO$_4$	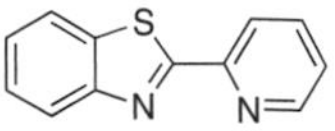	89
2	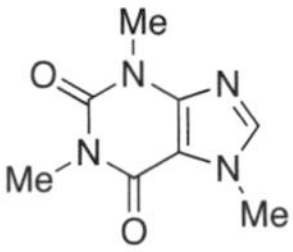	C$_6$H$_5$I/t-BuOLi	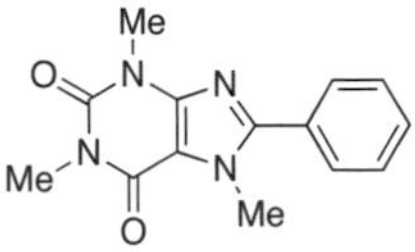	85
3	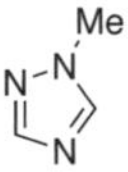	C$_6$H$_5$I/t-BuOLi	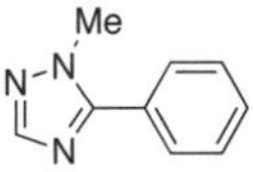	88
4	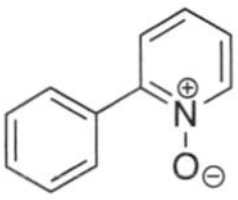	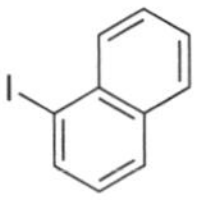 t-BuOLi	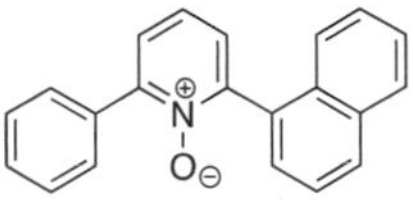	91
5	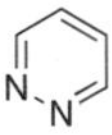	C$_6$H$_5$I/ Et$_3$COLi	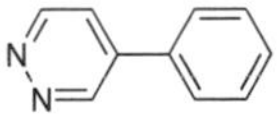	60
6	C$_6$F$_5$H	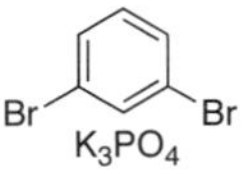 K$_3$PO$_4$	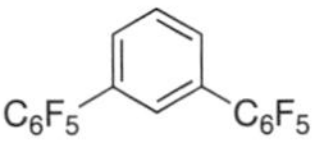 C$_6$F$_5$ ⟨⟩ C$_6$F$_5$	51
7	C$_6$F$_5$H	K$_3$PO$_4$	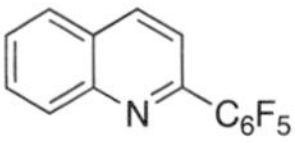	85
8	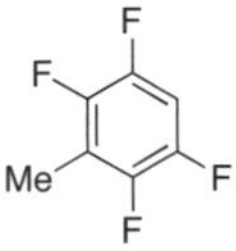	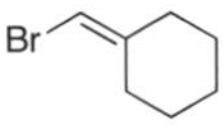 K$_3$PO$_4$	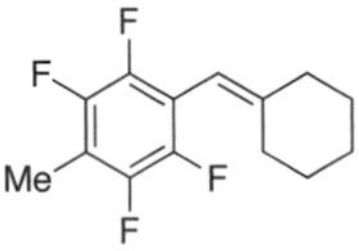	95
9	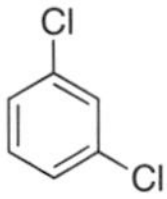	C$_6$H$_5$I/ Et$_3$COLi	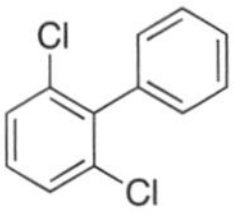	43
10	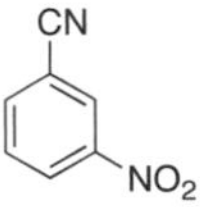	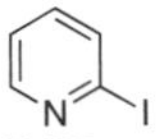 K$_3$PO$_4$	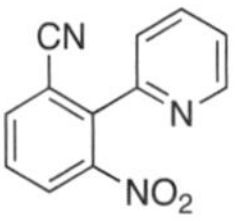	95

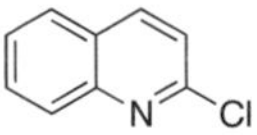

[a] From ref. 2b. Copper (I) iodide (0.1 mmol), phenanthroline (0.1 mmol), halide (1-2 mmol), arene (1-3 mmol), base (1.7-4 mmol), DMF, DMPU, or DMF/xylenes solvent.

1. University of Houston, Department of Chemistry, Houston, TX 77204-5003. Email: olafs@uh.edu.
2. (a) Do, H.-Q.; Daugulis, O. *J. Am. Chem. Soc.* **2008**, *130*, 1128-1129. (b) Do, H.-Q.; Khan, R. M. K.; Daugulis, O. *J. Am. Chem. Soc.* **2008**, *130*, 15185-15192.
3. (a) Shen, K.; Fu, Y.; Li, J.-N.; Liu, L.; Guo, Q.-X. *Tetrahedron* **2007**, *63*, 1568-1576. (b) Bordwell, F. G. *Acc. Chem. Res.* **1988**, *21*, 456-463.
4. Chiong, H. A.; Daugulis, O. *Org. Lett.*, **2007**, *9*, 1449-1451.
5. (a) Caron, L.; Campeau, L.-C.; Fagnou, K. *Org. Lett.* **2008**, *10*, 4533-4536. (b) Bellina, F.; Benelli, F.; Rossi, R. *J. Org. Chem.* **2008**, *73*, 5529-5535.
6. (a) Kiyomori, A.; Marcoux, J.-F.; Buchwald, S. L. *Tetrahedron Lett.* **1999**, *40*, 2657-2660. (b) Gujadhur, R. K.; Bates, C. G.; Venkataraman, D. *Org. Lett.* **2001**, *3*, 4315-4317.
7. Phipps, R. J.; Grimster, N. P.; Gaunt, M. J. *J. Am. Chem. Soc.* **2008**, *130*, 8172-8174.

Appendix
Chemical Abstracts Nomenclature; (Registry Number)

1-Iodo-3-methylbenzene; (625-95-6)

2-Chlorothiophene; (96-43-5)

Lithium t-butoxide; (1907-33-1)

1,10-Phenanthroline; (66-71-7)

Copper(I) iodide; (7681-65-4)

Tetrahydro-1,3-dimethyl-2(1H)-pyrimidinone; (7226-23-5)

2-Chloro-5-(3-methylphenyl)-thiophene; (1078144-58-7)

Olafs Daugulis was born in Riga, Latvia in 1968. He obtained his degree in chemical engineering from Riga Technical University in 1991. His PhD research was performed at the University of Wisconsin-Madison in the group of Prof. E. Vedejs. After obtaining the PhD in 1999 he joined the group of Prof. M. Brookhart at UNC-Chapel Hill as a postdoctoral associate. He is currently an Associate Professor of Chemistry at the University of Houston.

Joseph Alvarado was born in Houston, Texas in 1985. He entered University of Houston in 2004 where he is currently an undergraduate chemistry major. He plans to attend graduate school in chemistry.

Hien-Quang Do was born in 1978 in Trà Vinh, Vietnam. He received his B.S. in chemistry from the University of Natural Sciences at Ho Chi Minh City in Vietnam. A few years after obtaining a Masters degree from the same University in 2003, he came to the University of Houston where he currently is a fourth-year graduate student in Prof. Daugulis' group.

Peter Marsden was born in 1984 in Pasadena, CA. He obtained his B.A. in Chemistry from Pomona College in 2006. He is currently a 4th year graduate student working with Profs. Jonathan Ellman and Robert Bergman on creating commodity chemicals from biomass.

1,3,5-TRIACETYLBENZENE

Submitted by Peter J. Alaimo,[1] Amanda-Lynn Marshall, David M. Andrews and Joseph M. Langenhan.[1]
Checked by Liang Huang, Eric A. Bercot, and Margaret Faul.

1. Procedure

A single-necked, 250-mL round-bottomed flask equipped with a Teflon-coated magnetic stir bar (33 x 15 mm, egg-shaped) is charged with *trans*-4-methoxy-3-buten-2-one (15.3 mL, 15.0 g, 0.135 mol, 1.0 equiv), ethanol (120 mL) (Note 1), and deionized water (30 mL) (Note 1) under an atmosphere of air. Glacial acetic acid (17.5 M, 0.9 mL, 0.944 g, 0.015 mol, 0.11 equiv) (Note 1) is added via syringe to the reaction mixture. A reflux condenser is fitted to the round-bottomed flask and the resulting homogeneous pale yellow reaction mixture is heated gently to reflux, with stirring, for 48 h ensuring that the temperature of the oil bath is kept at 77 °C. Over this time, the reaction remains homogeneous, and turns orange in color (Note 2). The oil bath is removed, and the reaction flask is allowed to cool to room temperature, and then placed in a –20 °C freezer for 18 h to crystallize the product. The resulting mixture is then vacuum filtered while cold through a sintered-glass funnel (Note 3) to collect the yellow needles (Note 4), which are rinsed with pre-cooled (0 °C) anhydrous diethyl ether (4 x 50 mL) (Notes 1, 5 and 6). The crude product (6.4–6.6 g; 70–72%; mp = 161–165 °C) is recrystallized by dissolving the yellow needles in hot ethanol (200 mL, 78 °C), allowing the solution to cool to room temperature, then storing it in a –20 °C freezer for 18 h. The resulting mixture is then vacuum filtered while cold, through a sintered-glass funnel (Note 3), and the pale yellow needles are rinsed with pre-cooled (0 °C) anhydrous ether (4 x 50 mL) (Note 7). The 1,3,5-triacetylbenzene obtained is air dried overnight to provide 5.6–5.9 g of product (61–64%) as pale yellow needles (mp 163–165 °C) (Notes 8 and 9).

Org. Synth. **2010**, *87*, 192-200
Published on the Web 3/17/2010

2. Notes

1. The submitters purchased *trans*-4-methoxy-3-buten-2-one (tech. grade, 90%) and ethanol (200 proof; HPLC grade) from Aldrich and used as received. The submitters purchased glacial acetic acid (Baker; 99%, 17.5M) and diethyl ether (Baker; 99%, anhydrous) from VWR and used as received. The checkers used deionized water from a house system (municipal water is purified by reverse osmosis using a commercial system manufactured by US Filters). The checkers analyzed *trans*-4-methoxy-3-buten-2-one (tech. grade, 90%) from Aldrich by ^{1}H NMR versus phenylmethyl sulfone (99 % purity, Acros cat # 353040050) and found the material to be 90 wt %. All product yields are adjusted based on the assay of starting *trans*-4-methoxy-3-buten-2-one. The instrument details for ^{1}H NMR analysis are as follows: 400 MHz magnet, Bruker Avance II console, 5 mm BBFO probe, and a standard Bruker parameter set (zg30) with d1 set to 30 s.

2. Reaction progress is monitored by thin layer chromatography (TLC) analysis using Whatman silica gel 60 F_{254} analytical plates and 25% ethyl acetate/hexanes. Visualization is accomplished using 254 nm UV light and *p*-anisaldehyde stain [prepared by dissolving *p*-anisaldehyde (15g) in EtOH (250 mL) and slowly adding concentrated H_2SO_4 (2.5 mL)]. R_f (*trans*-4-methoxy-3-buten-2-one; tech. grade) = 0.26 (pink) and 0.35 (purple). R_f (1,3,5-triacetylbenzene) = 0.20 (light orange).

3. The funnel used is a 150 mL medium-porosity 6 cm sintered-glass funnel purchased from Chemglass.

4. The crude yellow needles are 1,3,5-triacetylbenzene.

5. The filtrate (the combined supernatant and ether wash) is concentrated by rotary evaporation (34 °C, 20 mmHg) and analyzed by ^{1}H NMR spectroscopy. Any 1,3,5-triacetylbenzene present may be recovered according to the same crystallization procedure described above to give 90–120 mg of additional 1,3,5-triacetylbenzene in approximately 80% purity (^{1}H NMR; diagnostic resonances at 8.70 and 2.71 ppm).

6. The checkers carried out an analysis of the crude isolated product by ^{1}H NMR and found the material to be 99 wt % versus phenylmethyl sulfone (99 % purity, Acros cat # 353040050) as an internal standard.

7. Alternatively, purification of the crude product can be accomplished by sublimation. The crude yellow needles are placed in a water-cooled vacuum sublimator, which is then evacuated (0.01 mmHg), and heated to

110 °C for 12 h to give 0.90–0.95 g of product (as pale yellow crystals) per gram of starting crude yellow needles. The material obtained is spectroscopically indistinguishable (^{1}H NMR, ^{13}C NMR, IR, GC-MS) from the pale yellow needles obtained from recrystallization from ethanol. Anal. calcd. for $C_{12}H_{12}O_3$: C, 70.56; H 5.93; found C, 70.50; H, 5.93. An advantage of the sublimation method is that it reduces the solvent waste by half, relative to recrystallization.

8. Analytical data for the product are as follows: pale yellow needles; mp 162–165 °C; IR (ATR): 3358 (m), 3086 (m), 3064 (m), 3003 (m), 2924 (m), 1845 (w), 1683 (s), 1586 (m), 1418 (s), 1359 (s), 1316 (m), 1221 (s), 1089 (w), 1020 (w), 976 (m), 957 (w), 941 (w), 905 (m) cm^{-1}; ^{1}H NMR (400 MHz, CDCl$_3$) δ: 2.72 (s, 9 H), 8.71 (s, 3 H); ^{13}C NMR (100 MHz, CDCl$_3$) δ: 26.9, 131.7, 137.9, 196.6; HRMS (ESI) m/z calcd. for $C_{12}H_{13}O_3$ ([M+H]$^+$) 205.0859; found 205.0857; GCMS (EI): 204 (M$^+$), 189 ([M–CH$_3$]$^+$), 161 ([M–C(O)CH$_3$]$^+$), 118 ([M–(C(O)CH$_3$)$_2$]$^+$), 75 (M–(C(O)CH$_3$)$_3$]$^+$); Anal. calcd. for $C_{12}H_{12}O_3$: C, 70.56; H 5.93; found: C, 70.45; H, 5.84. GCMS analysis was performed using an Agilent 6890 GC with 5973 MS detector. Helium was used as carrier gas at 1 mL/min and a column from Restek (Rtx-5MS, 30 m x 0.25 mm, 0.5 μm film thickness). Splitless injection; Injector: 250 °C; transfer line 280 °C; source 150 °C. Initial T: 50 °C for 0.5 min then ramp 15 °C/min to 300 °C, isothermal for 4 min. The retention time for the product was 13.86 min, and no additional signals were present. The checkers did not analyze the isolated product by GC/MS.

9. Performing this procedure on smaller scale using 0.5 g of *trans*-4-methoxy-3-buten-2-one consistently gives a 70–89% yield of recrystallized 1,3,5-triacetylbenzene; however, on a larger scale than reported herein reproducibly leads only to a 61–64% yield of recrystallized 1,3,5-triacetylbenzene, despite extensive optimization. The product obtained in the small scale reaction is spectroscopically indistinguishable (^{1}H NMR, ^{13}C NMR) from the pale yellow needles obtained in the preparation described.

194

All hazardous materials should be handled and disposed of in accordance with "Prudent Practices in the Laboratory"; National Academy Press; Washington, DC, 1995.

3. Discussion

1,3,5-Triacetylbenzene is a valuable material that has been used in the synthesis of materials such as dendrimers,[2] phenolic molecular glasses,[3] and cyanine dyes.[4] It has found use as a starting material for the synthesis of cyclophanes[5] and several ligands for transition metals;[6] and it is a commonly used probe for investigations of reaction mechanism and scope.[7]

Several syntheses of 1,3,5-triacetylbenzene have been documented, including one in *Organic Syntheses (Coll. Vol. 3, p.829 (1955); Vol. 27, p.91 (1947))*. The procedure previously published in *Organic Synthesis* involves the synthesis of the sodium salt of acetoacetaldehyde and its subsequent condensation in the presence of acetic acid to yield 1,3,5-triacetylbenzene in 30-38% yield (Scheme 1).[8] The method presented herein is preferred because it is more efficient (54-60% yield of analytically pure material), it is more convenient (reflux commercially available materials for 2 d), and it generates a small volume of less hazardous waste (250 mL of reaction waste that is a mixture of 120 mL of ethanol, 30 mL of distilled water, 0.9 mL of acetic acid, and 200 mL of diethyl ether). The recrystallization generates 400 mL of additional solvent waste consisting of 200 mL of ether and 200 mL of ethanol, but sublimation can be performed instead (Note 6). The approximate cost of running this reaction, excluding solvents (ethanol, water, acetic acid, diethyl ether), is $38.

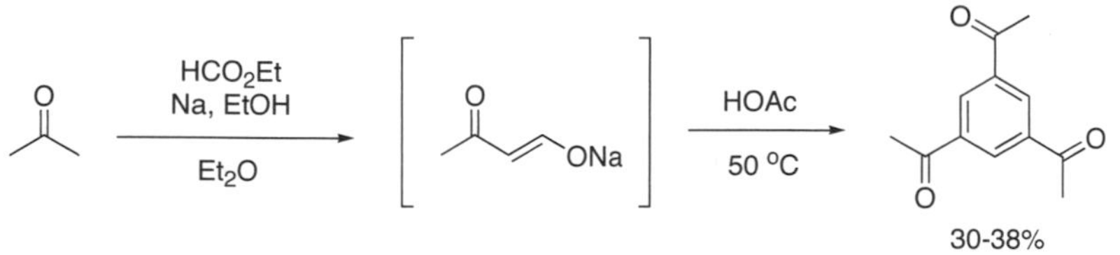

Scheme 1

Seven methods have been reported previously for the synthesis of 1,3,5-triacetylbenzene. Three of these methods use transition metal catalysts.

For example, the Eaton lab[9] employed cobalt(I) catalysts (Scheme 2A) in an overall [2+2+2] cyclotrimerization reaction of butyn-2-one to form 1,3,5-triacetylbenzene in 44% isolated yield. The Cadierno lab[10] used a ruthenium(IV) catalyst (Scheme 2B) in a similar cyclotrimerization reaction to generate a mixture (88:12) of 1,3,5-triacetylbenzene and 1,2,4-triacetylbenzene. Due to the expense of butyn-2-one (10 mL for $449.50; Aldrich), the method presented herein is preferable. Jiang and co-workers found that $PdCl_2$ catalyzes the cyclotrimerization of buten-2-one to give 1,3,5-triacetylbenzene in 58% yield (Scheme 2C).[11] Although this reaction yield is similar to the reaction reported herein, our procedure avoids the use of palladium, and is simpler to perform (the procedure reported by Jiang requires an autoclave for the reaction and chromatography for the subsequent separation).

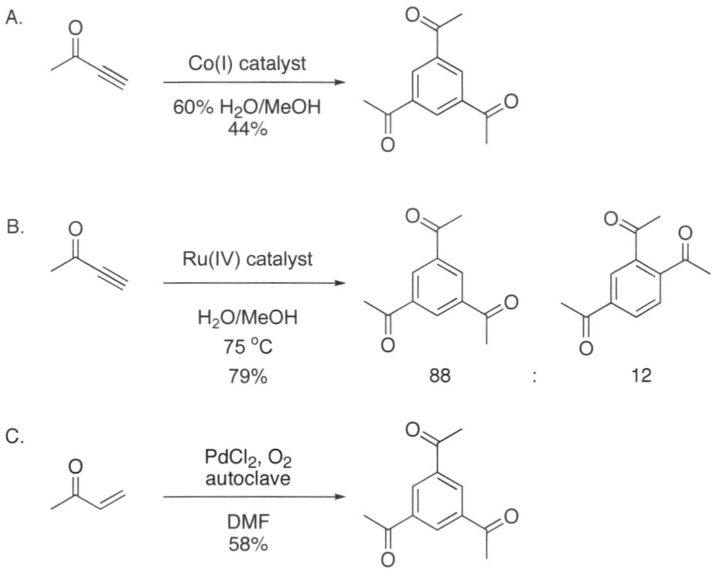

Scheme 2

Four methods that do not employ transition metal catalysts have also been reported. For example, Shi and co-workers describe the reaction of a tribenzimidazolium salt with Grignard reagents to afford 1,3,5-triacetylbenzene (Scheme 3).[12] This multi-step procedure has the disadvantages of poor atom economy and yield (36%). Xue and co-workers reported a cyclotrimerization reaction in which 1,3,5-triacetylbenzene is formed by reacting butyn-2-one in the presence of 2,4-pentanedione and

196

DMAP (Scheme 4A).[13] Although the reported yield is excellent (95%), the method requires chromatography and uses butyn-2-one, an expensive reagent (10 mL for $449.50; Aldrich), as the starting material. Lastly, the condensation of enaminones provides 1,3,5-triacetylbenzene, as described by Elnagdi and co-workers (Scheme 4B).[14] This procedure, which is very similar to the procedure presented herein, uses *trans*-4-dimethylamino-3-buten-2-one (25 g for $185; Aldrich), which is roughly twice as expensive as *trans*-4-methoxy-3-buten-2-one (50 g for $184; Aldrich).

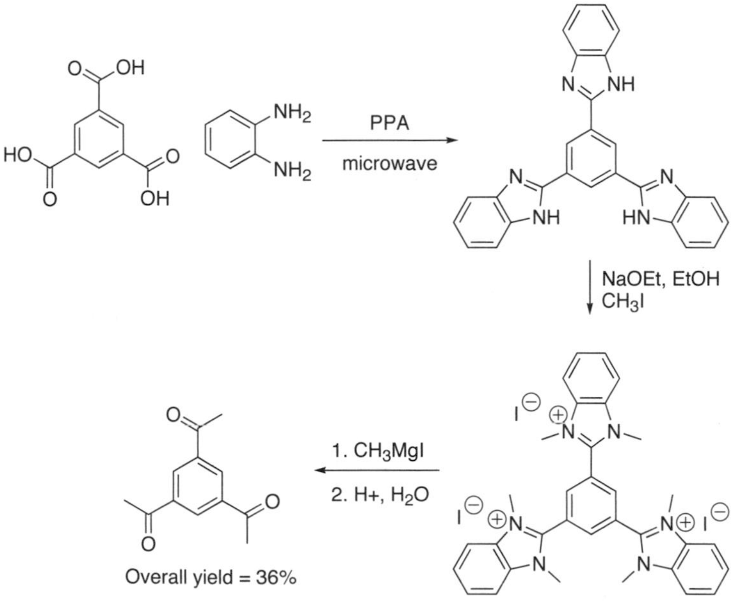

Scheme 3

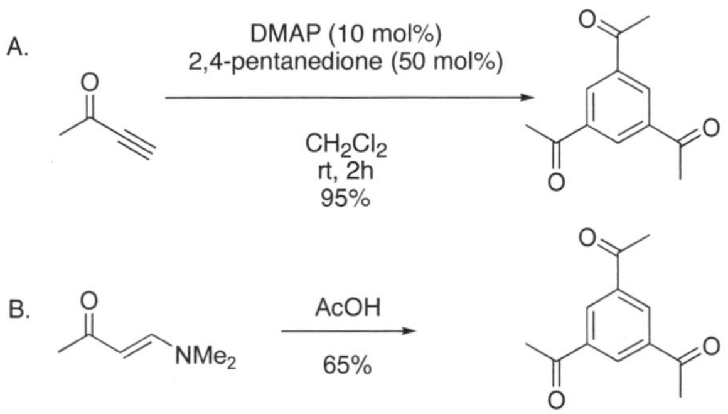

Scheme 4

In summary, the procedure presented herein is a simple process for the preparation and purification of 1,3,5-triacetylbenzene from *trans*-4-methoxy-3-buten-2-one. This method is advantageous relative to other synthetic strategies because the method is simple, requires no specialized equipment, is cost-effective, and generates a small quantity of waste that is relatively nontoxic.

1. 901 12th Avenue, Department of Chemistry, Seattle University, Seattle, WA 98122. alaimop@seattleu.edu; langenha@seattleu.edu

2. (a) Diez-Barra, E.; Gonzalez, R.; de la Hoz, A.; Rodriguez, A.; Sanchez-Verdu, P. *Tetrahedron Lett.* **1997**, *38*, 8557-8560. (b) L'abbé, G.; Haelterman, B.; Dehaen, W. *J. Chem. Soc., Perkin Trans. 1* **1994**, 2203-2204.

3. De Silva, A.; Lee, J.-K.; André, X.; Felix, N. M.; Cao, H. B.; Deng, H.; Ober, C. K. *Chem. Mater.* **2008**, *20*, 1606-1613.

4. Stadler, S.; Bräuchle, C; Brandl, S.; Gompper, R. *Chem. Mater.* **1996**, *8*, 676-678.

5. Karpuk, E.; Schollmeyer, D.; Meier, H. *Eur. J. Org. Chem.* **2007**, 1983-1990.

6. (a) Riesgo, E.C.; Jin, X.; Thummel, R. P. *J. Org. Chem.* **1996**, *61*, 3017-3022. (b) Gajardo, J.; Araya, J. C.; Moya, S. A.; Pardey, A. J.; Guerchais, V.; Le Bozec, H.; Aguirre, P. *Appl. Organomet. Chem.* **2006**, *20*, 272-276. (c) Rahman, A. F. M. M.; Jahng, Y. *Heteroat. Chem.* **2007**, *18*, 650-656.

7. (a) West, C. T.; Donnelly, S. J.; Kooistra, D. A.; Doyle, M. P. *J. Org. Chem.* **1973**, *38*, 2675-2681. (b) Makeiff, D. A.; Sherman, J. C. *Chem. Eur. J.* **2003**, *9*, 3253-3262. (c) Royles, B. J. L.; Smith, D. M. *J. Chem. Soc., Perkin Trans. 1* **1994**, 355-358. (d) Görlitz, G.; Hartmann, H. *Heteroat. Chem.* **1997**, *8*, 147-155. (e) Wolf, R. A.; Warakomski, J. M.; Staples, T. L.; Fazio, M. J. *J. Polym. Sci., Part A: Polym. Chem.* **2001**, *39*, 1391-1402.

8. Frank, R. L.; Varland, R. H. *Org. Synth.* **1947**, *3*, 829-830.

9. Sigman, M. S.; Fatland, A. W.; Eaton, B. E. *J. Am. Chem. Soc.* **1998**, *120*, 5130-5131.

10. Cadierno, V.; Garcia-Garrido, S. E.; Gimeno, J. *J. Am. Chem. Soc.* **2006**, *128*, 15094-15095.

11. Jiang, H.; Shen, Y.; Wang, Z. *Tetrahedron Lett.* **2007**, *48*, 7542-7545.

12. Zhang, B.; Gao, Y.; Li, J. L.; Shi, Z. *Chin. Chem. Lett.* **2006**, *17*, 1165-1168.

13. Zhou, Q.-F.; Yang, F.; Guo, Q.-X.; Xue, S. *Synlett* **2007**, 215-218.

14. Abdel-Khalik, M. M.; Elnagdi, M. H. *Synth. Commun.* **2002**, *32*, 159-164.

Appendix
Chemical Abstracts Nomenclature; (Registry Number)

trans-4-Methoxy-3-buten-2-one; (51731-17-0)
1,3,5-Triacetylbenzene; (779-90-8)

Peter J. Alaimo was born in 1968 in New York. He completed his undergraduate studies at the University of Michigan with Prof. Brian P. Coppola, and earned his Ph.D. at the University of California at Berkeley with Prof. Robert G. Bergman. He spent five years with Prof. Kevan M. Shokat (University of California at San Francisco) as a postdoctoral researcher. He began his independent career at Seattle University in 2004. His research interests include the development of new synthetic methods and asymmetric catalysts.

Joseph M. Langenhan completed his undergraduate degree at Allegheny College in 1998. He then pursued graduate studies with Professor Samuel H. Gellman at the University of Wisconsin-Madison's Department of Chemistry and earned a Ph.D. in Organic Chemistry in 2003. After two years of post-doctoral research with Professor Jon S. Thorson at the Laboratory for Biosynthetic Chemistry in the School of Pharmacy at the University of Wisconsin-Madison, he moved to Seattle University in 2005, where he is currently an Assistant Professor. He is currently developing chemoselective glycosylation methods for novel applications.

Amanda-Lynn Marshall was born in 1988 in Georgia. She completed her undergraduate studies at Seattle University where she earned her B.S. in General Science with concentrations in Chemistry and Biology. She was recently awarded a Rotary Ambassadorial Scholarship, which will allow her the opportunity to study abroad and obtain her M.Sc. in Clinical Microbiology/Immunology. After her studies abroad she plans to attend medical school.

David M. Andrews was born in 1988 in Utah. He is currently enrolled in Seattle University's undergraduate program working towards a B.S. in General Science. He is researching the enantioselective synthesis of 2,3-dihydropyridin-4-ones under the direction of Professor Alaimo. After graduation he plans to attend graduate school.

Liang Huang obtained his M.S. degree in Chemistry from Murray State University in 1997. He then began working at AlliedSignal as a synthetic organic chemist. In 1998 he moved to CB Research and Development, Inc. taking a position as Scientist. He has been working in Amgen's Chemical Process Research and Development group as a Process Chemist since 1999.

Eric A. Bercot completed his B.S. in Chemistry at the University of Oregon in 1999 while working in the laboratories of Michael M. Haley. He received his Ph.D. from Colorado State University in 2004 working with Professor Tomislav Rovis where he investigated the use of cyclic anhydrides as eletrophilic coupling partners in transition metal mediated cross-coupling reactions. In 2004, Eric joined the Chemical Process Research and Development group at Amgen where he is currently a Senior Scientist.

ORGANOCATALYTIC α-METHYLENATION OF ALDEHYDES: PREPARATION OF 3,7-DIMETHYL-2-METHYLENE-6-OCTENAL

Submitted by Meryem Benohoud, Anniina Erkkilä, and Petri M. Pihko.[1]
Checked by Yunus Turkmen and Viresh Rawal.

1. Procedure

A 100-mL, three-necked round-bottomed flask (Note 1), equipped with an 8 x 13 mm octagonal magnetic stir bar, a condenser, a rubber septum, and an internal thermometer, is charged with 4-(dimethylamino)benzoic acid (661 mg, 4.0 mmol) (Note 2). The system is evacuated, filled with nitrogen, and charged with dichloromethane (40 mL) (Notes 3-4) and pyrrolidine (334 µL, 4.0 mmol) (Notes 5-6), added sequentially. The resulting colorless solution is placed in an oil bath and heated to 35 °C (internal temperature). To this stirred solution is added via syringe (±)-citronellal ((±)-3,7-dimethyl-6-octenal) (7.20 mL, 40.0 mmol) (Note 7) and 37% aqueous formaldehyde solution (2.98 mL, 40.0 mmol) (Notes 8-9). The mixture is heated to reflux (bath temperature 50 °C) and maintained at reflux for 30 min (Note 10). The reaction mixture is cooled to room temperature, diluted with dichloromethane (20 mL) and transferred to a 250-mL separatory funnel. The reaction flask was rinsed with additional dichloromethane (3 x 7 mL). The organic phase is washed with water (40 mL) and then a saturated solution of NaHCO₃ (40 mL). The combined aqueous layers are extracted with dichloromethane (3 x 20 mL). The combined organic phases are dried over anhydrous Na₂SO₄ (35 g), filtered and concentrated by rotary evaporation (30 °C, 20 mmHg). The flask is then placed under a vacuum (1.5 mmHg) (Note 11) for ca. 30 min to afford 3,7-dimethyl-2-methylene-6-octenal as a light yellow liquid in quantitative yield

Org. Synth. **2010**, *87*, 201-208
Published on the Web 3/19/2010

(6.65 g). The resulting material is sufficiently pure for further reactions. Distillation of the product from a 25-mL flask, under reduced pressure, using a one-piece distillation apparatus equipped with a 6-cm Vigreux column, affords 3,7-dimethyl-2-methylene-6-octenal (5.44–5.93 g, 82–89%) as a colorless liquid (bp 68 °C/0.3 mmHg) (Notes 12-16).

2. Notes

1.   The checkers carried out the reaction under an inert atmosphere. The apparatus was oven-dried overnight and assembled while hot. It was then evacuated and filled with nitrogen three times and maintained under a positive pressure of nitrogen during the course of the reaction.

2.   4-(Dimethylamino)benzoic acid (98%) was purchased from Aldrich Chemical Co.

3.   A white suspension is observed at this point. 4-(Dimethylamino)benzoic acid is only sparingly soluble in dichloromethane.

4.   Dry dichloromethane was obtained from an MBraun Solvent Purification System. Checkers used dichloromethane (Optima®) purchased from Fisher Scientific, Inc. and purified through an Innovative Technology solvent purification system (activated alumina). The solvent was dried to ensure reproducible water content in the reaction mixture. However, the conditions are not strictly anhydrous as water is present in the formaldehyde solution.

5.   The submitters obtained pyrrolidine from Aldrich Chemical Co. and purified it before use by distillation under argon. Checkers purchased pyrrolidine (≥99.5%, redistilled, Sure/Seal™) from Aldrich Chemical Co. and used it as received.

6.   Most of the white suspension disappeared when pyrrolidine was added to the reaction mixture. The remaining few solid particles dissolved when the internal temperature reached 30 °C

7.   (±)-Citronellal (purum, >95.0% (GC)) was obtained from Aldrich Chemical Co. and used as received. This compound should be handled carefully as it is irritating to the eyes, skin and respiratory system.

8.   Formaldehyde (37 wt. % solution in water, A.C.S. reagent) was purchased from Aldrich Chemical Co. and used as received.

9.   The checkers used 1.05 equiv (42.0 mmol, 3.13 mL) of formaldehyde for the second run. However, the ^{1}H-NMR of the crude

product didn't show any improvement when compared with that of the first run.

10. The progress of the reaction can be followed by TLC (hexanes/ethyl acetate 9/1, anisaldehyde stain). (±)-Citronellal R_f = 0.51, product R_f = 0.60).

11. The submitters used 0.1 mmHg vacuum.

12. The submitters reported the boiling point as 94–97 °C at 0.1 mmHg.

13. The submitters reported obtaining 5.60 g (84%) of the product.

14. The checkers carried out the reaction twice at full scale. First run: 5.44 g, 82% (68 °C, 0.3 mmHg). Second run: Fraction 1 (0.72 g), fraction 2 (5.04 g), fraction 3 (obtained by heating the Vigreux column with a heat gun, 0.17 g); total yield 5.93 g, 89%; bp 73 °C at 0.4 mmHg. All three fractions were pure by NMR. ^{1}H and ^{13}C NMR, GC-MS, and elemental analysis data were obtained by using the 2nd fraction.

15. Spectroscopic data of (±)-3,7-dimethyl-2-methyl-6-octenal closely matched published data[2,3]: ^{1}H NMR (500 MHz, CDCl$_3$) δ: 1.04 (d, J = 7.0 Hz, 3 H), 1.32–1.40 (m, 1 H), 1.47–1.53 (m, 1 H), 1.54 (s, 3 H), 1.65 (s, 3 H), 1.85–1.96 (m, J = 8.0 Hz, 2 H), 2.68 (sext, J = 7.0 Hz, 1 H), 5.05 (m, J = 7.0, 1.5 Hz, 1 H), 5.97 (s, 1 H), 6.21 (s, 1 H), 9.51 (s, 1 H); ^{13}C NMR (125 MHz, CDCl$_3$) δ: 17.8, 19.7, 25.8, 25.9, 31.1, 35.8, 124.3, 131.8, 133.2, 155.7, 194.8; IR (film) 2966, 2926, 2857, 1696, 1454, 1377, 945 cm^{-1}; Anal. calcd. for C$_{11}$H$_{18}$O: C, 79.46; H, 10.91; O, 9.63; found: C, 79.66; H, 10.88; O, 9.46; GC-MS (column ZB-5MS 0.25 μm x 250 μm x 30 m, initial T = 50 °C for 1.0 min, ramp = 15.0 °C/min to 260 °C, hold for 10 min., MS: Source T = 150 °C, EI$^+$): R_t = 7.15 min., m/z 165.9, 151.0, 149.0, 123.0, 109.0, 81.0.

16. The product may polymerize during storage and it should be stored in the freezer (–20 °C).

Safety and Waste Disposal Information

All hazardous materials should be handled and disposed of in accordance with "Prudent Practices in the Laboratory"; National Academy Press; Washington, DC, 1995.

3. Discussion

Several methods for the preparation of α-substituted acroleins are available; however, typically, these require relatively drastic conditions, including high temperature and high pressure.[4,5] Moreover, only the preparation of simple α-substituted acroleins has been reported using these methods. More complex aldehydes have required milder, but non-catalytic, procedures[6,7] such as Horner-Wadsworth-Emmons reactions[8,9] or Mannich reactions[10] with Eschenmoser's salt. The present method allows the rapid synthesis of α-substituted-α,β-unsaturated aldehydes in good yields with simple reagents and tolerates a variety of functionalities (Table 1). The procedure described herein represents an improved and faster protocol when compared to our originally published method.[11] The products are potentially highly useful intermediates in organic synthesis,[12] given the range of possible transformations, such as 1,2-[13,14] and 1,4-additions,[15-18] Baylis-Hillman reactions,[19-20] Diels-Alder reactions,[21-24] and a number of organocatalyzed transformations.[25-27]

Table 1 Formation of α-substituted acroleins

Starting material	Product	Time [min]	Yield [%]
		180	85
		5	99
		25	> 99
		60	> 99
		10	85
		25	> 99
		20	> 99
		40	> 99
		15	98

1. Department of Chemistry, University of Jyväskylä, P. O. B: 35, FI-40014 JYU, Finland. E-mail: Petri.Pihko@jyu.fi
2. Erkkilä, A.; Pihko, P. M. *Eur. J. Org. Chem.* **2007**, 4205-4216.

3. White, J. D.; Amedio, J. C.; Gut, S.; Ohira, S.; Jayasinghe, L. R. *J. Org. Chem.* **1992**, *57*, 2270-2284.
4. Häusermann, M. *Helv. Chim. Acta* **1951**, *34*, 1482-1491.
5. Offenhauer, R. D.; Nelsen, S. F. *J. Org. Chem.* **1968**, *33*, 775-777.
6. Hon, Y.-S.; Chang, F.-J.; Lu, L. *J. Chem. Soc., Chem. Commun.* **1994**, 2041-2042.
7. Hon, Y.-S.; Chang, F.-J.; Lu, L.; Lin, W.-C. *Tetrahedron* **1998**, *54*, 5233-5246.
8. Boehm, H. M.; Handa, S.; Pattenden, G.; Roberts, L.; Blake, A. J.; Li, W.-S. *J. Chem. Soc., Perkin Trans. I* **2000**, 3522-3538.
9. Villiéras, J.; Rambaud, M. *Synthesis* **1984**, 406-408.
10. Kinast, G.; Tietze, L. F. *Angew. Chem. Int. Ed.* **1976**, *15*, 239-240.
11. Erkkilä, A.; Pihko, P. M. *J. Org. Chem.* **2006**, *71*, 2538-2541.
12. Evans, D. A.; Ratz, A. M.; Huff, B. E.; Sheppard, G. S. *J. Am. Chem. Soc.* **1995**, *117*, 3448-3467.
13. Ken, O.; Shigeru, M.; Ryosuke, G.; Masayuki, I. *Angew. Chem. Int. Ed.* **2010**, *49*, 329-332.
14. Mann, R. K.; Parsons, J. G.; Rizzacasa, M. A. *J. Chem. Soc., Perkin Trans. I* **1998**, 1283-1294.
15. Perlmutter, P. *Conjugate Addition Reactions in Organic Synthesis*; Pergamon: Oxford, 1992.
16. Rossiter, B. E.; Swingle, N. M. *Chem. Rev.* **1992**, *92*, 771-806.
17. Giorgi, G.; Miranda, S.; Lopez-Alvarado, P.; Avendano, C.; Rodriguez, J.; Menendez, J. C. *Org. Lett.* **2005**, *7*, 2197-2200.
18. Ho, C.-Y.; Schleicher; D., K.; Chan, C.-W.; Jamison, T. F. *Synlett* **2009**, 2565-2582.
19. Basavaiah, D.; Dharma Rao, P.; Suguna Hyma, R. *Tetrahedron* **1996**, *52*, 8001-8062.
20. Basavaiah, D.; Rao, A. J.; Satyanarayana, T. *Chem. Rev.* **2003**, *103*, 811-892.
21. Kagan, H. B.; Riant, O. *Chem. Rev.* **1992**, *92*, 1007-1019.
22. Li, J.; Li, X.; Zhou, P.; Zhang, L.; Luo, S.; Cheng, J.-P. *Eur. J. Org. Chem.* **2009**, 4486-4493.
23. Min, S.-J.; Jones, G. O.; Houk, K. N.; Danishefsky, S. J. *J. Am. Chem. Soc.* **2007**, *129*, 10078-10079.
24. Corey, E. J.; Guzman-Perez, A. *Angew. Chem. Int. Ed.* **1998**, *37*, 388-401.

25. Kano, T.; Tanaka, Y.; Osawa, K.; Yurino, T.; Maruoka, K. *Chem. Commun.* **2009**, 1956-1958.
26. Erkkilä, A.; Pihko, P. M.; Clarke, M.-R. *Adv. Synth. Catal.* **2007**, *349*, 802-806.
27. Ishihara, K.; Nakano, K. *J. Am. Chem. Soc.* **2005**, *127*, 10504-10505.

Appendix
Chemical Abstracts Nomenclature; (Registry Number)

4-(Dimethylamino)benzoic acid: (619-84-1)
Pyrrolidine (123-75-1)
(±)-Citronellal: ((±)-3,7-dimethyl-6-octenal); (106-23-0)
3,7-Dimethyl-2-methylene-6-octenal; (22418-66-2)
37% Aqueous formaldehyde; (50-00-0)

Petri Pihko was born in 1971 in Oulu, Finland. He became interested in chemistry several years before entering the University, assisted by inspiring teachers, Maija Aksela (currently a Professor of Chemical Education at the University of Helsinki) and Prof. Hans Krieger, who also taught him organic chemistry at the University of Oulu before his retirement. Petri Pihko then joined the research group of Professor Ari Koskinen, graduating with a Ph.D. in 1999. Between 1999 and 2001, he enjoyed nearly two years of a wonderful time as a postdoctoral associate with Professor K. C. Nicolaou at the Scripps Research Institute in La Jolla, California, USA. In 2001, he joined the faculty of Helsinki University of Technology (TKK), and in 2008, his research group moved to the University of Jyväskylä, Finland. His research interests include organocatalysis, catalyst design, and total synthesis of natural products.

Meryem Benohoud was born in 1981 in Antwerp, Belgium. She graduated from the Université Paris XI Orsay with a Ph.D. in 2008 realized under the supervision of Robert H. Dodd at the Institut de Chimie des Substances Naturelles in Gif-sur-Yvette, France. She is now a postdoctoral associate with Petri M. Pihko at the University of Jyväskylä, Finland, and she works on the development of oxidation reactions using organocatalysis and tertiary and quaternary centers formation.

Anniina Erkkilä was born in 1980 in Helsinki, Finland. She received her M.Sc. (Tech.) degree in Organic Chemistry from the Helsinki University of Technology in 2003. Subsequently, she joined the research group of Professor Petri Pihko working on developing new methodologies for organocatalytic reactions, earning her D.Sc. (Tech.) degree in 2007. In 2005, she worked as a visiting researcher in the laboratory of Professor Benjamin List at Max-Planck-Institut für Kohlenforschung in Germany for 3 months. In May 2007, she was awarded the Publication Prize of the Division of Synthetic Chemistry, Association of the Finnish Chemical Societies.

Yunus E. Turkmen was born in Bursa, Turkey in 1983. He received his B.S. and M.S. in chemistry from Middle East Technical University, Turkey and in 2006, he started his PhD study at the University of Chicago. He's currently a graduate student in Prof. Viresh Rawal group, working on hydrogen-bond catalysis.

RUTHENIUM-CATALYZED ARYLATION OF ORTHO C-H BOND IN AN AROMATIC WITH AN ARYLBORONATE: 8-PHENYL-1-TETRALONE

Submitted by Kentaroh Kitazawa, Takuya Kochi, and Fumitoshi Kakiuchi.[1]
Checked by Somenath Chowdhury and Jonathan A. Ellman.

1. Procedure

8-Phenyl-1-tetralone (Note 1). All glassware and a magnetic stirring bar are oven-dried for about 0.5 h prior to use. A 100-mL three-necked round-bottomed flask is equipped with a reflux condenser connected to a vacuum/N_2 line, a thermometer, a rubber septum, and a Teflon coated octagonal magnetic stirring bar of dimension 8mm x 13mm. The flask is evacuated and backfilled with nitrogen three times. While the rubber septum is temporarily removed and a positive flow of nitrogen is maintained, the flask is charged with 5,5-dimethyl-2-phenyl[1,3,2]dioxaborinane, $PhB(OCH_2CMe_2CH_2O)$, (8.55 g, 45.0 mmol, 1.50 equiv) (Note 2) and carbonyldihydridotris(triphenylphosphine)ruthenium(II) ($RuH_2(CO)(PPh_3)_3$) (1.38 g, 1.50 mmol, 0.050 equiv) (Note 3). The flask is again evacuated and backfilled with nitrogen three times. 3,3-Dimethylbutan-2-one (pinacolone) (15 mL) (Note 4) and 1-tetralone (4.39 g, 4.00 mL, 30.0 mmol, 1.00 equiv) (Note 5) are added to the flask through the rubber septum via syringes in the order indicated at room temperature to give a white suspension (Note 6). The resulting mixture is heated at reflux in an oil bath (oil bath temperature 125 °C) (Note 7). The reaction mixture becomes a dark red solution within 5 min. After heating for 12 h, the mixture is cooled to room temperature (Notes 8 and 9). The resulting dark red solution is transferred to a 250-mL round-bottomed flask and concentrated by rotary evaporation (15 mmHg). The three-necked flask is rinsed with dichloromethane (50 mL) in two portions, and the combined rinse is transferred to the 250-mL flask, followed by concentration by rotary evaporation. The volatile materials are further removed under vacuum (1 mmHg) for 30 min. The residue is dissolved in

Org. Synth. **2010**, *87*, 209-217
Published on the Web 4/1/2010

dichloromethane (50 mL) and transferred to a 1-L Erlenmeyer flask. The round-bottomed flask is further rinsed with dichloromethane (2 x 125 mL) and the rinse is also transferred to the Erlenmeyer flask (Note 10). Basic alumina (250 g) (Note 11) is added to the solution, and the resulting mixture is stirred at room temperature for 30 min (Note 12). The mixture is filtered through Celite (20 g) (Note 13), and the filter cake is washed with dichloromethane (500 mL). After concentrating by rotary evaporation (15 mmHg), the resulting material is transferred to a 200-mL recovery flask using dichloromethane (50 mL). SiO_2 (9.0 g) (Note 14) is added to the dark red solution. The solvent is removed by rotary evaporation (15 mmHg) and then under vacuum (1 mmHg) for 2 h. Flash column chromatography is performed by charging the resulting solid onto the top of the column of SiO_2 (120 g) (Note 14) and eluting with 5% ethyl acetate in hexane (Note 15). The combined fractions containing the phenylated product are concentrated by rotary evaporation (15 mmHg). The resulting material is dissolved in a single-neck, open to air 1-L round-bottomed flask using methanol (90 mL) (Note 16) by heating in an oil bath with the bath temperature maintained at 90 °C. Distilled water (300 mL) is slowly added to the mixture to form a white precipitate, and then the oil bath is removed. After cooling the mixture to room temperature, water (600 mL) is slowly added. The precipitated white solid is collected by suction filtration on a Büchner funnel (Note 17) and washed with water (200 mL). Drying the solid under vacuum (1 mmHg) for 20 h gives 5.39-5.72 g (81-86% yield) of 8-phenyl-1-tetralone as a white solid (Note 18).

2. Notes

1. This procedure is a modification of that published by the submitters.[2]
2. The phenylboronate, 5,5-dimethyl-2-phenyl[1,3,2]dioxaborinane, can be synthesized according to the literature method.[3,4] This reagent is commercially available from chemical suppliers including Aldrich Chemical Company, Inc. and Wako Pure Chemical Industries, Ltd. The checkers purchased this reagent from Sigma-Aldrich.
3. The ruthenium complex, carbonyldihydridotris(triphenyl-phosphine)ruthenium(II), can be synthesized according to the literature method.[5,6] This complex is commercially available from various chemical suppliers including Aldrich Chemical Company, Inc., Strem Chemicals Inc.,

Org. Synth. **2010**, *87*, 209-217

Acros Organics, Wako Pure Chemical Industries, Ltd., and Tokyo Chemical Industry Co., Ltd. The checkers purchased the reagent from Sigma-Aldrich.

4.   Pinacolone was purchased from Sigma-Aldrich by the checkers and from Tokyo Chemical Industry Co., Ltd by the submitters. The reagent was dried over $CaSO_4$ and distilled (bp = 106 °C/760 mmHg).

5.   1-Tetralone was purchased from Sigma-Aldrich, dried over $CaSO_4$, and distilled under reduced pressure (bp = 117 °C/5 mmHg).

6.   The ruthenium complex was only slightly soluble in pinacolone at room temperature.

7.   Maintaining the oil bath temperature at 120–130 °C (preferentially around 125 °C) is necessary to effect the catalytic reaction with good reproducibility. Oil bath temperature was set to 125 °C and the internal temperature varied from 117–119°C.

8.   The progress of the reaction can be followed by TLC analysis on silica gel (with 20% EtOAc-hexane as eluent and visualization with UV light (254 nm)). 1-Tetralone has an R_f = 0.51 and the phenylation product has an R_f = 0.44. TLC analysis was performed on Merck silica gel $60F_{254}$ pre-coated glass plates.

9.   After the reaction, a yellow precipitate is sometimes observed. In this case, after cooling to room temperature, the reaction mixture is filtered through Celite. The Celite pad is then washed with diethyl ether (10 mL x 3) and the filtrates are combined. The checkers did not observe any yellow precipitate.

10.   Dichloromethane was purchased from Fisher Scientific and used as received.

11.   Merck aluminum oxide 90 active basic (0.063-0.200 mm, activity stage I) was used.

12.   Basic alumina was added to remove the unreacted phenylboronate from the mixture, because the boronic ester exhibits significant tailing on silica gel column chromatography.

13.   Celite® 545 was purchased from Kanto Chemical Co., Inc. and used as received.

14.   Merck silica gel 60 (mesh 230-400) was added to the flask, which is the same batch of silica gel that was used to perform the chromatography.

15.   The silica gel column (33 x 230 mm) was eluted with 3.5 L of 5% ethyl acetate in hexane at a flow rate of approximately 40 mL/min. 50 mL fractions were collected, and fractions 19 to 57, for which the product was observed by TLC analysis, were combined. In some cases, a small amount of

1-tetralone was also found in the early fractions that contained the product. These fractions can be combined with pure ones, which only contain the product, because 1-tetralone can be removed by a reprecipitation procedure following chromatography. Ethyl acetate (>99.0%) and hexane (>95.0%) were purchased from Fisher Scientific by the checkers and from Wako Pure Chemical Industries, Ltd. by the submitters and were used as received.

16. Methanol (>99.5%) was purchased from Fisher Scientific by the Checkers and from Kanto Chemical Co., Inc. by the submitters and was used as received.

17. Whatman grade 202 filter paper was used for the filtration.

18. The product exhibited the following physicochemical properties: Mp 97.0–97.5 °C; IR (ATR): 1678 (C=O) cm^{-1}; ^{1}H NMR (600 MHz, CDCl$_3$) δ: 2.14 (tt, J = 6.6, 6.0 Hz, 2 H), 2.62 (t, J = 6.6 Hz, 2 H), 3.01 (t, J = 6.3 Hz, 2 H), 7.13 (d, J = 7.2 Hz, 1H), 7.21–7.22 (m, 2 H), 7.24–7.26 (d, J = 7.8 Hz, 1 H), 7.31–7.33 (m, 1 H), 7.35–7.39 (m, 2 H), 7.43 (t, J = 7.8 Hz, 1 H); ^{13}C NMR (150 MHz, CDCl$_3$) δ: 23.2, 30.8, 40.6, 126.7, 127.9, 128.2, 128.2, 130.4, 131.3, 131.9, 143.0, 144.1, 145.7, 198.5; HRMS (ESI) m/z calc. for C$_{16}$H$_{15}$O ([M+H]$^+$) 223.1117; found 223.1110; Anal. Calcd. for C$_{16}$H$_{14}$O: C, 86.45; H, 6.35; found: C, 86.15; H, 6.52.

Safety and Waste Disposal Information

All hazardous materials should be handled and disposed of in accordance with "Prudent Practices in the Laboratory"; National Academy Press; Washington, DC, 1995.

3. Discussion

Transition metal-catalyzed functionalizations of C-H bonds are highly attractive research subjects in current organic synthesis.[7,8] Among these reactions, heteroatom-directed direct C-H arylations of aromatic compounds have been extensively studied. Almost all of these arylations involve electrophilic substitution at aromatic C-H bonds[9] or base-assisted proton abstraction from aromatic C-H bonds.[10] In these processes, relatively high valent transition metals, such as, Pd(II), Ru(II), and Rh(III), participate in the C-H bond cleavage step, and the hydrogen of the C-H bond is removed as a proton from the aromatic ring. In the C-H arylation described here, however, an ortho C-H bond in an aromatic ketone is cleaved via oxidative

212

addition to a ruthenium(0) center and an (aryl)(H)Ru species is formed.[2,11] One of the key steps in the catalytic cycle is generation of a ruthenium-alkoxide intermediate which is essential for transmetalation with an arylboronate. The use of pinacolone as a solvent effectively generates an (aryl)(alkoxy)ruthenium intermediate without sacrificing much of the aromatic ketone substrate. Thus, pinacolone functions not only as a solvent but also as a hydride acceptor from the Ru-H species. Transmetalation between the ruthenium-alkoxide and the organoboronate provides a diarylruthenium intermediate, and subsequent reductive elimination yields the corresponding ortho-arylated aromatic ketone and regenerates the catalytically active species.

An important feature of the present ruthenium-catalyzed C-H arylation is that many functional groups, except for the ortho C-H bond, are tolerated in the reaction. A variety of aromatic ketones bearing an electron-donating or electron-withdrawing group can be used for this ruthenium-catalyzed arylation with arylboronates. Representative results concerning the $RuH_2(CO)(PPh_3)_3$-catalyzed C-H arylation are listed in the Table.

Table 1. Ruthenium-catalyzed C-H arylation of aromatic ketones with arylboronates.[a,b]

Entry	Ketone	Arylboronate	Time	Product	Yield[c]
1[d]		**1**	1 h		89%
2		**1**	1 h		76%
3		**1**	2 h		61%
4		**1**	1 h		81%
5		**1**	1 h		78%
6		**1**	3 h		78%
7[e]	**2**	**1**	4 h	**2**	98%
		3		**4**	
8[e]	**2**	**3a:** Ar = 4-MeOC$_6$H$_4$	1 h	**4a**	88%
9[e]	**2**	**3b:** Ar = 4-Me$_2$NC$_6$H$_4$	1 h	**4b**	84%
10[e]	**2**	**3c:** Ar = 4-FC$_6$H$_4$	1 h	**4c**	75%
11[e]	**2**	**3d:** Ar = 4-CF$_3$C$_6$H$_4$	1 h	**4d**	84%
12[e]	**2**	**3e:** Ar = 2-MeC$_6$H$_4$	1 h	**4e**	96%
13[e]	**2**	**3f:** Ar = 1-naphthyl	1 h	**4f**	92%

[a]See ref. [b]Reaction conditions: aromatic ketone (1 mmol), phenylboronate (1.1 mmol), RuH$_2$(CO)(PPh$_3$)$_3$ (0.02 mmol), pinacolone (1.0 mL, 8 mmol), reflux. [c]Isolated yield. [d]**1** (2.2 mmol), [e]**1** (1.2 mmol), pinacolone (0.5 mL).

1. Department of Chemistry, Faculty of Science and Technology, Keio University, 3-14-1 Hiyoshi, Kohoku-ku, Yokohama, Kanagawa 223-8522, Japan. E-mail: kakiuchi@chem.keio.ac.jp.

2. Kakiuchi, F.; Matsuura, Y.; Kan, S.; Chatani, N. *J. Am. Chem. Soc.* **2005**, *127*, 5936-5945.

3. Blakemore, P. R.; Marsden, S. P.; Vater, H. D. *Org. Lett.* **2006**, *8*, 773-776.

4. Chaumeil, H.; Signorella, S.; Le Drian, C. *Tetrahedron* **2000**, *56*, 9655-9662.

5. Kakiuchi, F.; Sekine, S.; Tanaka, Y.; Kamatani, A.; Sonoda, M.; Chatani, N.; Murai, S. *Bull. Chem. Soc. Jpn.* **1995**, *68*, 62-83.

6. Ahmad, N.; Levison, J. J.; Robinson, S. D.; Uttley, M. F. *Inorg. Synth.* **1974**, *15*, 45-64.

7. Alberico, D.; Scott, M. E.; Lautens, M. *Chem. Rev.*, **2007**, *107*, 174-238.

8. Kakiuchi, F.; Kochi, T. *Synthesis* **2008**, 3013-3039.

9. Ackermann, L. *Synlett* **2007**, 507-526.

10. García-Cuadrado, D.; Braga, A. A. C.; Maseras, F.; Echavarren, A. M. *J. Am. Chem. Soc.* **2006**, *128*, 1066-1067.

11. Kakiuchi, F.; Kan, S.; Igi, K.; Chatani, N.; Murai, S. *J. Am. Chem. Soc.* **2003**, *125*, 1698-1699.

Appendix
Chemical Abstracts Nomenclature; (Registry Number)

RuH$_2$(CO)(PPh$_3$)$_3$: Ruthenium, carbonyldihydrotris(triphenylphosphine); (25360-32-1)

1-Tetralone: 1(2H)-Naphthalenone, 3,4-dihydro-; (529-34-0)

pinacolone: 2-Butanone, 3,3-dimethyl-; (75-97-8)

Phenylboronate: 1,3,2-Dioxaborinane, 5,5-dimethyl-2-phenyl-; (5123-13-7)

Fumitoshi Kakiuchi was born in Hyogo, Japan, in 1965 and received his B.Sc. in 1988 and Ph.D. in 1993 from Osaka University under the guidance of Professor Shinji Murai. He was appointed as an Assistant Professor at Osaka University in 1993. He did his postdoctoral work with Prof. E. N. Jacobsen at Harvard University in 1996-1997. In 2000, he was promoted to an Associate Professor at Osaka University. In 2005, he moved to Keio University as a Professor. His research interests include the development of new transition metal-catalyzed reactions.

Kentaroh Kitazawa was born in Nagano, Japan in 1983. B. Sc. in 2006 and M.Sc. in 2008 from Keio University, working with Professor Fumitoshi Kakiuchi. He is currently a Ph.D. student in Professor Kakiuchi's group at Keio University. His research interest is catalytic C-H functionalization.

Takuya Kochi was born in Tokyo, Japan in 1975. He received his undergraduate and master's degree from the University of Tokyo, working with Professors Masanobu Hidai and Youichi Ishii, and his Ph.D. in chemistry from the University of California at Berkeley, working with Professor Jonathan A. Ellman. After carrying out postdoctoral research with Professor Kyoko Nozaki at the University of Tokyo, he joined the group of Professor Fumitoshi Kakiuchi at Keio University as an Assistant Professor in 2007. His research interests include development of new reactions and their application to the synthesis of a wide range of organic molecules.

Somenath Chowdhury grew up in Majirdanga, India. He obtained his M.Sc. degree in Chemistry from the Indian Institute of Technology, Kharagpur, India and his Ph. D. degree from the University of Saskatchewan, Canada. In his Ph.D. research he worked on the design and synthesis of peptide beta sheet mimics under the supervision of Professor Heinz-Berhard Kraatz. After finishing his Ph.D. he carried out research with Professor Giuseppe Melacini's group at McMaster University, Canada for one and a half years where he studied protein ligand interactions by NMR spectroscopy. He then joined Professor Jonathan Ellman's lab at the University of California, Berkeley in Sept 2008 as an NSERC postdoctoral fellow. His current research focuses on developing small molecule protease inhibitors.

AMIDE FORMATION BY DECARBOXYLATIVE CONDENSATION OF HYDROXYLAMINES AND α-KETOACIDS: N-[(1S)-1 PHENYLETHYL]-BENZENEACETAMIDE

Submitted by Lei Ju and Jeffrey W. Bode.[1]
Checked by Tatsuya Toma and Tohru Fukuyama.

1. Procedure

N-[(1S)-1-Phenylethyl]-benzeneacetamide. A 500-mL, single-necked, round-bottomed flask equipped with an 8.0 × 30 mm, octagon-shaped Teflon coated-magnetic stir bar, a reflux condenser, a rubber septum and nitrogen gas inlet is charged with phenylpyruvic acid (4.75 g, 28.9 mmol, 1.0 equiv) (Note 1) and *N,N*-dimethylformamide (289 mL) (Note 2). After stirring for 5 min, *N*-hydroxy-(*S*)-1-phenylethylamine oxalate (9.20 g, 40.5 mmol, 1.4 equiv) (Note 3) is added to the homogeneous solution (Note 4) as a solid in one portion at ambient temperature (23 °C). The reaction mixture is warmed to 40 °C in an oil bath and stirred at that temperature under a nitrogen atmosphere until completion (Note 5). The reaction is subsequently concentrated by rotary evaporation (50 °C, 10 mmHg) to approximately 20 mL. The resulting slightly yellow solution is allowed to cool to ambient temperature (23 °C) over the course of 30 min and diluted with ethyl ether (Et₂O) (200 mL) (Note 6). The solution is carefully poured into a 2-L separatory funnel, to which 1 N aqueous hydrochloric acid solution (200 mL) is added. The organic layer is separated and extracted with an additional portion of 1 N aqueous hydrochloric acid solution (200 mL), and the combined aqueous phase is extracted with Et₂O (3 × 200 mL). To the combined organic layers is added saturated sodium bicarbonate (NaHCO₃) solution (200 mL) and the mixture is partitioned. The basic aqueous solution is back-extracted with Et₂O (2 × 200 mL). The combined organic phase is washed with brine (400 mL) (Note 7), dried over anhydrous sodium sulfate (Na₂SO₄) (Note 8), and filtered. The filtrate is concentrated on a rotary evaporator (40 °C, 20 mmHg), and dried under reduced pressure (2 mmHg)

Org. Synth. **2010**, *87*, 218-225
Published on the Web 4/1/2010

overnight to afford a yellow, viscous oil (6.55–6.92 g, 95–100%) (Note 9). The resulting crude product is dissolved in dichloromethane (5 mL) and loaded on a column (6.0 cm i.d. × 20 cm) of SiO_2 (280 g) (Note 10, 11). After elution with 400 mL of 30% ethyl acetate in hexanes, the product is obtained by collecting 450 mL (18 × 25 mL fractions) of the eluent (Note 12). The combined fractions are concentrated by rotary evaporation (40 °C, 20 mmHg) followed by high vacuum (2 mmHg) to provide N-[(1S)-1 phenylethyl]-benzeneacetamide (5.92–5.96 g, 85–86%) as a white solid. The product is dissolved in hot Et_2O (300 mL) (50 °C, at reflux) and cooled to 4 °C overnight. The resulting crystals are collected by filtration on a Büchner funnel and washed with ice-cold Et_2O (50 mL). The crystals are then transferred to a 100-mL round-bottomed flask and dried overnight at 2.0 mmHg to afford a spindle-like solid (5.06–5.08 g, 73%) (Notes 13, 14, and 15).

2. Notes

1.   Phenylpyruvic acid (98%) was purchased from Aldrich Chemical Co., Inc and was recrystallized from hot benzene before use. CAUTION: Benzene is carcinogenic, and must be handled with care.

2.   N,N-Dimethylformamide (99.8%), purchased by the submitters from EMD Biosciences, Inc. and by the checkers from Wako Pure Chemical Industries, Ltd., was passed over activated molecular sieves 4A under an argon atmosphere before use.

3.   N-Hydroxy-(S)-1-phenylethylamine oxalate was prepared from (S)-1-phenylethylamine following a reported procedure.[2] The oxalate salt of the hydroxylamine was recrystallized from hot ethanol and washed with ethyl ether. The oxalate salt form of the hydroxylamine is bench stable and more efficient in the ligation reaction. CAUTION: Free hydroxylamines may cause explosions under certain conditions. Careful handling is required when they are heated.

4.   Phenylpyruvic acid was allowed to dissolve completely in N,N-dimethylformamide as a 0.1 M solution. White precipitate persisted if phenylpyruvic acid was used without recrystallization.

5.   The progress of the reaction was monitored on reverse phase HPLC by following the disappearance of phenylpyruvic acid. Analytical conditions were: Column: Shiseido Capcell Pac C18; Eluent: 0.1% TFA in H_2O/MeCN; Flow rate: 1 mL/min; Detection: 216 nm, 235 nm, 288 nm, 221

nm; Gradient: 0–95% MeCN over 30 min. R_T of phenylpyruvic acid = 19.04 min; R_T of product = 21.56 min. The reaction typically takes 26–30 h to complete, while trace amounts of phenylpyruvic acid could still be observed at 288 nm.

6.   Ethyl ether (Et_2O) (99.9%), unstabilized, was purchased by the submitters from EMD Biosciences, Inc. and by the checkers from Wako Pure Chemical Industries, Ltd., and used without further purification.

7.   Sodium chloride (NaCl), crystalline, was purchased by the submitters from EMD Biosciences, Inc. and by the checkers from Wako Pure Chemical Industries, Ltd.

8.   Sodium sulfate (Na_2SO_4) anhydrous, crystalline, was purchased by the submitters from EMD Biosciences, Inc. and by the checkers from Wako Pure Chemical Industries, Ltd.

9. ^{1}H NMR analysis by the submitters indicated a high purity of the crude product, which was contaminated by 3 wt. % of residual N,N-dimethylformamide. No residual N,N-dimethylformamide was detected by the checkers and the crude product was obtained as a yellow solid.

10.   The column was packed by the submitters with EMD Silica Gel 60 (230-400 Mesh, Art 7747) and by the checkers with Silica Gel 60 purchased from Kanto Chemical Co, Inc.

11.   Thin layer chromatography (TLC) was performed by the submitters on EMD precoated plates (silica gel 60 F_{254}, Art 5715, 0.25 mm) and by the checkers on Merck precoated plates (silica gel 60 F_{254}, 0.25 mm). TLC analysis of the crude product (with elution of 30% EtOAc/Hexanes) was visualized with a 254-nm UV lamp and phosphomolybdic acid stain. The crude mixture contained desired product with R_f = 0.22 (blue), and a trace spot on the baseline.

12.   The checkers found that the column chromatographic purification could be omitted. Direct recrystallization from hot Et_2O afforded the ligation product in a similar yield (5.05-5.09 g, 73%).

13.   The pure product exhibits the following physical and spectroscopic properties: mp 117.4–118.6 °C; $[\alpha]_D^{20}$ +0.97 (c 1.00, $CHCl_3$); IR (KBr, thin film): ν 3306 (s, NH), 3062, 3029, 5975, 1645 (s, C=O, amide), 1541, 1495, 1446, 1413, 1357, 1337, 1247, 1206, 1128, 1018, 760 cm^{-1}; ^{1}H NMR (500 MHz, $CDCl_3$) δ: 1.40 (d, J = 7.2 Hz, 3 H), 3.58 (s, 2 H), 5.12 (quintet, J = 7.2 Hz, 1 H), 5.60 (br s, 1 H), 7.14–7.40 (m, 10 H); ^{13}C NMR (125 MHz, $CDCl_3$) δ: 21.8, 43.9, 48.7, 125.9, 127.2, 127.3, 128.6, 129.0, 129.3, 134.8, 143.0, 170.0; HRMS (ESI) m/z calcd. for $C_{16}H_{18}NONa$ ([M+Na]$^+$) 262.1208;

220

found 262.1201; Anal. calcd. for $C_{16}H_{17}NO$: C, 80.30; H, 7.16; N, 5.85; found: C, 80.14; H, 7.16; N, 5.85.

14. When 1.0 equiv of (S)-N-1-phenylethylhydroxylamine oxalate was used instead of 1.4 equiv relative to phenylpyruvic acid, the yield of the product was reduced to 65% due to decomposition of the hydroxylamine upon heating. The reaction rate was observed by HPLC to be slower when 10:1 DMF:H_2O was used as solvent.

15. The enantiomeric purity was shown by the submitters to be > 99% ee by SFC analysis on a chiral stationary phase. Analytical conditions were: Column: Chiralcel OJ-H; Eluent: MeOH and super critical CO_2; Flow rate: 1 mL/min; Detection: 220 nm, 254 nm; Gradient: 5% MeOH hold 0.1 min, 5–80% MeOH 13 min. R_T of racemic standard = 4.81 min, 5.40 min; R_T of ligation product = 5.40 min. The checker determined the enantiomeric purity by HPLC analysis. Analytical conditions were: Column: Chiralcel OD-H; Eluent: 15% i-PrOH/n-hexane; Flow rate: 1 mL/min; Detection: 220 nm, 254 nm; R_T of racemic standard = 25.32 min, 26.57 min; R_T of ligation product = 26.57 min.

Waste Disposal Information

All toxic materials were disposed of in accordance with "Prudent Practices in the Laboratory"; National Academy Press; Washington, DC, 1995.

3. Discussion

The construction of an amide bond generally involves activation of carboxylic acids as acyl halides, acyl azides, anhydrides, or esters followed by reaction with an amine.[3] Employing common activating reagents complicates the amide formation with racemization, difficult purification, protection of reactive functionalities and deprotection steps. Therefore, chemoselective amide-bond forming reactions are in demand.[4,5] We have developed a novel approach to amide synthesis[6] by decarboxylative condensation of N-alkyl hydroxylamines[7] and α-ketoacids.[8]

Decarboxylative condensation of N-hydroxy-(S)-1-phenylethylamine oxalate and phenylpyruvic acid illustrates the amide bond-forming protocol under mild, simple condition. This method is reagent/catalyst-free, proceeds cleanly by simply mixing the two ligation partners in polar solvents with gentle heating, and produces only innocuous, volatile byproducts. A variety of functionalities have been demonstrated to be compatible with the ligation conditions including carboxylic acids, azides, amines, alcohols, and heterocyclic substrates (Table 1).[9] Benzoyl-protected hydroxylamine undergoes the ligation reaction with a similar outcome to the use of the free hydroxylamine oxalate salt. The steric environment due to substituents adjacent to the ligation centers affects coupling efficiency. A more hindered junction (Entry 8) gave the ligation product in lower yield even when heated to 60 °C.

The ligation between α-ketoacids and hydroxylamines is a powerful, chemoselective strategy that proceeds in the presence of reactive functional groups without activating reagents or catalysts. This method should find application in circumstances that require convergent synthesis of amides in the presence of unprotected functionalities.

Table 1. Decarboxylative Condensation of α-Ketoacids and *N*-Alkyl Hydroxylamines

Entry	α-Ketoacid	Hydroxylamine	Ligation product	Yield[a] (%)
1		•(COOH)₂		90
2[b]		•(COOH)₂		63[c]
3		•(COOH)₂		81
4		•(COOH)₂		78
5				80
6		•(COOH)₂		56
7		•(COOH)₂		43[d]
8[c]		•(COOH)₂		27

[a] Isolated yield, entries are average of two experiments on a 0.2 mmol scale. [b] 5-azido 2-oxopentanoic acid was prepared from the corresponding phosphorous ylide via DMDO oxidation.[10] [c] Yield was calculated from the starting phosphorous ylide over two steps. [d] *N*-Benzylidene-(*S*)-±-α-methylbenzylamine *N*-oxide was isolated as a major byproduct in 33% yield. [e] Reaction was carried out at 60 °C.

1. Roy and Diana Vagelos Laboratories, Department of Chemistry, University of Pennsylvania, Philadelphia, PA 19104. Current address: Laboratorium für Organische Chemie, ETH-Zürich, CH-8093, Switzerland. Email: bode@org.chem.ethz.ch

2. Tokuyama, H.; Kuboyama, T.; Fukuyama, T. *Org. Synth.* **2003**, *80*, 207-212.

3. Montalbetti, C. A. G. N.; Falque, V. *Tetrahedron* **2005**, *61*, 10827-10852.

4. For other strategies in chemoselective amide-forming ligation reactions see: (a) Dawson, P. E.; Muir, T. W.; Clark-Lewis, I.; Kent, S. B. H. *Science* **1994**, *266*, 776-779. (b) Nilsson, B. L.; Kiessling, L. L.; Raines, R. T. *Org. Lett.* **2001**, *3*, 9-12. (c) Saxon, E.; Armstrong, J. I.; Bertozzi, C. R. *Org. Lett.* **2000**, *2*, 2141-2143. (d) Crich, D.; Sana, K.; Guo, S. *Org. Lett.* **2007**, *9*, 4423-4426.

5. Shangguan, N.; Katukojvala, S.; Greenberg, R.; Williams, L. J. *J. Am. Chem. Soc.* **2003**, *125*, 7754-7755.

6. Bode, J. W.; Fox, R. M.; Baucom, K. D. *Angew. Chem. Int. Ed.* **2006**, *45*, 1248-1252.

7. For a review on hydroxylamines see: Ottenheijm, H. C. J.; Herscheid, J. D. M. *Chem. Rev.* **1986**, *86*, 697-707.

8. For a review on ketoacids see: Cooper, A. J. L.; Ginos, J. Z.; Meister, A. *Chem. Rev.* **1983**, *83*, 321-358.

9. For the application of this reaction to the synthesis of peptides containing unprotected side chains, see (a) Ju, L.; Lippert, A. R.; Bode, J. W. *J. Am. Chem. Soc.* **2008**, *130*, 4253-4255. (b) Ju, L.; Bode, J. W. *Org. Biomol. Chem.* **2009**, *7*, 2259-2264.

10. Wasserman, H. H.; Peterson, A. K. *J. Org. Chem.* **1997**, *62*, 8972-8973.

Appendix
Chemical Abstracts Nomenclature; (Registry Number)

Phenylpyruvic acid: α-Oxobenzenepropanoic acid, 2-Oxo-3-phenylpropionic acid; (156-06-9)

N, N-Dimethylformamide: DMF; (68-12-2)

(*S*)-(α-Methylbenzyl)hydroxylamine oxalate salt: (α*R*)-*N*-Hydroxy-α-methyl-benzenemethanamine ethanedioate salt; (78798-33-1)

Benzeneacetamide, *N*-[(1*S*)-1-phenylethyl]-: 2-phenyl-*N*-(1-phenylethyl)acetamide; (17194-90-0)

Jeffrey W. Bode was born in California in 1974 and studied chemistry and philosophy at Trinity University in San Antonio, Texas. He received his Dok. Nat. Sci. from the Eidgenössicsche Technische Hochschule (ETH) in Zürich, Switzerland with Prof. Erick M. Carreira in 2001. Following a JSPS Postdoctoral Fellowship with Prof. Keisuke Suzuki at the Tokyo Institution of Technology, he joined the faculty of the University of California, Santa Barbara as an Assistant Professor in 2003. In 2007, he joined the University of Pennsylvania in Philadelphia, Pennsylvania as an Associate Professor of Chemistry and in 2010 returned to ETH-Zürich as Professor of Chemistry. His research interests include the development of new synthetic methods, catalysis, peptide synthesis, and bioorganic chemistry.

Lei Ju was born in Dandong, China in 1982. In 2005, she received her B.S degree in chemistry/biochemistry from University of California, San Diego, where she conducted research with Yoshihisa Kobayashi. She subsequently started her graduate studies at University of California, Santa Barbara under the supervision of professor Jeffrey W. Bode. Her research efforts focused on the amide bond-forming ligations between a-ketoacids and hydroxylamines. In the summer of 2007, she moved to University of Pennsylvania in Philadelphia, PA to continue with her research.

Tatsuya Toma was born in 1984 in Saitama, Japan. He graduated in 2007 and received his M. S. degree in 2009 from the University of Tokyo. The same year he started his Ph. D. study under the supervision of Professor Tohru Fukuyama. His current interest is enantioselective total synthesis of complex natural products.

ONE-POT DIAZOTIZATION AND HECK REACTION OF METHYL ANTHRANILATE: 2-(3-OXOPROPYL)BENZOIC ACID METHYL ESTER

Submitted by Florencio Zaragoza.[1]

Checked by Fumiki Kawagishi and Tohru Fukuyama.

1. Procedure

2-(3-Oxopropyl)benzoic acid methyl ester (**3**). To a 1-L, three-necked, round-bottomed flask equipped with a magnetic stirring bar (round, 9 x 50 mm) and an internal thermometer are added methyl anthranilate (15.1 g, 0.100 mol, 1.00 equiv) (Note 1), MeCN (150 mL) and a mixture of sulfuric acid (11.3 mL, 20.7 g, 0.210 mol, 2.10 equiv) and water (100 mL) at room temperature. Allyl alcohol (12.8 g, 0.220 mol, 2.20 equiv) and a solution of $PdCl_2$ (112 mg, 0.630 mmol, 0.00630 equiv) in MeCN (50 mL) (Note 2) are then added. Finally, while stirring vigorously, a solution of $NaNO_2$ (8.48 g, 0.123 mol, 1.23 equiv) in water (20 mL) is added in one portion (Note 3). After an induction period of 0.5 h to 2 h, a slightly exothermic reaction ensues. The temperature of the mixture is kept below 40 °C by cooling with a water bath (Note 4).

The mixture is stirred at room temperature for 8 h, diluted with water (0.5 L), and extracted (3 × 50 mL EtOAc). The combined organic extracts are washed once with brine (135 g), dried over $MgSO_4$ (25 g) for 15 min, filtered, and concentrated under reduced pressure (final pressure/temperature: 20 mmHg, 35 °C) to yield 18.67 g of an oil (Note 5). Short path vacuum distillation (30 mmHg, 105 °C) of this oil yielded 11.73 g (61.0%) of the title aldehyde as slightly yellow oil (Note 6).

2. Notes

1. The submitter purchased methyl anthranilate (> 98.0%), allyl alcohol (> 98.0%), $PdCl_2$ (anhydrous, 60% Pd), and $NaNO_2$ (> 98.0%) from

226

Fluka, MeCN (> 99.0%) and H_2SO_4 (95-97%) from Merck KGaA. The checkers purchased methyl anthranilate (99+%) from Aldrich Chemical Co., allyl alcohol (> 99%) from Tokyo Chemical Industry Co., $PdCl_2$ (anhydrous, 60% Pd) from Fluka, $NaNO_2$ (> 98.5%) from Kanto Chemical Co., MeCN (> 99.0%) and H_2SO_4 (> 95.0%) from Wako Pure Chemical Industries respectively. No inert atmosphere is required.

2.   This solution is prepared by stirring $PdCl_2$ and MeCN at 80 °C for 18 h. No undissolved $PdCl_2$ should remain visible.

3.   During the addition of $NaNO_2$ the mixture turns red but finally a yellow-orange solution should result. The diazotization leads to an immediate temperature increase to approximately 40 °C, which wears off within 15–30 min, depending on the scale of the preparation. Two liquid phases are usually formed, and a gentle gas evolution takes place.

4.   Unsurprisingly, the larger the scale, the more pronounced the exotherm. Cooling becomes usually necessary when more than 50 g of methyl anthranilate are employed. The current procedure has been performed successfully with up to 60 g (0.4 mol) of methyl anthranilate. When the exotherm ceases and the reaction mixture reaches room temperature no more product is formed.

5.   According to the submitter, this oil contains approximately 62% (weight) of the title compound based on ^{1}H NMR with an internal standard (4-nitrobenzaldehyde). The crude product may contain variable amounts of the corresp onding hydrate (2-(3,3-dihydroxypropyl)benzoic acid methyl ester): ^{1}H NMR (400 MHz, CDCl$_3$) δ: 2.02 (m, 2 H), 3.08 (br t, J = 7 Hz, 2 H), 4.89 (t, J = 6 Hz, 1 H), 7.28 (m, 2 H), 7.45 (t, J = 7 Hz, 1 H), 7.89 (d, J = 7 Hz, 1 H). The EtOAc-extract may be used directly for ensuing synthetic operations. Thus, treatment of this extract with $NaBH_4$, followed by saponification, yielded 2-(3-hydroxypropyl)benzoic acid in 53% overall yield (three steps). Analytical data were as follows: Mp (toluene) 68–69 °C (lit.[2] Mp (benzene) 70 °C); ^{1}H NMR (300 MHz, d$_6$-DMSO) δ: 1.68 (quintet, J = 7 Hz, 2 H), 2.91 (t, J = 7 Hz, 2 H), 3.39 (t, J = 6 Hz, 2 H), 4.45 (s, br, 1 H), 7.28 (m, 2 H), 7.43 (d, J = 7 Hz, 1 H), 7.73 (d, J = 7 Hz, 1 Hz, 1 H), 12.80 (br s, 1 H).

6.   A small forerun, mainly methyl benzoate, was collected and discarded. Analytical data: IR (film): 2952, 1720, 1435, 1259, 1087 cm^{-1}; ^{1}H NMR (400 MHz, CDCl$_3$) δ: 2.81 (td, J = 7.8, 1.4 Hz, 2 H), 3.28 (t, J = 7.8 Hz, 2 H), 3.90 (s, 3 H), 7.26-7.31 (m, 2 H), 7.44 (td, J = 7.8, 1.4 Hz, 1 H), 7.93 (dd, J = 7.8, 1.4 Hz, 1 H), 9.82 (t, J = 1.4 Hz, 1 H); ^{13}C NMR (100

MHz, CDCl$_3$) δ: 27.3, 45.5, 52.0, 126.5, 129.2, 131.0, 131.2, 132.4, 142.6, 167.6, 201.7; Anal. calcd. for C$_{11}$H$_{12}$O$_3$: C, 68.74; H, 6.29; found: C, 68.58; H, 6.28. The product may undergo aldol addition and condensation upon prolonged storage. According to the submitter, ^{1}H NMR with an internal standard (4-nitrobenzaldehyde) indicated that the distilled product contained 96% (weight) of the title compound. The purity of the product was 97% by GC analysis. Anal. calcd. for C$_{11}$H$_{12}$O$_3$: C, 68.74; H, 6.29; found: C, 68.58; H, 6.28.

Safety and Waste Disposal Information

All hazardous materials should be handled and disposed of in accordance with "Prudent Practices in the Laboratory"; National Academy Press; Washington, DC, 1995.

3. Discussion

Aryldiazonium salts are cheap, readily accessible intermediates for Pd-catalyzed C-C bond forming reactions.[3] Because aryldiazonium salts are potentially explosive and often carcinogenic, one-pot procedures, which obviate the isolation of these salts, are more convenient than procedures based on isolated diazonium salts,[4] and may even be applicable to large-scale preparations.

3-Arylpropanals are useful synthetic intermediates, and are widely used as fragrances in cosmetics, perfumes, and numerous household products. The present procedure offers a practical one-step conversion of anilines into 3-arylpropanals. It is based on readily available reagents, requires only 0.5-1.0% of PdCl$_2$, and does without anhydrous solvents or heating or cooling. Other allylic alcohols than allyl alcohol, such as 3-buten-2-ol, can also be used in the present procedure, to yield the corresponding 1-aryl-3-butanones. Particularly well suited for this protocol are anilines ortho-substituted with electron-withdrawing groups (alkoxycarbonyl, cyano, acetyl, trifluoromethyl, halogens). Ortho-unsubstituted anilines yield mixtures of 3-arylpropanals and 2-arylpropanals, which can, however, be separated by bisulfite adduct formation.[5]

Although diazotizations are often performed with only one equivalent of nitrite, in the present procedure an excess of nitrite actually results in higher yields and clean crude products. Thus, if the title procedure is

228

performed with only 1.0 equivalent of sodium nitrite, a product mixture containing only small amounts of the desired aldehyde is obtained. Although alcohols readily react with HNO_2 to yield alkylnitrites, this potential side reaction appears not to interfere with the diazotization or the Heck reaction. 2-(3-Oxopropyl)benzoates have previously been prepared by oxidative cleavage of tetralones[6] and by oxidation of 2-(3-hydroxypropyl)benzoates.[7]

1. Lonza AG, CH-3930 Visp, Switzerland; florencio.zaragoza@lonza.com. I gladly acknowledge the skillful technical assistance by Verena Heinze and the GC-analysis by Simon Gaul.
2. Rieche, A.; Gross, H. *Chem. Ber.* **1962**, *95*, 91-95.
3. Roglans, A.; Pla-Quintana, A.; Moreno-Mañas, M. *Chem. Rev.* **2006**, *106*, 4622-4643.
4. Beller, M.; Fischer, H.; Kühlein, K. *Tetrahedron Lett.* **1994**, *35*, 8773-8776.
5. Kjell, D. P.; Slattery, B. J.; Semo, M. J. *J. Org. Chem.* **1999**, *64*, 5722-5724.
6. Nishinaga, A.; Yamazaki, S.; Matsuura, T. *Tetrahedron Lett.* **1986**, *27*, 2649-2652. Wrobel, J.; Dietrich, A.; Gorham, B. J.; Sestanj, K. *J. Org. Chem.* **1990**, *55*, 2694-2702.
7. Hashizume, H.; Ito, H.; Yamada, K.; Nagashima, H.; Kanao, M. *Chem. Pharm. Bull.* **1994**, *42*, 512-520. John, V.; Maillard, M.; Tucker, J.; Aquino, J.; Jagodzinska, B.; Brogley, L.; Tung, J.; Bowers, S.; Dressen, D.; Probst, G.; Shah, N. WO 2005/087751.

Appendix
Chemical Abstracts Nomenclature; (Registry Number)

Methyl anthranilate; (134-20-3)
Allyl alcohol; (107-18-6)
Methyl 2-(3-oxopropyl)benzoate; (106515-77-9)

Florencio Zaragoza was born 1964 in Hamburg (Germany). He studied chemistry in Göttingen (Germany), and obtained his doctorate 1990 under the guidance of Professor Franck-Neumann in Strasbourg (France). After postdoctoral training with Professor Pfaltz in Basel (Switzerland) and Professor A. P. Marchand in Denton (TX, USA) he initiated his habilitation in Dresden (Germany). From 1994 until 2007 he worked as medicinal chemist at Novo Nordisk A/S (Måløv, Denmark), and since 2007 as process development chemist at Lonza AG in Switzerland.

Fumiki Kawagishi was born in 1986 in Shizuoka, Japan and received B.S. in 2009 from the University of Tokyo. In 2009 he began his graduate studies at the Graduate School of Pharmaceutical Sciences, the University of Tokyo, under the guidance of Professor Tohru Fukuyama. His research interests are in the area of total synthesis of natural products.

230

SYNTHESIS OF YNAMIDES BY COPPER-MEDIATED COUPLING OF 1,1-DIBROMO-1-ALKENES WITH NITROGEN NUCLEOPHILES. PREPARATION OF 4-METHYL-*N*-(2-PHENYLETHYNYL)-*N*-(PHENYLMETHYL)BENZENESULFONAMIDE

Submitted by Alexis Coste, François Couty and Gwilherm Evano.[1]
Checked by Thomas P. Willumstad and Rick L. Danheiser.

1. Procedure

A. 4-Methyl-N-(phenylmethyl)benzenesulfonamide (**2**). An oven dried, 1-L, three-necked, round-bottomed flask equipped with a magnetic stirring bar (5 cm Teflon coated ovoid-shaped), a 250-mL pressure-equalizing addition funnel fitted with an argon inlet, two rubber septa, and a thermocouple probe is sequentially charged with benzylamine (11.5 mL, 11.3 g, 105.3 mmol, 1.05 equiv), triethylamine (15.5 mL, 11.3 g, 111.2 mmol, 1.1 equiv), and 250 mL of dichloromethane (Notes 1 and 2). The solution is cooled to 2–4 °C (internal temperature) in an ice bath and a solution of *p*-toluenesulfonyl chloride (19.1 g, 100.2 mmol) (Note 3) in 100 mL of dichloromethane is added dropwise via the addition funnel over 45

Org. Synth. **2010**, *87*, 231-244
Published on the Web 4/16/2010

min. During the addition, the internal temperature of the flask is maintained below 5 °C. The mixture is stirred for 30 min at 2-4 °C. The ice bath is then removed and the resulting solution is allowed to warm to 21 °C, stirred at this temperature for 13 h, and then quenched by addition of 1 M HCl (500 mL). The resulting mixture is transferred to a 1-L separatory funnel and the organic layer is separated and washed with 1 M HCl (500 mL), 1 M NaOH (500 mL), saturated aqueous NaCl solution (500 mL), dried over MgSO$_4$ (15 g), filtered, and concentrated by rotary evaporation (25 °C, 25 mmHg). The resulting solid is dried under vacuum (0.2 mmHg, 6 h) to afford 25.6 g (98%) of 4-methyl-*N*-(phenylmethyl)benzenesulfonamide as a white solid (Note 4).

    B.   (2,2-Dibromoethenyl)benzene (**4**). An oven dried 1-L, three-necked, round-bottomed flask equipped with a magnetic stirring bar (5 cm Teflon coated ovoid-shaped), 250-mL pressure-equalizing addition funnel fitted with an argon inlet, two rubber septa, and a thermocouple probe is sequentially charged with benzaldehyde (10.0 mL, 10.4 g, 98.5 mmol), carbon tetrabromide (49.7 g, 149.9 mmol, 1.5 equiv), and dichloromethane (350 mL) (Notes 5 and 6). The solution is cooled to 2-4 °C (internal temperature) with an ice-bath, and a solution of triisopropyl phosphite (54.0 mL, 45.6 g, 218.9 mmol, 2.2 equiv) (Note 7) in dichloromethane (70 mL) is added dropwise via the addition funnel over 90 min. During the addition, the internal temperature of the flask is maintained below 5 °C. The reaction mixture is stirred for 30 min at 2-4 °C and then transferred to a 1-L, single-necked, round-bottomed flask and concentrated on a rotary evaporator (25 °C, 25 mmHg). The flask is equipped with a magnetic stirring bar (5 cm Teflon coated ovoid-shaped) and a water-cooled condenser. 12 M HCl solution (110 mL) and glacial acetic acid (110 mL) (Note 8) are successively added, the condenser is capped with a two-tap Schlenk adaptor connected to a bubbler and an argon/vacuum manifold, and the reaction mixture is heated at reflux in an oil bath (120 °C external temperature) for 14 h. The resulting mixture is then allowed to cool to room temperature, diluted with 100 mL of dichloromethane, and transferred to a 1-L separatory funnel. The organic layer is separated and successively washed with distilled water (3 x 500 mL) and then transferred to a 1-L, single-necked, round-bottomed flask equipped with a teflon coated ovoid shaped magnetic stirring bar. Distilled water (400 mL) is added, followed by portionwise addition of sodium carbonate (50 g, 0.47 mol) (Note 9). The contents of the flask are transferred to a 1-L separatory funnel, and the organic layer is separated and washed with saturated aqueous NaCl solution (200 mL), dried over MgSO$_4$ (7 g), filtered, and concentrated on a rotary evaporator (25 °C, 25 mmHg). The resulting yellow liquid is transferred to a 50-mL single-necked, round-bottomed flask. The flask is equipped with a magnetic stirring bar and fitted with a Vigreux column (8 cm, 1-cm diameter). The yellow liquid is purified by fractional distillation (80–111 °C, 7 mmHg) (Note 10) to give 20.8–22.4 g (81–87%) of (2,2-dibromoethenyl)benzene as a pale yellow liquid (Note 11).

C. *4-Methyl-N-(2-phenylethynyl)-N-(phenylmethyl)* *benzene-sulfonamide* (**5**). An oven-dried, 100-mL, single-necked, round-bottomed flask equipped with a magnetic stirring bar (Note 12) is successively charged with *N*-4-toluenesulfonylbenzylamine **2** (6.3 g, 24.1 mmol), cesium carbonate (31.0 g, 95.1 mmol, 4 equiv) (Note 13), copper(I) iodide (570 mg, 3.0 mmol, 0.12 equiv) (Note 14), and (2,2-dibromoethenyl)-benzene (9.5 g, 36.3 mmol, 1.5 equiv). The flask is equipped with a water-cooled condenser capped with an argon gas inlet, and evacuated (8 mmHg) and backfilled with argon three times. The gas inlet is replaced by a rubber septum with an argon inlet needle. *N,N'*-Dimethylethylenediamine (0.45 mL, 0.37 g, 4.2 mmol, 0.18 equiv) (Note 15) and 1,4-dioxane (70 mL) (Note 16) are successively added through the septum. The resulting green-brown slurry is heated in a pre-heated oil bath at 70 °C (bath temperature) and vigorously stirred for 24 h (Note 17). The reaction mixture is cooled to room temperature and filtered into a 250-mL, single-necked, round-bottomed flask through a 125-mL fritted filtration funnel containing a pad of silica gel (Note 18). The filtrate is washed three times with 50 mL of ethyl acetate and then concentrated on a rotary evaporator (25 °C, 25 mmHg). The resulting oily residue is dissolved in diethyl ether (50 mL) and pentane (50 mL), concentrated by rotary evaporation at 25 °C, 25 mmHg) and then at 0.2 mmHg for 1 h, and then placed at –20 °C in a freezer overnight where it slowly solidifies (Note 19). A stir bar (1.5 cm Teflon coated ovoid-shaped) is added to the flask, absolute ethanol (25 mL) is added, and the flask is equipped with a water-cooled condenser and heated with stirring in an oil bath at 90 °C (bath temperature) until all of the material dissolves. The flask is then allowed to cool to room temperature (Note 20) and then cooled at – 20 °C in a freezer for 8 h. The product is isolated by vacuum filtration through a fritted glass funnel (3 cm diameter, porosity 3) and washed with pentane (2 x 20 mL), and then dried under vacuum (0.2 mmHg, 4 h) to afford 6.3 g (72%) of 4-methyl-*N*-(2-phenylethynyl)-*N*-(phenylmethyl)benzenesulfonamide as a pale beige solid (Notes 21 and 22).

2. Notes

1.   Benzylamine (99%) was purchased from Acros and used as received. Triethylamine (99%) was obtained from Acros by the Submitters and from J. T. Baker by the Checkers and used as received.

2.   Dichloromethane (> 99.5%) was purchased from Carlo Erba by the Submitters and from Mallinckrodt by the Checkers and used as received.

3.   *p*-Toluenesulfonyl chloride (98 %) was purchased from Alfa Aesar and used as received.

4.   In a run carried out on half-scale, 12.4 g (95%) of 4-methyl-*N*-(phenylmethyl)-benzenesulfonamide was obtained. 4-Methyl-*N*-(phenylmethyl)benzenesulfonamide has the following physical properties: Mp 114–115 °C; IR (KBr) 3270, 1598, 1496, 1454, 1326, 1162, 1059, 876, 743 cm^{-1}; ^{1}H NMR (500 MHz, CDCl$_3$) δ: 2.45 (s, 3 H), 4.12 (d, *J* = 6.0 Hz, 2 H), 4.81 (t, *J* = 6.0 Hz, 1 H), 7.18–7.22 (m, 2 H), 7.24–7.33 (m, 5 H), 7.77 (d, *J* = 8.5 Hz, 2 H); ^{13}C NMR (125 MHz, CDCl$_3$) δ: 21.7, 47.4, 127.4, 128.06, 128.08, 128.9, 129.9, 136.5, 137.0, 143.7; HRMS-DART-ESI *m/z* calcd. for C$_{14}$H$_{14}$NO$_2$S ([M-H]$^-$) 260.0751, found 260.0748.

5.   Benzaldehyde (99%) was purchased from Aldrich and used as received.

6.   Carbon tetrabromide (98%) was purchased from Alfa Aesar. It was dissolved in dichloromethane (60 g in 300 mL), dried over MgSO$_4$ (20 g), filtered, concentrated on a rotary evaporator (30 °C, 20 mmHg), and dried under high vacuum (0.5 mmHg, 1 h) prior to use.

7.   Triisopropyl phosphite (90%) was purchased from Alfa Aesar and used as received.

8.   Concentrated hydrochloric acid and glacial acetic acid were purchased from Alfa Aesar and Mallinckrodt respectively and used as received.

9.   Sodium carbonate was added in 10 portions.

10.   A forerun of approximately 3 mL is collected and discarded.

11.   (2,2-Dibromoethenyl)benzene has the following physical properties: IR (film) 3057, 304, 1703, 1595, 1493, 1445, 1270, 1075, 920, 863 cm^{-1}; ^{1}H NMR (500 MHz, CDCl$_3$) δ: 7.34–7.43 (m, 3 H), 7.51 (s, 1 H), 7.54–7.58 (m, 2 H); ^{13}C NMR (125 MHz, CDCl$_3$) δ: 89.8, 128.60, 128.63, 128.8, 135.5, 137.1; HRMS-DART-ESI *m/z* calcd. for C$_8$H$_6$Br$_2$ (M$^+$) 259.8831, found 259.8827.

12.   The reaction mixture being heterogeneous, an ovoid-shaped

magnetic stirring bar longer than 2 cm is preferred. The use of a smaller magnetic bar did result in a failure to stir the reaction mixture properly in some cases.

13. Cesium carbonate (99.9%) was purchased from Alfa Aesar and used as received. Use of less than 4 equivalents of cesium carbonate led to reduced yields and to the formation of small amounts of unidentified by-products.

14. Copper iodide (99.999%) was purchased from Aldrich and used as received. The Submitters report that the yield was not significantly different when 98% copper iodide (purchased from Aldrich and used without purification) was employed for the reaction.

15. N,N'-Dimethylethylenediamine (99%) was purchased from Aldrich and used as received.

16. 1,4-Dioxane (>99%) was purchased from Mallinckrodt and purified by distillation over sodium/benzophenone ketyl and then degassed by three successive freeze-pump-thaw cycles.

17. The progress of the reaction was monitored by thin layer chromatography. TLC analysis was conducted using silica gel plates (elution with 10% EtOAc/hexanes, visualization by UV and KMnO$_4$); R$_f$ values: vinyl dibromide **4** R$_f$ = 0.97, sulfonamide **2** R$_f$ = 0.22, ynamide **5** R$_f$ = 0.58.

18. The pad is prepared by slurrying silica gel (30 g) with ethyl acetate in the fritted funnel (125 mL, 8 cm diameter, porosity 3) and filtering off the excess ethyl acetate.

19. The Submitters report that in some cases the crude residue crystallized after concentration from the diethyl ether/pentane solution. It can then be immediately recrystallized.

20. The Checkers found that crystallization was best initiated at room temperature by scratching with a glass rod prior to cooling at –20 °C.

21. In a run carried out on half-scale, 3.20 g (73%) of 4-methyl-N-(2-phenylethynyl)-N-(phenylmethyl)benzenesulfonamide was obtained. This compound has the following physical properties: Mp 82–84 °C; IR (KBr) 2232, 1597, 1495, 1361, 1169, 1090, 939, 772 cm^{-1}; ^{1}H NMR (500 MHz, CDCl$_3$) δ: 2.46 (s, 3 H), 4.60 (s, 2 H), 7.25 (app. s, 5 H), 7.31–7.37 (m, 7 H), 7.81 (d, J = 8.0 Hz, 2 H); ^{13}C NMR (125 MHz, CDCl$_3$) δ: 21.9, 55.9, 71.5, 82.8, 123.0, 127.9, 127.9, 128.4, 128.5, 128.7, 129.1, 129.9, 131.3, 134.6, 134.8, 144.8. Anal. calcd. for C$_{22}$H$_{19}$NO$_2$S: C: 73.10; H: 5.30; N: 3.88; found: C: 73.29; H: 5.25; N: 3.93.

22. In solid form, 4-methyl-*N*-(2-phenylethynyl)-*N*-(phenylmethyl) benzenesulfonamide has been found to be stable to storage for months even at room temperature.

Waste Disposal Information

All hazardous materials should be handled and disposed of in accordance with "Prudent Practices in the Laboratory"; National Academy Press; Washington, DC, 1995.

3. Discussion

Due to their superior stability compared to ynamines, ynamides are clearly emerging as key and versatile synthetic intermediates and have received considerable attention over the past decade.[2] They have been shown only quite recently to be excellent substrates for RCM,[3] palladium-mediated coupling reactions or rearrangements,[4] cycloisomerizations[5] and [2 + 2] cycloadditions,[6] radical transformations,[7] Pauson-Khand cyclizations,[8] diastereoselective sigmatropic rearrangements,[9] yne-carbonyl metathesis,[10] carbometalations,[11] and many other useful transformations. In addition, they have been used as efficient synthons for the preparation of a wide array of molecules such as, for example, keto-imides,[12] amino-indoles,[13] oxazolones,[14] indolines and carbazoles,[15] triazoles[16] or benzofurans (Figure 1).[17]

In addition to a number of traditionally employed approaches that often lack generality,[2] several strategies for the synthesis of ynamides have recently emerged. Although isomerization of propargyl amides or bromination/dehydrohalogenation of enamides represent facile entries to ynamides, they have been shown to be either substrate-specific or of too narrow scope. The first efficient procedure was based on the use of hypervalent alkynyliodonium salts which have been shown to cleanly react with metalated amides.[5b,8] This method is, however, limited to the preparation of ynamides substituted with either a trialkylsilyl or an aromatic group.

Figure 1. Utility of ynamides in organic synthesis

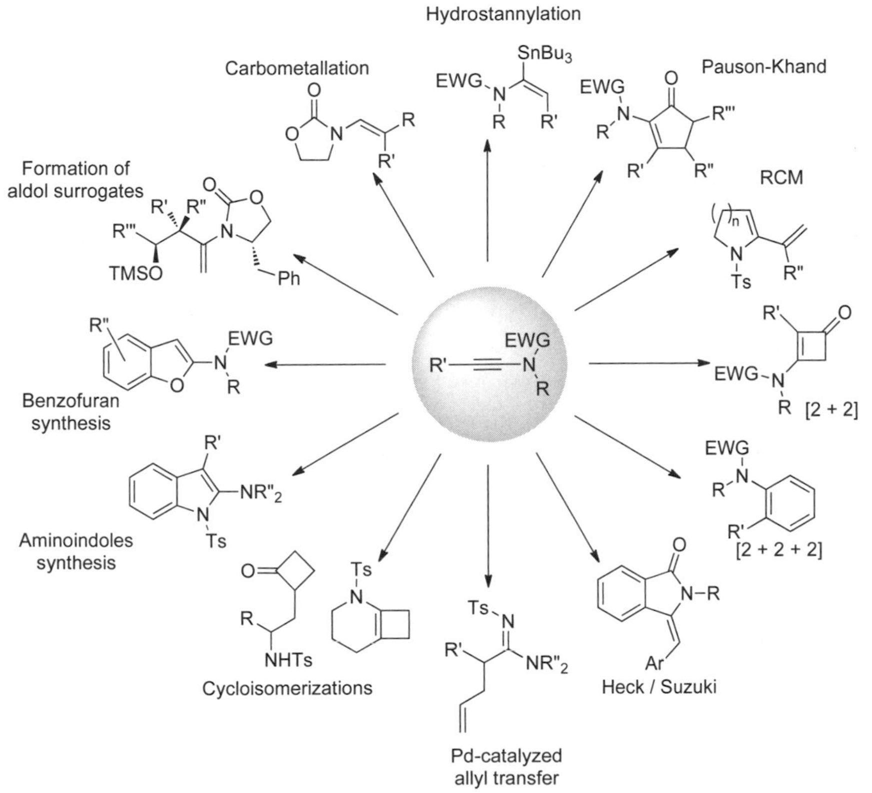

Recent advances in copper-catalyzed transformations provided the basis for new perspectives in ynamide synthesis. Inspired by the arylation of amides, Hsung and co-workers first developed a copper-catalyzed coupling between alkynyl bromides and amides, which provided an improved synthetic access to ynamides over existing protocols.[18] However, severe limitations remained, such as the use of high temperature and low substrate scope: although oxazolidinones were useful in the coupling, amides were mostly poor and sulfonamides were not suitable at all. In addressing this limitation, Danheiser and co-workers reported a useful solution using stoichiometric amounts of copper iodide along with KHMDS: this protocol allows reactions to proceed at room temperature with carbamates and sulfonamides but still requires the use of a strong base.[19] A general and quite mild procedure was finally published in 2004 by Hsung, who re-examined his coupling protocol by screening a variety of copper sources and ligands. The use of copper sulphate pentahydrate in combination with 1,10-phenanthroline proved to be especially successful, allowing the reaction to occur at temperatures ranging from 60 to 95 °C with potassium phosphate as base.[20]

Another option relies on the copper-catalyzed oxidative coupling between terminal alkynes and nitrogen nucleophiles recently reported by the Stahl group.[21] While good to excellent yields of various ynamides were obtained with this reaction, a major limitation is the use of five equivalents of the nucleophile, which can be prohibitive in some cases.

We have recently reported a new synthesis of ynamides based on the copper-mediated coupling between nitrogen nucleophiles and 1,1-dibromoalkenes.[22] This reaction has been shown to be general and provides a straightforward entry to ynamides that is complementary to previously reported synthetic routes. Ynamides are usually obtained in good yields and it is note worthy that this copper-catalyzed alkynylation of amides, sulfonamides, carbamates and oxazolidinones can be easily carried out on a gram or multigram scale (Figure 2). This method has the advantageous feature that it starts from readily available 1,1-dibromo-1-alkenes, which are attractive alkynylating agents. They can be prepared either using the classical Ramirez olefination[23] or the remarkably efficient Lautens modification,[24] which was used for the preparation of the vinyl-dibromide used in this procedure.

From a mechanistic point of view, we were able to demonstrate that the reaction proceeds through a regioselective monocoupling followed by dehydrobromination of the intermediate α-bromo-enamide which could be isolated in some cases by lowering the reaction temperature and cleanly transformed to the ynamide upon treatment with cesium carbonate.[25]

Figure 2. Scope of the copper-mediated coupling between nitrogen nucleophiles and 1,1-dibromoalkenes

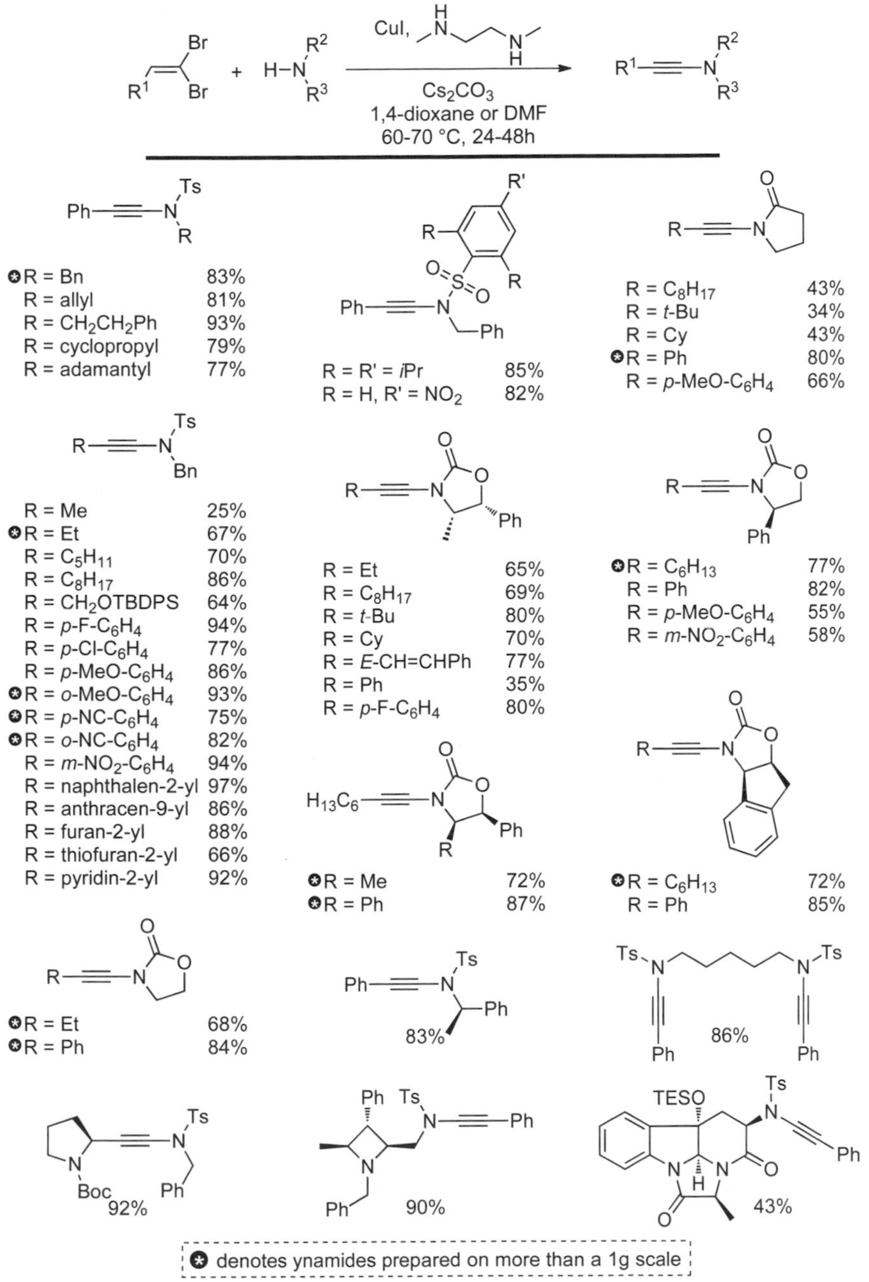

✪R = Bn	83%
R = allyl	81%
R = CH₂CH₂Ph	93%
R = cyclopropyl	79%
R = adamantyl	77%

R = R' = iPr	85%
R = H, R' = NO₂	82%

R = C₈H₁₇	43%
R = t-Bu	34%
R = Cy	43%
✪R = Ph	80%
R = p-MeO-C₆H₄	66%

R = Me	25%
✪R = Et	67%
R = C₅H₁₁	70%
R = C₈H₁₇	86%
R = CH₂OTBDPS	64%
R = p-F-C₆H₄	94%
R = p-Cl-C₆H₄	77%
R = p-MeO-C₆H₄	86%
✪R = o-MeO-C₆H₄	93%
✪R = p-NC-C₆H₄	75%
✪R = o-NC-C₆H₄	82%
R = m-NO₂-C₆H₄	94%
R = naphthalen-2-yl	97%
R = anthracen-9-yl	86%
R = furan-2-yl	88%
R = thiofuran-2-yl	66%
R = pyridin-2-yl	92%

R = Et	65%
R = C₈H₁₇	69%
R = t-Bu	80%
R = Cy	70%
R = E-CH=CHPh	77%
R = Ph	35%
R = p-F-C₆H₄	80%

✪R = C₆H₁₃	77%
R = Ph	82%
R = p-MeO-C₆H₄	55%
R = m-NO₂-C₆H₄	58%

✪R = Me	72%
✪R = Ph	87%

✪R = C₆H₁₃	72%
R = Ph	85%

✪R = Et	68%
✪R = Ph	84%

83%

86%

92%

90%

43%

─────────────────────────────
✪ denotes ynamides prepared on more than a 1g scale
─────────────────────────────

1. Institut Lavoisier de Versailles, UMR CNRS 8180, Université de Versailles Saint-Quentin en Yvelines, 45, avenue des Etats-Unis, 78035 Versailles Cedex, France, evano@chimie.uvsq.fr

2. For reviews, see: (a) Zificsak, C. A.; Mulder, J. A.; Hsung, R. P.; Rameshkumar, C.; Wei, L.-L. *Tetrahedron* **2001**, *57*, 7575-7606. (b) Mulder, J. A.; Kurtz, K. C. M.; Hsung, R. P. *Synlett* **2003**, 1379-1390. (c) Katritzky, A. R.; Jiang, R.; Singh S. K. *Heterocycles* **2004**, *63*, 1455-1475. (d) Tracey, M. R.; Hsung, R. P.; Antoline, J.; Kurtz, K. C. M.; Shen, L.; Slafer, B. W.; Zhang Y. in *Science of Synthesis, Houben-Weyl Methods of Molecular Transformations*, S. M. Weinreb, Ed. Georg Thieme Verlag KG, Stuttgart: Chapter 21.4.2, **2005**. (e) Witulski, B.; Alayrac C. in *Science of Synthesis, Houben-Weyl Methods of Molecular Transformations*, A. de Meijere Ed. Georg Thieme Verlag KG, Stuttgart. Chapter 24.4.4.2, **2005**. (f) Evano, G.; Coste, A.; Jouvin, K. *Angew. Chem. Int. Ed.* **2010**, *49*, 2840-2859.

3. (a) Saito, N.; Sato, Y.; Mori, M. *Org. Lett.* **2002**, *4*, 803-805. (b) Huang, J.; Xiong, H.; Hsung, R. P.; Rameshkumar, C.; Mulder, J. A.; Grebe, T. P. *Org. Lett.* **2002**, *4*, 2417-2420. (c) Mori, M.; Wakamatsu, H.; Saito, N.; Sato, Y.; Narita, R.; Sato, Y.; Fujita, R. *Tetrahedron* **2006**, *62*, 3872-3881.

4. (a) Couty, S.; Liegault, B.; Meyer, C.; Cossy, J. *Tetrahedron* **2006**, *62*, 3882-3895. (b) Zhang, Y.; DeKorver, K. A.; Lohse, A. G.; Zhang, Y.-S.; Huang, J.; Hsung, R. P. *Org. Lett.* **2009**, *11*, 899-902.

5. (a) Couty, S.; Liegault, B.; Meyer, C.; Cossy, J. *Tetrahedron* **2009**, *65*, 1809-1832. (b) Marion, F.; Coulomb, J.; Courillon, C.; Fensterbank, L.; Malacria, M. *Org. Lett.* **2004**, *6*, 1509-1511. (c) Couty, S.; Meyer, C.; Cossy, J. *Angew. Chem. Int. Ed.* **2006**, *45*, 6726-6730. (d) Movassaghi, M.; Hill, M. D.; Ahmad, O. K. *J. Am. Chem. Soc.* **2007**, *129*, 10096-10097. (e) Tanaka, K.; Takeishi, K. *Synthesis* **2007**, 2920-2923. (f) Witulski, B.; Stengel, T. *Angew. Chem., Int. Ed.* **1998**, *37*, 489-492. (g) Tanaka, K.; Takeishi, K.; Noguchi, K. *J. Am. Chem. Soc.* **2006**, *128*, 4586-4587.

6. (a) Kohnen, A. L.; Mak, X. Y.; Lam, T. Y.; Dunetz, J. R.; Danheiser, R. L. *Tetrahedron* **2006**, *62*, 3815-3822. (b) Villeneuve, K.; Riddell, N.; Tam, W. *Tetrahedron* **2006**, *62*, 3823-3836.

7. (a) Marion, F.; Courillon, C.; Malacria, M. *Org. Lett.* **2003**, *5*, 5095-5097. (b) Sato, A.; Yorimitsu, H.; Oshima, K. *Synlett* **2009**, 28-31.

8. (a) Rainier, J. D.; Imbriglio, J. E. *J. Org. Chem.* **2000**, *65*, 7272-7276.

240

(b) Witulski, B.; Stengel, T. *Angew. Chem., Int. Ed.* **1998**, *37*, 489-492. (c) Witulski, B.; Gössmann, M. *Chem. Commun.* **1999**, 1879-1880. (d) Witulski, B.; Gössmann, M. *Synlett* **2000**, 1793-1797. (e) Witulski, B.; Stengel, T.; Fernández-Hernández, J. M. *Chem. Commun.* **2000**, 1965-1966.

9. (a) Mulder, J. A.; Hsung, R. P.; Frederick, M. O.; Tracey, M. R.; Zificsak, C. A. *Org. Lett.* **2002**, *4*, 1383-1386. (b) Kurtz, K. C. M.; Frederick, M. O.; Lambeth, R. H.; Mulder, J. A.; Tracey, M. R.; Hsung, R. P. *Tetrahedron* **2006**, *62*, 3928-3938.

10. (a) Kurtz, K. C. M.; Hsung, R. P.; Zhang, Y. *Org. Lett.* **2006**, *8*, 231-234. (b) Saito, N.; Katayama, T.; Sato, Y. *Org. Lett.* **2008**, *10*, 3829-3832.

11. (a) Chechik-Lankin, H.; Livshin, S.; Marek, I. *Synlett* **2005**, 2098-2100. (b) Yasui, H.; Yorimitsu, H.; Oshima, H. *Bull. Chem. Soc. Jpn.* **2008**, *81*, 373-379. (c) Gourdet, B.; Lam, H. W. *J. Am. Chem. Soc.* **2009**, *131*, 3802-3803. (d) Das, J. P.; Chechik, H.; Marek, I. *Nature Chem.* **2009**, *1*, 128-132.

12. Al-Rashid, Z. F.; Johnson, W. L.; Hsung, R. P.; Wei, Y.; Yao, P.-Y.; Liu, R.; Zhao, K. *J. Org. Chem.* **2008**, *73*, 8780-8784.

13. (a) Witulski, B.; Alayrac, C.; Tevzadze-Saeftel, L. *Angew. Chem., Int. Ed.* **2003**, *42*, 4257-4260. (b) Dooleweerdt, K.; Ruhland, T.; Skrydstrup, T. *Org. Lett.* **2009**, *11*, 221-224.

14. (a) Hashmi, A. S. K.; Salathé, R.; Frey, W. *Synlett* **2007**, 1763-1766. (b) Istrate, F. M.; Buzas, A. K.; Jurberg, I. D.; Odabachian, Y.; Gagosz, F. *Org. Lett.* **2008**, *10*, 925-928.

15. (a) Dunetz, J. R.; Danheiser, R. L. *J. Am. Chem. Soc.* **2005**, *127*, 5776-5777. (b) Martínez-Esperón, M. F.; Rodríguez, D.; Castedo, L.; Saá, C. *Tetrahedron* **2008**, *64*, 3674-3686.

16. (a) Zhang, X.; Li, H.; You, L.; Tang, Y.; Hsung, R. P. *Adv. Syn. Catal.* **2006**, *348*, 2437-2442. (b) IJsselstijn, M.; Cintrat, J.-C. *Tetrahedron* **2006**, *62*, 3837-3842.

17. Oppenheimer, J.; Johnson, W. L.; Tracey, M. R.; Hsung, R. P.; Yao, P.-Y.; Liu, R.; Zhao, K. *Org. Lett.* **2007**, *9*, 2361-2364.

18. Frederick, M. O.; Mulder, J. A.; Tracey, M. R.; Hsung, R. P.; Huang, J.; Kurtz, K. C. M.; Shen, L.; Douglas, C. J. *J. Am. Chem. Soc.* **2003**, *125*, 2368-2369.

19. (a) Dunetz, J. R.; Danheiser, R. L. *Org. Lett.* **2003**, *5*, 4011-4014. (b) Kohnen, A. L.; Dunetz, J. R.; Danheiser, R. L. *Org. Synth.* **2007**, *84*,

88-101.

20. (a) Zhang, Y.; Hsung, R. P.; Tracey, M. R.; Kurtz, K. C. M.; Vera, E. L. *Org. Lett.* **2004**, *6*, 1151-1154. (b) Zhang, X.; Zhang, Y.; Huang, J.; Hsung, R. P.; Kurtz, K. C. M.; Oppenheimer, J.; Petersen, M. E.; Sagamanova, I. K.; Shen, L.; Tracey, M. R. *J. Org. Chem.* **2006**, *71*, 4170-4177. (c) Sagamanova, I. K.; Kurtz, K. C. M.; Hsung R. P. *Org. Synth.* **2007**, *84*, 359-367.

21. Hamada, T.; Ye, X.; Stahl, S. S. *J. Am. Chem. Soc.* **2008**, *130*, 833-835.

22. Coste, A.; Karthikeyan, G.; Couty, F.; Evano, G. *Angew. Chem., Int. Ed.* **2009**, *48*, 4381-4385.

23. Ramirez, F.; Desal, N. B.; McKelvie, N. *J. Am. Chem. Soc.* **1962**, *84*, 1745-1747.

24. (a) Fang, Y.-Q.; Lifchits, O.; Lautens, M. *Synlett* **2008**, 413-417. (b) Bryan, C.; Aurregi, V.; Lautens, M. *Org. Synth.* **2009**, *86*, 36-46.

25. Another possible mechanism would proceed through elimination to the alkynyl bromide and its further copper-catalyzed cross coupling. We were however not able to characterize even trace amounts of bromoalkynes from crude reaction mixtures, even at lower temperatures.

Appendix
Chemical Abstracts Nomenclature; (Registry Number)

4-Methyl-*N*-(phenylmethyl)benzenesulfonamide; (1576-37-0) Benzylamine: Benzenemethanamine; (100-46-9)

p-Toluenesulfonyl chloride: Benzenesulfonyl chloride, 4-methyl-; (98-59-9)

Triethylamine: Ethanamine, *N,N*-diethyl-; (121-44-8)

(2,2-Dibromoethenyl)benzene; (7436-90-0)

Carbon tetrabromide: Methane, tetrabromo-; (558-13-4)

Triisopropyl phosphite: Phosphorous acid, tris(1-methylethyl) ester; (116-17-6)

Benzaldehyde; (100-52-7)

4-Methyl-*N*-(2-phenylethynyl)-*N*-(phenylmethyl)benzenesulfonamide: (609769-63-3)

Cesium carbonate; (534-17-8)

Copper(I) iodide; (1335-23-5)

N,N'-Dimethylethylenediamine: 1,2-Ethanediamine, *N*1,*N*2-dimethyl-; (110-70-3)

Gwilherm Evano was born in Paris in 1977 and studied chemistry at the Ecole Normale Supérieure in Paris and received his Ph.D from Université Pierre et Marie Curie in 2002 under the supervision of François Couty and Claude Agami. After postdoctoral study with James S. Panek at Boston University, he joined the CNRS as Chargé de Recherche at the University of Versailles in 2004. His research interests focus on asymmetric synthesis and reactivity of nitrogen heterocycles, copper-catalyzed transformations, and the total synthesis of natural and/or biologically relevant products.

Alexis Coste was born in 1982 and studied chemistry at the Ecole Supérieure de Chimie Organique et Minérale. Since 2007, he is engaged in Ph.D. research under the supervision of François Couty and Gwilherm Evano at the University of Versailles as a National Cancer Institute Fellow. His work focuses on the development of copper-catalyzed transformations with application in natural product synthesis and on the development of new proteasome inhibitors in a tumor targeting approach.

François Couty was born in Caen (France) in 1963. He studied chemistry at the University Pierre et Marie Curie in Paris and completed his Ph.D. with Claude Agami in 1991. The same year, he was appointed assistant professor in this University. He spent a year in Namur (Belgium) as a postdoctoral fellow with Prof. A. Krief. After having completed his Habilitation in 1999, he was promoted full Professor at the University of Versailles in 2001. His research interests are in the field of asymmetric synthesis, including the development of new synthetic methodologies for nitrogen heterocycles, the total synthesis of natural products and projects at the chemistry-biology interface.

 Thomas Willumstad was born in 1985 in Pueblo, Colorado. In 2004, he received his B.A. degree in chemistry from the University of Colorado at Boulder. While at CU he worked on methods directed toward the synthesis of diazonamide A in the lab of Professor Tarek Sammakia. Tom is currently pursuing graduate studies at the Massachusetts Institute of Technology under the guidance of Professor Rick Danheiser. His research focuses on the synthesis of highly substituted quinolines via a tandem benzannulation/iodocylization strategy.

244

DIRECT FLUORINATION OF THE CARBONYL GROUP OF BENZOPHENONES USING DEOXO-FLUOR®: PREPARATION OF BIS(4-FLUOROPHENYL)DIFLUOROMETHANE

Submitted by Ying Chang, Hyelee Lee, Chulsung Bae.[1]
Checked by David Hughes.

1. Procedure

Caution! Deoxo-Fluor® reacts rapidly and exothermally with water, generating HF. It is volatile and a respiratory hazard and must be handled in a well-ventilated hood.

Bis(4-fluorophenyl)difluoromethane. An oven-dried 50-mL round-bottomed flask equipped with an oval Teflon-coated magnetic stirring bar (2 cm) is charged with 4,4'-difluorobenzophenone (**1**) (5.11 g, 23.4 mmol, 1.0 equiv) (Note 1). Deoxo-Fluor® (Note 2) (13 mL, 15.7 g, 71 mmol, 3 equiv) is added via a disposable graduated pipette. The flask is fitted with a reflux condenser equipped with a gas inlet adapter connected to a nitrogen line and a gas bubbler. The reaction solution is stirred in a preheated 90 °C oil bath (Notes 3, 4, and 5) for 24 h under nitrogen to give a dark red mixture with yellow solids. The flask is removed from the oil bath and cooled to room temperature. Dichloromethane (30 mL) is added and the mixture is transferred to a 250-mL separatory funnel, followed by additional dichloromethane (2 × 30 mL) to rinse the flask. The organic solution is washed with water (2 × 50 mL) (Note 6), then saturated aqueous NaHCO₃ solution (50 mL) (*caution*: Due to the generation of CO_2, the separatory funnel should be shaken carefully and the pressure released frequently). A final wash is carried out with saturated aqueous NaCl solution (50 mL), then the organic layer is vacuum-filtered through a bed of Na_2SO_4 (20 g) in a 150-mL medium-porosity sintered-glass funnel. The cake is washed with dichloromethane (3 × 25 mL) until colorless. The filtrate is concentrated by rotary evaporation (40 °C bath, 100 mmHg initial vacuum, lowered to 20 mmHg) to afford the crude product as a red oil (6.9 g). Purification using column chromatography (Note 7) on SiO$_2$ affords 4.42–4.87 g of bis(4-fluorophenyl)difluoromethane (**2**) as a colorless oil (78–86% yield) (Notes 8 and 9).

2. Notes

1.   Reagents and solvents were used as received and sourced as follows: 4,4'-difluorobenzophenone   (Acros,   99%),   bis(2-methoxyethyl)aminosulfur   trifluoride   (Deoxo-Fluor®)   (Sigma-Aldrich), dichloromethane (Sigma-Aldrich, ACS reagent, 99.5%), silica gel (Sigma-Aldrich, 230-400 mesh, 60 Å), ethyl acetate (Sigma-Aldrich, ACS reagent, >99.5%), hexanes (Sigma-Aldrich, ACS reagent, >98.5%), and sodium sulfate (Fisher Scientific).

2.   Deoxo-Fluor® is volatile and a respiratory hazard and must be handled in a well-ventilated hood. The checker pipetted Deoxo-Fluor® in the hood and weighed the stoppered flask before and after addition.

3.   Deoxo-Fluor® is known to decompose initiating at 140 °C.[2] Thus, the oil bath temperature must be kept below 110 °C for safety concerns. A 90 °C oil bath temperature provides efficient fluorination; raising the oil bath temperature to 100 °C does not improve the yield. For substrates with high molecular weight, such as 4,4'-dibromobenzophenone, the stirring may be difficult at the beginning, but this does not affect the reaction yield as the stirring gradually becomes more efficient during the course of the reaction.

4.   In a separate experiment, a 2.5 g reaction was carried out in 2-necked, 50-mL flask with a thermocouple thermometer inserted through a septum. The internal temperature was monitored (84 °C) using a J-Kem Gemini digital thermometer with a Teflon-coated T-Type thermocouple probe (12-inch length, 1/8 inch outer diameter, temperature range –200 to +250 °C). This reaction proceeded to 84% conversion with a 78% isolated yield.

5.   The reaction was monitored by ^{1}H NMR as follows. A drop of the reaction mixture was added to 1 mL of CDCl$_3$ and 1 mL of sat. NaHCO$_3$. The layers were mixed, then the bottom layer was filtered into an NMR tube through Na$_2$SO$_4$ and a cotton plug. Multiplets at 7.20 and 7.85 ppm from the starting material were integrated relative to the product resonances at 7.1 and 7.5 ppm to assess conversion. The reaction proceeded to 84–90% conversion

246

for experiments carried out at the 2.5-5 g scale. Additional reaction time did not result in increased conversion. Addition of one equiv of Deoxo-Fluor® at the end of the reaction and heating for a further 24 h only increased conversion by 2-3%.

6. Slight foams are produced due to the remaining Deoxo-Fluor® or its decomposition products.

7. A 5-cm glass column is wet-packed (2.5% hexanes/EtOAc) with SiO$_2$ (200 g) topped with 0.5 cm sand. The crude reaction product is loaded neat onto the column and eluted as follows: 2.5% EtOAc/hexanes (600 mL), 3% EtOAc/hexanes (750 mL), collecting 50-mL fractions. TLC (UV visualization) is used to follow the chromatography. The R$_f$ value of the title compound is 0.5 (2.5% EtOAc/hexanes), the starting material has an R$_f$ of 0.05. Fractions 8-21 are concentrated by rotary evaporation (40 °C bath, 20 mmHg), then vacuum dried (20 mmHg) at 22 °C for 2 h to constant weight (4.42–4.87 g, 78–86% yield). The product contains <0.1 wt % EtOAc and hexanes by ^{1}H NMR analysis. Fractions 25-29 are combined and concentrated to afford 0.50–0.8 g (10–16%) of unreacted starting material **1**.

8. The checker obtained the following yields from three experiments:

Scale (1)	Isolated product 2 (%)	Unreacted starting material (1) (%)
5.11 g	4.86 g (86%)	0.50 g (10%)
3.88 g	3.56 g (83%)	0.49 (13%)
2.56 g	2.19 g (78%)	0.41 (16%)

9. *Bis(4-fluorophenyl)difluoromethane* (**2**) has the following physical and spectroscopic data: ^{1}H NMR (400 MHz, CDCl$_3$) δ: 7.09-7.14 (m, 4 H), 7.46-7.51 (m, 4 H); ^{19}F NMR (376 MHz, CDCl$_3$) δ: –110.7 (t, $^6J_{FF}$ = 3 Hz, 2 F, *Ar-F*), –86.3 (t, $^6J_{FF}$ = 3 Hz, 2 F, *CF$_2$*); ^{13}C NMR (100 MHz, CDCl$_3$) δ: 115.5 (d, $^2J_{CF}$ = 22 Hz), 120.1 (t, $^1J_{CF}$ = 242 Hz), 128.1 (dt, $^3J_{CF}$ = 9.0 Hz, $^3J_{CF}$ = 5.2 Hz), 133.5 (td, $^2J_{CF}$ = 29 Hz, $^4J_{CF}$ = 3 Hz), 163.6 (dt, $^1J_{CF}$ = 249 Hz, $^5J_{CF}$ = 2 Hz); MS (EI) *m/z* 240 (M$^+$, 48), 221 (16), 145 (100), 126 (16), 95 (18), 75 (17). GC/MS (Shimadzu QP2010S equipped with a 30 m × 0.25 mm SHR-XLB GC column and an EI ionization MS detector) indicated product purity >99%. An analytical sample was prepared by dissolving 100 mg of product in 3 mL of hexanes, filtering through a 0.45 micron PTFE syringe filter, and vacuum concentration for 3 h at ambient temperature. Anal. calcd. for C$_{13}$H$_8$F$_4$: C, 65.01; H, 3.36; found: C, 64.77; H, 3.32.

Safety and Waste Disposal Information

All hazardous materials should be handled and disposed of in accordance with "Prudent Practices in the Laboratory"; National Academy Press; Washington, DC, 1995.

3. Discussion

Selective fluorination of carbonyl groups to the *gem*-difluorides is a useful transformation that has been traditionally achieved by the use of gaseous sulfur tetrafluoride[3] and diethylaminosulfur trifluoride (DAST)[4]; however, the harsh reaction conditions and the toxicity of sulfur tetrafluoride and the thermal instability of DAST have limited their use in large-scale reactions. Deoxo-Fluor® is known to be more thermally stable than DAST.[2,5]

Compared with the well-studied fluorination of the carbonyl group of aldehydes and alkyl ketones, the carbonyl group of diaryl ketones is much less reactive under the general fluorination conditions and requires harsh conditions, which causes safety concerns.[6] Thus, two-step procedures which involve conversion of the carbonyl group of benzophenones to more reactive thioketones or thiolanes were developed for the *gem*-difluorination of benzophenones.[6-7] The direct fluorination procedure of benzophenones with Deoxo-Fluor® described here is a modified procedure of our previously reported method.[8] In this work, the fluorination reactions are conducted in a flask equipped with a reflux condenser under a nitrogen atmosphere rather than in a closed system, such as a pressure tube and screw-capped vial. The implementation of this more general experimental procedure allows a more convenient, safer, and scalable preparation of *gem*-difluoride compounds from the corresponding benzophenones.

The fluorination of 4,4'-difluorobenzophenone (1 g scale) with Deoxo-Fluor® was monitored by GC/MS and the conversion was improved from 56% to 75% by increasing the reaction time from 4 to 24 h. Considering the high cost of Deoxo-Fluor® and the limited solubility of some of the substrates in Deoxo-Fluor® under neat conditions, we choose 24 h as the standard reaction time. The effect of the fluorinating reagent ratio on the conversion was investigated using benzophenone (1 g scale) as substrate. When 1.4 equiv of Deoxo-Fluor® was used, the corresponding *gem*-difluoride was isolated in 27% yield. Increasing the amount of Deoxo-Fluor®

248

Table 1. Conversion of Diaryl Ketones to Diaryldifluoromethanes[a]

Entry	Substrate	Product	Yield[b]
1			63
2			20
3			75
4			83
5			77
6			95
7			68
8			85
9			61
10			74, 86[c]
11			13

[a] Reactions were conducted in an oven-dried 10 mL one-necked round-bottomed flask using 1.0 g of benzophenone substrate and 3 equiv of Deoxo-Fluor®. [b] Isolated yield; remainder of mass balance is unreacted starting material. [c] Yield from a 5.0 g scale reaction.

to 3 equiv improved the yield to 63%. No significant additional improvement was obtained by a further increase of Deoxo-Fluor® to 4 equiv.

The scope of the direct conversion of the carbonyl groups of benzophenones to the *gem*-difluorides was examined with substrates containing halogen, alkoxy, and nitro substituents on one or both aromatic rings (Table 1). The effect of steric hindrance was observed in the *gem*-difluorination of 2-chlorobenzophenone, as only a 20% yield was obtained (entry 2). The electronic effect also plays an important role in the reaction yields: substrates bearing an electron-withdrawing group formed the *gem*-difluorinated products in good to high yields (61–95%, entries 1 and 3–10) while low yields were obtained for benzophenones bearing an electron-donating group such as methoxy (13%, entry 11).

1. Department of Chemistry, University of Nevada Las Vegas, 4505 Maryland Parkway, Box 454003, Las Vegas, Nevada 89154-4003.
2. Lal, G. S.; Pez, G. P.; Pesaresi, R. J.; Prozonic, F. M. *Chem. Commun.* **1999**, 215–216.
3. (a) Uneyama, K. Organofluorine Chemistry; Blackwell: Oxford, 2006 (b) Smith, W. C.; Tullock, C. W.; Muetterties, E. L.; Hasek, W. R.; Fawcett, F. S.; Engelhardt, V. A.; Coffman, D. D. *J. Am. Chem. Soc.* **1959**, *81*, 3165–3166. (c) Hasek, W. R.; Smith, W. C.; Gngelhardt, V. A. *J. Am. Chem. Soc.* **1960**, *82*, 543–551. (d) Boswell, G. G.; Ripka, W. C.; Scribner, R. M.; Tullock, C. W. *Org. React.* (N. Y.) **1974**, *21*, 20–30. (e) Gerstenburger, M. R. C.; Haas, A. *Angew. Chem., Int. Ed. Engl.* **1981**, *20*, 647–667. (d) Dmowski, W.; Kaminski, M. *J. Fluorine Chem.* **1983**, *23*, 219–228.
4. (a) Messina, P. A.; Mange, K. C.; Middleton, W. W. *J. Fluorine Chem.* **1989**, *42*, 137–143. (b) McCarthy, J. R.; Peet, N. P.; LeTourneau, M. E.; Inbasekaran, M. *J. Am. Chem. Soc.* **1985**, *107*, 735–737. (c) Manandhar, S.; Singh, R. P.; Eggers, G. V.; Shreeve, M. M. *J. Org. Chem.* **2002**, *67*, 6415–6420. (d) Nishizono, N.; Sugo, M.; Machica, M.; Oda, K. *Tetrahedron* **2007**, *63*, 11622–11625. (e) Robins, M. J.; Wnuk, S. F. *J. Org. Chem.* **1993**, *58*, 3800–3801. (f) Middleton, W. J. *J. Org. Chem.* **1975**, *40*, 574–578. (g) Singh, R. P.; Shreeve, J. M. *J. Org. Chem.* **2003**, *68*, 6063–6065. (h) Negi, D. S.; Köppling, L.; Lovis, K.; Abdallah, R.; Geisler, J.; Budde, U. *Org. Process Res. Dev.* **2008**, *12*, 345–348.

5. Lal, G. S.; Pez, G. P.; Pesaresi, R. J.; Prozonic, F. M.; Cheng, H. *J. Org. Chem.* **1999**, *64*, 7048–7054.
6. Lal, G. S.; Lobach, E.; Evans, A. *J. Org. Chem.* **2000**, *65*, 4830–4832.
7. (a) Sondej, S. C.; Katzenellenbogen, J. A. *J. Org. Chem.* **1986**, *51*, 3508–3513. (b) Chambers, R. D.; Sandford, G.; Sparrowhawk, M. E.; Atherton, M. J. *J. Chem. Soc., Perkin Trans. 1* **1996**, 1941–1944. (c) Motherwell, W. B.; Wilkinson, J. A. *Synlett* **1991**, 191–192. (d) Prakash, G. K. S.; Hoole, D.; Reddy, V. P.; Olah, G. A. *Synlett* **1993**, 691–693. (e) Kuroboshi, M.; Hiyama, T. *Synlett* **1991**, 909–910. (f) Reddy, V. P.; Alleti, R.; Perambudrur, M. K.; Welz-Biermann, U.; Buchholz, H.; Prakash, G. K. S. *Chem. Commun.* **2005**, 654–656.
8. Chang, Y.; Tewari, A.; Adi, A.-I.; Bae, C. *Tetrahedron* **2008**, *64*, 9837–9842.

Appendix
Chemical Abstracts Nomenclature (Registry Number)

Deoxo-Fluor: Bis(2-methoxyethyl)aminosulfur trifluoride: Ethanamine, 2-ethoxy-N-(2-ethoxyethyl)-N-(trifluorothio)-; (202289-38-1)
Benzophenone: Diphenylmethanone: Benzoylbenzene; (119-61-9)
Bis(4-fluorophenyl) ketone: 4,4'-Difluorobenzophenone; (345-92-6)

Chulsung Bae received his B.S. degree from Inha University and M.S. degrees from Pohang University of Science & Technology in Korea and University of Massachusetts Lowell. He received a Ph.D. in chemistry from University of Southern California in 2002 under the guidance of G. K. Surya Prakash and Nobel Laureate George A. Olah. After postdoctoral research at Yale University with John F. Hartwig in 2002–2004, he joined the faculty in the Chemistry Department at University of Nevada Las Vegas. His current research interests include organic chemistry, organofluorine chemistry, catalysis, green chemistry, polymer science for renewable energy applications.

Ying Chang was born in Anhui Province, China in 1981. He obtained his Ph.D. degree in Applied Chemistry from Nanjing University of Science and Technology, China, in 2006 under the supervision of Professor Chun Cai. After carrying out a postdoctoral research work with Professor William R. Dolbier, Jr. at the University of Florida (2006-2007) on synthetic organofluorine chemistry, he joined Professor Chulsung Bae's group at the University of Nevada Las Vegas as a postdoctoral scholar and now works in the fields of synthetic organofluorine chemistry, functional polymers synthesis for renewable energy applications.

Hyelee Lee was born in Incheon, South Korea in 1984. She received her B.S. degree in Chemistry from the Chung-Ang University in 2008. After working as a research assistant in the Drug Evaluation Department, Korea Food & Drug Administration until summer of 2009, she joined Professor Chulsung Bae's group at University of Nevada Las Vegas as a Ph.D. student. Her current interest is development of new organic synthetic method for the applications in proton exchange membrane fuel cells.

PREPARATION AND [2+2] CYCLOADDITION OF 1-TRIISOPROPYLSILOXY-1-HEXYNE WITH METHYL CROTONATE: 3-BUTYL-4-METHYL-2-TRIISOPROPYLSILOXY-CYCLOBUT-2-ENECARBOXYLIC ACID METHYL ESTER

Submitted by Valeriy Shubinets, Michael P. Schramm, and Sergey A. Kozmin.[1]

Checked by John Frederick B. Briones and Huw M. L. Davies.

1. Procedure

A. 1-Triisopropylsiloxy-1-hexyne (**2**). A three-necked, flame-dried, 500-mL, round-bottomed flask equipped with an egg-shaped 1 1/4 x 5/8 in magnetic stir bar, rubber septum fitted with argon inlet needle, and rubber stopper fitted with thermometer is flushed with argon and charged with THF (100 mL) (Note 1) and 1-hexyne (3.44 mL, 30 mmol, 1.0 equiv) (Note 2) by syringe through a septum. The solution is cooled in a dry ice acetone bath to –78 °C and treated dropwise over 10 min with anhydrous t-BuOOH (6.30 mL of 5.26 M solution in nonane, 33 mmol, 1.1 equiv) (Note 3) by means of a syringe pump (Note 4). The resulting mixture is treated dropwise with lithium bis(trimethylsilyl)-amide (69 mL of 1.0 M solution in THF, 69 mmol, 2.3 equiv) (Note 5) over 30 min by means of a syringe pump (Note 6). The flask is then placed in a 0 °C ice water bath and allowed to stir for 2 h. The reaction mixture is cooled in a dry ice acetone bath to –78 °C and treated dropwise over 10 min with triisopropylsilyl trifluoromethanesulfonate (8.90 mL, 33 mmol, 1.1 equiv) (Note 7) by means of a syringe pump (Note 8). The reaction mixture is stirred for 5 min, then placed in a 0 °C ice water bath and allowed to stir for 40 min. The reaction mixture is then diluted with hexanes (200 mL) (Note 9) and transferred into a 1-L separatory funnel containing saturated aqueous $NaHCO_3$ solution (125 mL), using an additional 100 mL of hexanes to wash the flask. The layers are separated. The aqueous layer is extracted with hexanes (75 mL). The combined organic layers are washed with saturated aqueous $Na_2S_2O_3$ (100 mL) (Note 10), 100 mL of brine, and dried over anhydrous $MgSO_4$. The solution is filtered through a fritted glass funnel with fine porosity. The solvent is removed on a rotary evaporator (Note 11). The crude product is transferred to a 50-mL round-bottomed flask using two hexanes washings. Hexane solvent is removed using a rotary evaporator, and the product is purified by fractional distillation (Notes 12 and 13) to afford 6.66 g (87%) of 1-triisopropylsiloxy-1-hexyne **2** (Note 14).

B.   *3-Butyl-4-methyl-2-triisopropylsilanyloxy-cyclobut-2-enecarboxylicacid methyl ester* (**4**). A single-necked, dry, 250-mL, round-bottomed flask equipped with an egg-shaped 1 x 1/2 in. magnetic stirbar is charged with 1-triisopropylsiloxy-1-hexyne **2** (5.12 g, 20.16 mmol, 1.2 equiv), dry methylene chloride (140 mL) (Note 15), and methyl crotonate (1.78 mL, 16.8 mmol, 1.0 equiv) (Note 16) at 23 °C. Silver trifluoromethylsulfonimide[4] (326 mg, 0.84 mmol, 0.05 equiv) (Note 17) is added in one portion and the reaction mixture is stirred for 15 min at which time TLC (Note 18) indicates complete consumption of starting material and the product formation is evident from a new spot with $R_f = 0.53$ (10:1 hexanes:ether), corresponding to siloxy cyclobutene **4**. The reaction mixture is washed with brine (50 mL). The aqueous phase is then extracted with methylene chloride (25 mL). The combined organic layers are dried over anhydrous $MgSO_4$ and filtered through a fritted glass funnel with fine porosity. The solvent is removed on a rotary evaporator and the crude product is purified by column chromatography (Note 19) using 85 g of SiO_2 (elution with 10:1, hexanes:ether). The fractions containing the product are concentrated to afford 4.57–4.79 g (77–80% yield) of 3-butyl-4-methyl-2-triisopropylsilanyloxy-cyclobut-2-enecarboxylicacid methyl ester **4** as a colorless oil (Note 20).

254

2. Notes

1. THF (HPLC grade, H_2O = 0.003%) was purchased from Fisher Scientific Company and distilled from sodium benzophenone ketyl.

2. 1-Hexyne (97%) was purchased from Aldrich Chemical Company, Inc. and used without further purification.

3. The solution of t-BuOOH (5.26 M in nonane) was purchased from Aldrich Chemical Company, Inc. (Fluka). It was sold as an approximately 5.5 M solution, but the exact concentration (up to one-hundredth) of a particular batch can be obtained at the Aldrich website.

4. During the addition of t-BuOOH the internal temperature was maintained between –65 and –60 °C.

5. The solution of lithium bis(trimethylsilyl)-amide (1.0 M in THF) was purchased from Aldrich Chemical Company, Inc.

6. During the addition of LiHMDS, the internal temperature was kept steady at –60 °C.

7. Triisopropylsilyl trifluoromethanesulfonate (97%) was purchased from Aldrich Chemical Company, Inc. and fractionally distilled over calcium hydride at 3 mmHg.

8. During the addition of TIPSOTf, the internal temperature was maintained between –65 and –60 °C

9. Hexanes (ACS grade) were purchased from Fisher Scientific Company and used without further purification.

10. This operation is intended to remove any residual t-BuOOH. In the past, however, when this operation was omitted from the procedure, the NMR of the crude product did not show any presence of t-BuOOH. Nevertheless, washing the organic phase with saturated aqueous $Na_2S_2O_3$ is recommended as a safety precaution.

11. Most of the nonane can be removed on a rotary evaporator at 40 °C using a diaphragm pump (ca. 20 mmHg).

12. Fractional distillation was done using a vigreux column. 1-Triisopropylsiloxy-1-hexyne was distilled at 102–105 °C (ca. 3 mmHg). Care must be taken at the beginning of the distillation process as the crude siloxy alkyne has a tendency to foam when placed under high vacuum (most likely as a result of the leftover nonane). A small amount of triisopropyl silanol (ca. 0.3-0.4 mL) is removed at 70–90 °C.

13. Alternatively, one could distill the siloxy alkyne using bulb-to-bulb distillation (first removing the triisopropyl silanol and then distill the siloxy alkyne itself).

14. Occasionally, a small amount of impurities (such as triisopropyl silanol) may be observed. The exact amount of these impurities depends on how well the distillation is performed. 1-Triisopropylsiloxy-1-hexyne **2** displays the following spectral data: IR (neat) 2945, 2869, 2278, 1463, 1254 cm^{-1}; ^{1}H NMR (400 MHz, C$_6$D$_6$) δ: 0.77 (t, 3 H, J = 6.8 Hz), 1.01–1.04 (m, 18 H), 1.07–1.13 (m, 3 H), 1.33–1.38 (m, 4 H), 2.06 (t, 2 H, J = 7.2 Hz); ^{13}C NMR (100 MHz, C$_6$D$_6$) δ: 12.0, 13.6, 17.2, 17.3, 22.1, 30.7, 32.4, 87.1; MS (APCI) m/z calcd. for C$_{15}$H$_{31}$OSi ([M+H]$^+$) 255.2139; found 255.2144; Anal. calcd. for C$_{15}$H$_{30}$OSi: C, 70.79; H, 11.88; found: C, 70.49; H, 11.97.

15. Methylene chloride (HPLC grade) was purchased from Fisher Scientific Company and distilled over calcium hydride.

16. Methyl crotonate (98%) was purchased from Aldrich Chemical Company, Inc. and used without further purification.

17. Silver trifluoromethylsulfonimide was prepared according to a literature procedure:[4] Trifluoromethanesulfonimide (2.81 g, 10 mmol, 95% from Aldrich Chemical Company, Inc.) was added to a 100-mL round-bottomed flask, which was wrapped in aluminum foil. To this flask, 20 mL of distilled water was added, followed by silver carbonate (1.5 g, 5.4 mmol, 99% from Aldrich Chemical Company, Inc.). The flask was warmed to 65 °C in an oil bath and stirred for 4 h, after which the water was removed using a rotary evaporator (10 mmHg). The flask was placed on a vacuum line (1 mmHg) and heated to 80 °C for 14 h. It was removed from the vacuum line, brought to room temperature, and treated with 30 mL of diethyl ether (anhydrous grade, Fisher Scientific Company), stirred for 2 h, and filtered through a glass sintered frit (fine grade) to remove some small black impurities. The filtrate was concentrated on a rotary evaporator to give a powder. The powder was placed under vacuum (1 mmHg) at room temperature for 12 h to remove any remaining water and ether to give 3.31 g of silver trifluoromethylsulfonimide (85 % yield).

18. TLC was performed using Whatman precoated 60 Å silica gel plates with fluorescent indicator.

19. Flash column chromatography was performed over Silicycles Inc. ultra pure 230–400 mesh silica gel, pH 7.0, H$_2$O content = 6%. Fractions of ca 15 mL each were collected using 16 x 150 mm test tubes. A total of 26

fractions were collected and fractions 12 to 18 were pooled and concentrated to give the desired product.

20. Cyclobutene **4** displays the following spectral data: IR (neat) 2947, 2867, 1737, 1702 cm^{-1}; ^{1}H NMR (400 MHz, CDCl$_3$) δ: 0.86 (t, 3 H, J = 7.2 Hz), 1.02–1.07 (m, 21 H), 1.10 (d, 3 H, J = 6.8 Hz), 1.20–1.40 (m, 4 H), 1.85–1.92 (m, 1 H), 2.01–2.09 (m, 1 H), 2.41 (q, 1 H, J = 6.8 Hz), 3.00 (br s, 1 H), 3.63 (s, 3 H); ^{13}C NMR (100 MHz, CDCl$_3$) δ: 12.8, 14.0, 17.80, 17.9, 18.1, 22.9, 24.8, 29.4, 35.1, 51.7, 56.6, 125.6, 138.6, 173.6; MS (APCl) m/z calcd. for C$_{20}$H$_{39}$O$_3$Si ([M+H]$^+$) 355.2663; found 355.2670; Anal. calcd. for C$_{20}$H$_{38}$O$_3$Si: C, 67.74; H, 10.80; found: C, 67.44; H, 10.87.

Safety and Waste Disposal Information

All hazardous materials should be handled and disposed of in accordance with "Prudent Practices in the Laboratory"; National Academy Press; Washington, DC, 1995.

3. Discussion

Siloxy alkynes represent a general class of organosilicon compounds with significant potential for the development of an arsenal of new synthetic methods having both practical and conceptual untility in organic synthesis.[3,5-9] Among several existing methods for preparation of siloxy alkynes, the protocol developed by Julia in 1993[2] represents perhaps the most simple, versatile and efficient approach to these compounds. The Julia method entails the generation of an acetylide anion from a terminal alkyne, followed by addition of lithium tert-butyl hydroperoxide, which at 0 °C triggers a facile oxidation of the lithium acetylide to give lithium ynolate and lithium tert-butoxide. The modification of the original procedure reported herein is based on the use of triisopropylsilyl triflate instead of the silyl chloride, which was utilized in the initial report by Julia. This modification enables the highly efficient silylation of the ynolate anion to furnish the corresponding siloxy alkyne in excellent yield. Importantly, only 1 equivalent of silyl trilfate is sufficient for the silylation step. Presumably, the competitive silylation of lithium tert-butoxide is significantly slower.

Scheme 1

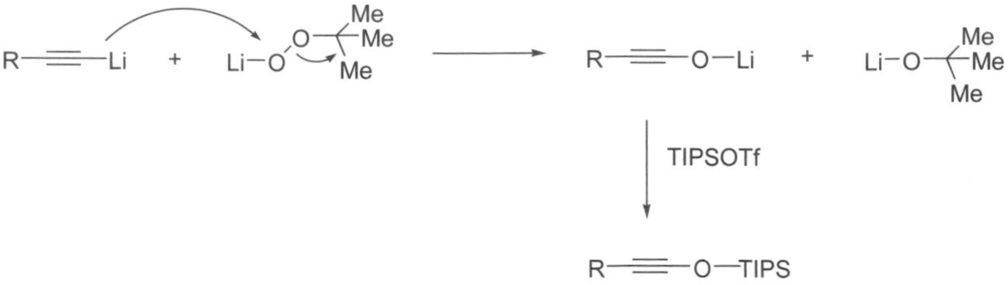

This simple protocol was utilized to convert a range of terminal alkynes to the corresponding 1-siloxy-1-alkynes. Several representative examples are depicted in Table 1. We found that the use of bulky siloxy groups, i.e. TIPS, was required to improve the hydrolytic stability of the resulting alkynes, which can be readily purified via distillation or chromatography on SiO_2. In many instances, the alkynes were obtained in sufficiently pure form for direct use in the next transformation.

Table 1 Yields of various 1-siloxy-1-alkynes

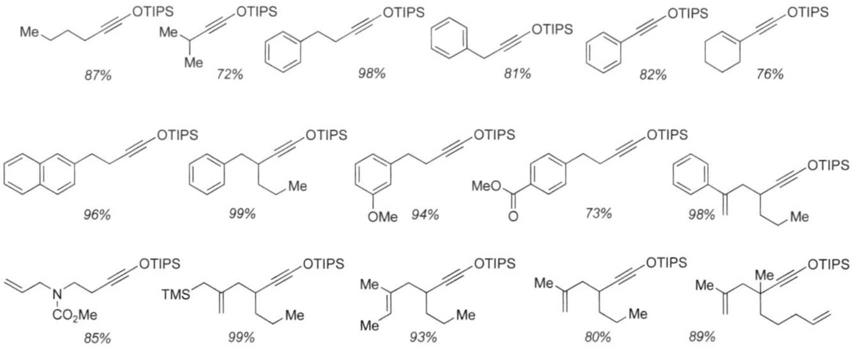

We found that a catalytic amount of silver trifluoromethylsulfonamide can efficiently promote the [2+2] cycloadditions of siloxy alkynes with simple unsaturated ketones, esters, and nitriles.[3] This transformation provided an efficient method for the assembly of a range of highly functionalized siloxy cyclobutenes. The investigation of the scope of this method revealed that many unsaturated carbonyl compounds are capable of efficient participation in this reaction with a range of siloxy alkynes (Table 2). In most cases, these reactions proceeded to completion in 5 minutes at

258

Scheme 2

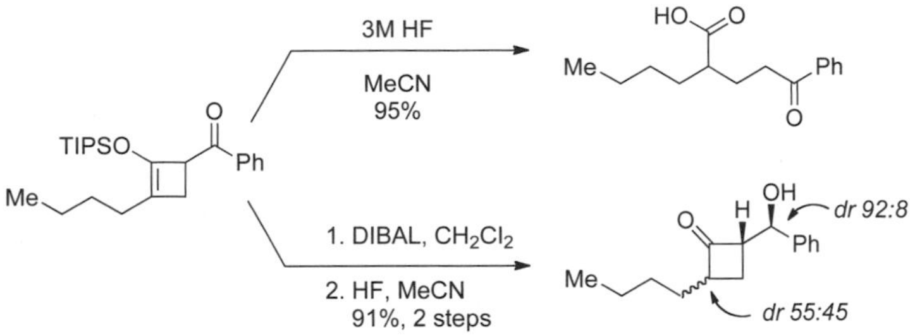

Table 2 Selected [2+2] cycloadditions of siloxy alkynes

Entry	Siloxyalkyne	Alkene	Product	Yield (%)
1	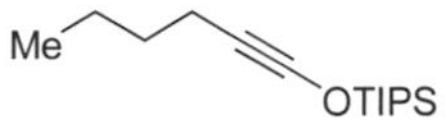	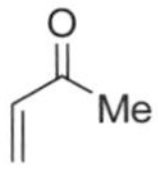	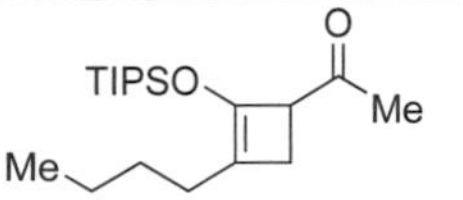	77
2	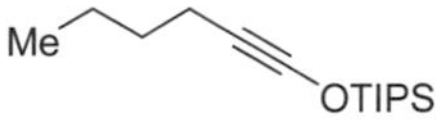	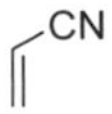	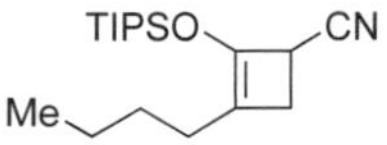	69
3	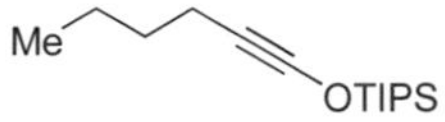	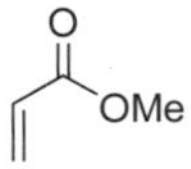	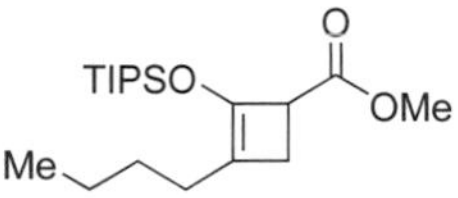	78
4	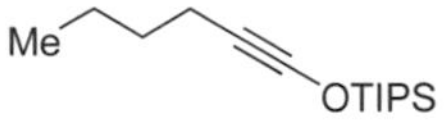	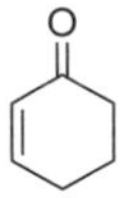	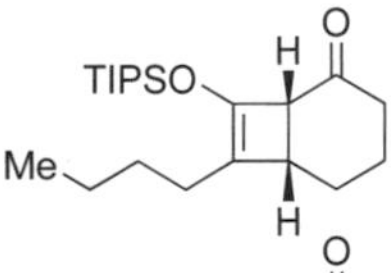	73
5	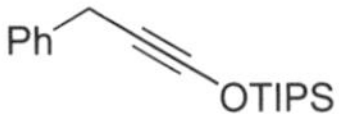	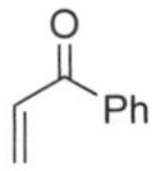	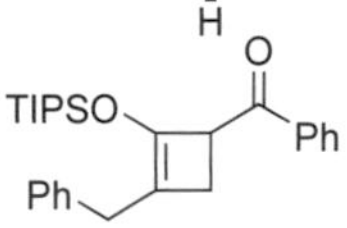	83
6	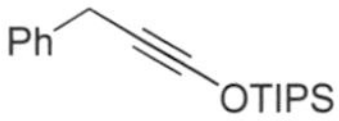	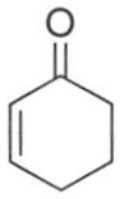	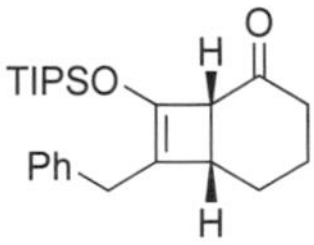	68
7	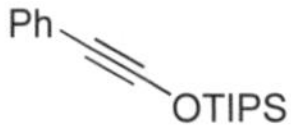	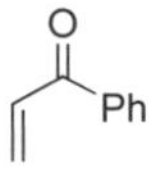	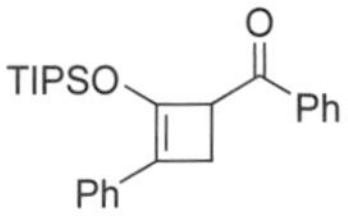	75

room temperature, demonstrating the mild and general nature of this catalytic protocol. Two representative examples of subsequent transformations of siloxy cyclobutenes are depicted in Scheme 2. Direct protodesilylation of silyl enol ether afforded keto acid in 95% yield, arising from the fragmentation of the initially produced acyl cyclobutanone. Alternatively, a highly diastereoselective DIBAL reduction (dr 92:8), followed by protodesilylation furnished the cyclobutenone (dr 55:45) without any detectable ring-opening products.

1. Department of Chemistry, The University of Chicago, 5735 S. Ellis Avenue, Chicago, IL 60637.
2. Preparation of siloxy alkyne represents a modification of the Julia method: Julia, M.; Saint-Jalmes, V.P.; Verpeaux, J. N. *Synlett* **1993**, 233–234.
3. Sweis, R. F.; Schramm, M. P.; Kozmin, S. A. *J. Am. Chem. Soc.* **2004**, *126*, 7442–7443.
4. A. Vij, Y.Y. Zheng, R. I. Kirchmeier, J. M. Shreeve, *Inorg. Chem.* **1994**, *33*, 3281–3288.
5. Zhang, L.; Kozmin, S. A. *J. Am. Chem. Soc.* **2004**, *126*, 10204-10205.
6. Zhang, L.; Kozmin, S. A. *J. Am. Chem. Soc.* **2004**, *126*, 11806–11807.
7. Sun, J.; Kozmin, S.A. *J. Am. Chem. Soc.* **2005**, *127*, 13512–13513.
8. Sun, J.; Kozmin, S.A. *Angew. Chem. Int. Ed.* **2006**, *45*, 4991–4993.
9. Zhang, L.; Sun, J.; Kozmin, S. A. *Tetrahedron* **2006**, *62*, 11371–11380.

Appendix
Chemical Abstracts Nomenclature; (Registry Number)

Methyl crotonate; (623-43-8)
Triisopropylsilyl trifluoromethanesulfonate; (80522-42-5)
Trifluoromethanesulfonimide; (82113-65-3)
Silver carbonate; (534-16-7)

Sergey A. Kozmin received his Undergraduate Diploma at the Moscow State University in 1993. He obtained his Ph.D. in 1998 at the University of Chicago with Viresh H. Rawal, and completed postdoctoral studies at the University of Pennsylvania with Amos B. Smith, III in 2000. Kozmin is currently an Associate Professor of Chemistry at the University of Chicago. His research program combines several complementary efforts, including (a) development of new catalytic reactions of conceptual and practical utility; (b) synthesis of complex natural products of notable biological significance; and (c) generation of highly diverse chemical libraries, featuring the complexity of natural metabolites.

Valeriy Shubinets was born in 1985 in Ivano-Frankivsk, Ukraine. He obtained his B.S. in Chemistry and Biochemistry and his B.A. in Biology in 2007 from the University of Chicago, where he performed his undergraduate research with Professor Sergey Kozmin in the area of new method development and natural product synthesis. Valeriy is currently a student at the Harvard Medical School.

Michael P. Schramm was born in Syracuse, New York in 1975. He received his Ph.D. under the direction of Professor Sergey A. Kozmin from The University of Chicago in 2005. He then completed two years of postdoctoral work under the supervision of Professor Julius Rebek, Jr. at The Scripps Research Institute in La Jolla, California. In 2007 he began his independent career as Assistant Professor in the Department of Chemistry and Biochemistry at California State University Long Beach. His research efforts utilize principles of molecular recognition to design and prepare modular alpha-helicial peptidomimetic libraries. Additionally, these principles are being directed at the preparation and study of synthetic small molecule membrane transporters.

John Frederick Briones was born in 1982 in Laguna, Philippines. He earned his B.S. degree in Chemistry from the University of the Philippines, Los Banos in 2003 and later on pursued his Master's degree at the University of the Philippines, Diliman. He joined the research lab of Prof. Huw Davies in 2007 and currently his research project focuses on Rh(II) catalyzed enantioselective cyclopropenation of alkynes.

Pd(0)-CATALYZED ASYMMETRIC ALLYLIC AND HOMOALLYLIC DIAMINATION OF 4-PHENYL-1-BUTENE WITH DI-*TERT*-BUTYLDIAZIRIDINONE

A.

B.

Submitted by Baoguo Zhao, Haifeng Du, Renzhong Fu, and Yian Shi.*[1]
Checked by Eric E. Buck and Peter Wipf.[2]

1. Procedure

A.   (S)-1-(8,9,10,11,12,13,14,15-Octahydro-3,5-dioxa-4-phosphacyclohepta[2,1-a;3,4-a']dinaphthalen-4-yl)-2,2,6,6-tetramethylpiperidine [(S)-1]. A 250-mL, two-necked, round-bottomed flask equipped with a Teflon-coated magnetic stir bar (length: 3.5 cm), internal thermometer, nitrogen inlet, and rubber septum is charged with 2,2,6,6-tetramethylpiperidine (5.13 mL, 29.5 mmol, 1.24 equiv) and anhydrous THF (47.2 mL) (Notes 1, 2, and 3). The solution is cooled to 5 °C in an ice bath, treated with *n*-BuLi (19.5 mL, 29.8 mmol, 1.53 M in hexanes, 1.25 equiv) by syringe over 5 min, and stirred at 0 °C for 30 min (Notes 4 and 5). To this solution is added phosphorus trichloride (PCl₃) (7.75 mL, 88.8 mmol, 3.73 equiv) at such a rate as to keep the internal temperature below 10 °C (*ca.* 30 min) (Note 6). After complete addition of PCl₃, the ice bath is removed, and the brown reaction mixture is stirred for 1 h and concentrated by rotary evaporation (25 °C, 55 mmHg) (Notes 7 and 8). The resulting residue is dissolved in THF (71.0 mL) and cooled to 5 °C in an ice bath. A solution of (S)-H₈-BINOL (7.08 g, 23.8 mmol, 1.00 equiv) and Et₃N (10.1 mL, 71.4 mmol, 3.00 equiv) in THF (17.8 mL) is added by

Org. Synth. **2010**, *87*, 263-274
Published on the Web 4/23/2010

syringe pump over 15 min (Notes 9 and 10). The ice bath is removed. The reaction mixture is stirred for 20 h and concentrated by rotary evaporation (25 °C, 18 mmHg) (Note 11). The resulting residue is dissolved in toluene (60 mL) and the mixture is vacuum filtered (water aspirator, 45 mmHg) through a Büchner funnel (150 mL, 40–60 astm) (Note 12). The solid is washed with toluene (3 x 20 mL), the combined filtrates are concentrated by rotary evaporation (40 °C, 18 mmHg), and the residue is diluted with a toluene/dichloromethane/Et$_3$N solution (25 mL, 2:1:1.5). The solution is loaded onto a wet-packed neutral Al$_2$O$_3$ column (250 g Al$_2$O$_3$, diameter: 4.5 cm, height: 18 cm, pretreated with toluene/Et$_3$N, 20:1). The product is eluted with toluene/Et$_3$N (20:1) to give (S)-1 as a pale yellow solid (9.94 g) (Note 13). The solid is recrystallized from ethyl acetate (90 mL) and dried at 85 °C under vacuum for 4 h (87.0 mmHg) to give (S)-1 as colorless crystals (7.79 g, 70%) (Notes 14, 15, and 16).

B.   *(4S,5S)-1,3-Di-tert-butyl-4-phenyl-5-vinylimidazolidin-2-one*
(3). A 100-mL, three-necked, round-bottomed flask equipped with a Teflon-coated magnetic stir bar (length: 2.5 cm), argon inlet, and rubber septum is charged with Pd$_2$(dba)$_3$ (1.27 g, 1.37 mmol, 0.05 equiv) and ligand (S)-1 (2.81 g, 6.03 mmol, 0.22 equiv) (Note 17). The flask is evacuated and back-filled with argon three times; benzene (2.94 mL) is added, and the mixture is placed in a preheated oil bath at 65 °C and stirred for 40 min (Note 18). To the heated mixture is added 4-phenyl-1-butene (4.20 mL, 27.4 mmol, 1.00 equiv) followed by the addition of di-*tert*-butyldiaziridinone (2) (9.81 g, 57.6 mmol, 2.1 equiv) by syringe pump at a rate of 0.7 mL/h (17 h) (Notes 19 and 20). After complete addition, the reaction mixture is stirred for an additional 3 h, cooled to room temperature, and diluted with hexanes (30 mL) (Note 21). The mixture is vacuum filtered (water aspirator, 45 mmHg) through a Büchner funnel (150 mL, 40–60 astm), and the solid is washed with hexanes (5 x 20 mL). The combined filtrate is concentrated and the residue is diluted with hexanes/diethyl ether (6:1, 5 mL) (Note 22). The solution is loaded onto a wet-packed SiO$_2$ column (206 g SiO$_2$, diameter: 5.5 cm, height: 22.5 cm, pretreated with hexanes/diethyl ether, 6:1) (Note 23). The product is eluted with hexanes/diethyl ether (6:1) to give diamination product 3 along with a small amount of dibenzylideneacetone (dba). The resulting residue is diluted with hexanes/diethyl ether (6:1, 5 mL). The solution is loaded onto a wet-packed SiO$_2$ column (206 g SiO$_2$, diameter: 5.5 cm, height: 22.5 cm, pretreated with hexanes/diethyl ether, 6:1). The

product is eluted with hexanes/diethyl ether (6:1) to give 6.02 g (73%) of diamination product **3** as a pale yellow oil (90% ee) (Notes 24, 25 and 26).

2. Notes

1. All glassware was flame dried under a nitrogen atmosphere.
2. The checkers purchased 2,2,6,6-tetramethylpiperidine (98+%) from Alfa Aesar and used it as received.
3. The checkers purchased tetrahydrofuran (ACS grade) from Fisher Chemicals and dried it by distillation over sodium/benzophenone ketyl under an argon atmosphere.
4. The checkers purchased *n*-butyllithium (1.6 M solution in hexanes) from Sigma-Aldrich and used it as received.
5. *n*-Butyllithium was titrated before each use using the following protocol: A 25-mL, one-necked, round-bottomed flask equipped with a rubber septum and a nitrogen inlet was charged with (*L*)-menthol (*ca* 200 mg) and dipyridine (*ca* 5 mg). The flask was cooled to 5 °C in an ice bath and charged with THF (10 mL) followed by a known volume of *n*-BuLi (1.6 M solution in hexanes).
6. The checkers purchased phosphorus trichloride (99%) from Sigma-Aldrich and used it as received.
7. Due to the toxicity of phosphorus trichloride, the mixture was concentrated in a fume hood.
8. The remaining trace amount of PCl_3 was removed under vacuum (25 °C, 0.05 mmHg) for 4 h.
9. The checkers purchased (*S*)-H_8-BINOL (99%) from TCI America and used it as received.
10. The checkers purchased triethylamine (99%) from Fisher Chemicals, dried by distillation over CaH_2, and stored it over potassium hydroxide.
11. The reaction can be monitored by TLC on precoated Al_2O_3 150 F_{254} plates (purchased from Merck, Darmstadt, Germany) (toluene/Et_3N, 20:1 ($R_f = 0.9$)). The TLC plate was pretreated with toluene/Et_3N, 20:1.
12. The checkers purchased toluene (ACS grade) from Fisher Chemicals and dried it by passing through activated alumina.
13. Aluminum oxide, activated, neutral, Brockmann I, standard grade, ~150 mesh, 58 Å, Sigma-Aldrich. The fractions containing the product ($R_f = 0.9$, toluene/Et_3N, 20:1) were combined and concentrated.

14. The checkers purchased ethyl acetate (ACS grade) from Mallinckrodt Chemicals and distilled it under argon.

15. Working at 50% scale, the checkers obtained 4.01 g (72.4%).

16. The product has the following physicochemical properties: Colorless crystal, Mp 199–201 °C; $[\alpha]^{20}_D$ +153.0 (c 1.0, CHCl$_3$); ^{1}H NMR (300 MHz, benzene-d_6) δ: 1.16–1.60 (m, 26 H), 2.18–2.38 (m, 2 H), 2.48–2.74 (m, 6 H), 6.92 (d, 1 H, J = 8.4 Hz,), 6.98 (d, 1 H, J = 8.4 Hz), 7.10 (d, 1 H, J = 8.1 Hz), 7.12 (d, 1 H, J = 8.1 Hz); ^{13}C NMR (150 MHz, CDCl$_3$) δ: 17.4, 22.8, 23.0 (2), 27.9, 28.2, 29.2, 29.5, 32.2 (2), 43.0, 56.2 (2), 119.2 (2), 119.5, 128.1 (2), 128.8, 129.7, 130.0 (2), 132.2, 134.1, 137.9, 138.3 (2), 149.7, 149.8, 149.9; ^{31}P NMR (121 MHz, benzene-d_6) δ: 164.5; IR (ATR) 1579, 1467, 1232 cm^{-1}; HRMS (EI+) m/z calcd. for C$_{29}$H$_{40}$NO$_2$P 465.2797; found 465.2787; Anal. calcd. for C$_{29}$H$_{38}$NO$_2$P: C, 75.13; H, 8.26; N, 3.02; found: C, 74.80; H, 8.26; N, 2.97.

17. The checkers purchased tris(dibenzylideneacetone)dipalladium(0) from Strem and used it as received.

18. The checkers purchased benzene from EMD Chemicals and dried it by distillation over sodium/benzophenone ketyl under a nitrogen atmosphere.

19. The checkers purchased 4-phenyl-1-butene (98%) from TCI America and used it as received.

20. Di-tert-butyldiaziridinone (2) was prepared according to a recent Organic Syntheses procedure: Du, H.; Zhao, B.; and Shi, Y. Org. Synth. 2009, 86, 315-323.

21. The checkers purchased hexanes (ACS grade) from EMD Chemicals and used it as received.

22. The checkers purchased diethyl ether (ACS grade) from EMD Chemicals and used it as received.

23. Silica gel 40-63 μm (60 Å) (Silicycle, Quebec City, Canada) was used. The fractions containing the product (R_f = 0.4, hexanes/Et$_2$O, 6:1) were combined and concentrated.

24. Working at 50% scale, the checkers obtained 2.95 g (71.7 %).

25. The product has the following physicochemical properties: Pale yellow oil; $[\alpha]^{20}_D$ -30.5 (c 1.0, CHCl$_3$) (90% ee); ^{1}H NMR (300 MHz, CDCl$_3$) δ: 1.28 (s, 9 H), 1.33 (s, 9 H), 3.65 (d, 1 H, J = 8.1 Hz), 4.16 (d, 1 H, J = 1.2 Hz), 5.16 (d, 1 H, J = 9.6 Hz), 5.20 (d, 1 H, J = 15.9 Hz), 6.03 (ddd, 1 H, J = 8.1, 9.9, 17.1 Hz), 7.37–7.28 (m, 5 H); ^{13}C NMR (75 MHz, CDCl$_3$) δ: 28.9, 29.0, 53.4, 53.7, 63.3, 64.9, 115.8, 125.9, 127.9, 128.9, 140.9, 144.1,

159.2; IR (ATR) 1683 cm^{-1}; HRMS (ES+) m/z calcd. for $C_{19}H_{28}N_2ONa$ ([M+Na]$^+$) 323.2099; found 323.2079; Anal. calcd. for $C_{19}H_{28}N_2O$: C, 75.96; H, 9.39; N, 9.32; found: C, 75.84; H, 9.45; N, 9.38.

26. The enantiomeric excess is determined by chiral HPLC (Chiralpak AD column, flow rate: 0.7 mL/min, eluent: hexanes/iPrOH 4:1, t_+ = 8.0 min, t_- = 11.2 min) after removal of the two *tert*-butyl groups. The deprotection is carried out with the following procedure:[8a] A 5-mL screw-capped vial is charged with diamination product **3** (30.0 mg, 0.0999 mmol) and CF_3CO_2H (0.50 mL) (Note 27). The solution is placed into a preheated oil bath at 75–80 °C for 2 h, cooled to room temperature, and concentrated by rotary evaporation (25 °C, 18 mmHg). The residue is loaded onto a wet-packed SiO_2 column (diameter: 2.0 cm, height: 14.0 cm, pretreated with hexanes/diethyl ether, 4:1). The product is eluted with hexanes/diethyl ether, 4:1) to give 15.1 mg (93%) of the deprotected diamination product as a white powder (90% ee; Mp 204 °C (dec.)) (Notes 21, 22 and 23). The deprotected product is analyzed by chiral HPLC.

27. The checkers purchased trifluoroacetic acid (99%) from Alfa Aesar and used it as received.

Safety and Waste Disposal Information

All hazardous materials should be handled and disposed of in accordance with "Prudent Practices in the Laboratory"; National Academy Press; Washington, DC, 1995.

3. Discussion

Metal-mediated or catalyzed diamination of olefins is an attractive strategy for the construction of vicinal diamines and their derivatives.[3,4,5] Recently, we found that di-*tert*-butyldiaziridinone (**2**)[6,7] is an efficient diamination reagent for olefins with Pd(0) and Cu(I) catalysts.[8,9] When Pd(0) is used as the catalyst, a variety of dienes **4** can be regioselectively diaminated on the internal double bond to give compounds **5** in high yields under mild conditions (Scheme 1).[8a,b,d,e] Further studies show that simple terminal olefins **6** can be diaminated at the allylic and homoallylic carbons to give similar diamination products **7** (Scheme 1).[8c,f]

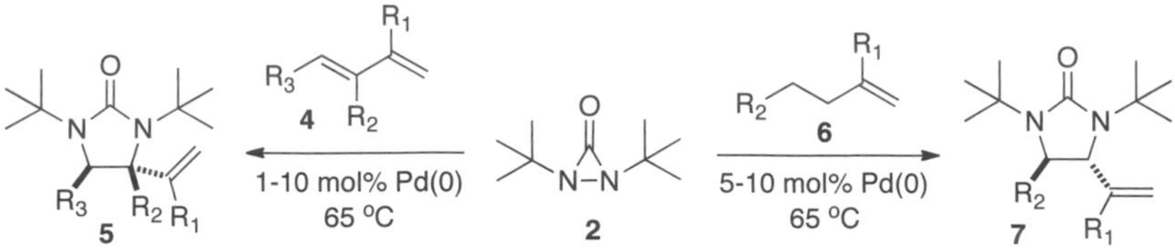

Scheme 1

A possible mechanism for this C-H diamination is shown in Scheme 2.[8c] The Pd(0) complex initially inserts into the N-N bond of di-*tert*-butyldiaziridinone (**2**) to form four-membered Pd(II) species **8**, which coordinates with olefin **6** and forms π-allyl Pd(II) complex **10** upon allylic H extraction. β-H elimination of **10** gives diene **11** and regenerates the Pd(0) catalyst. The formed diene (**11**) coordinates with Pd(II) species **8** and generates π-allyl Pd(II) complex **14** upon aminopalladation to the internal double bond. Pd(II) species **14** undergoes reductive elimination to give diamination product **7** and regenerates the Pd(0) catalyst.

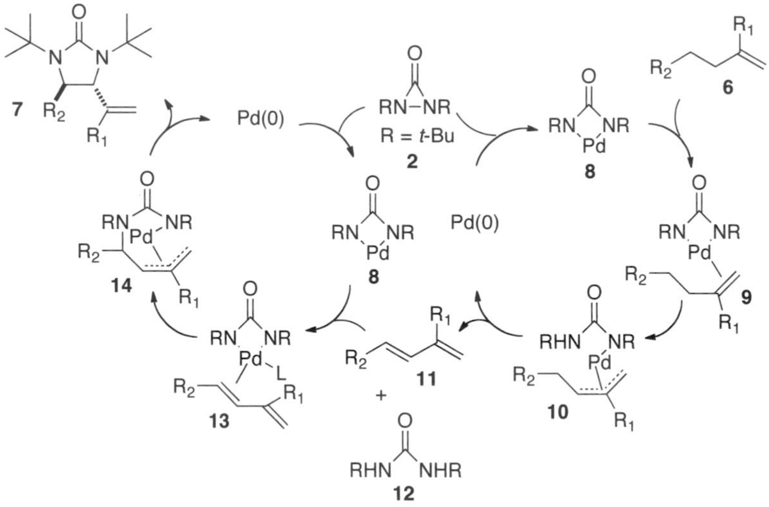

Scheme 2

When the chiral phosphorus ligand (*R*)-**1** is used, asymmetric C-H diamination can be achieved under mild conditions as shown in Scheme 3.[8f,g] A variety of terminal olefins are efficiently C-H diaminated in good yields and high enantioselectivities (Table 1). When 1,9-decadiene and 1,7-

268

octadiene are used, four C-H bonds can be replaced with four C-N bonds in an enantioselective fashion.

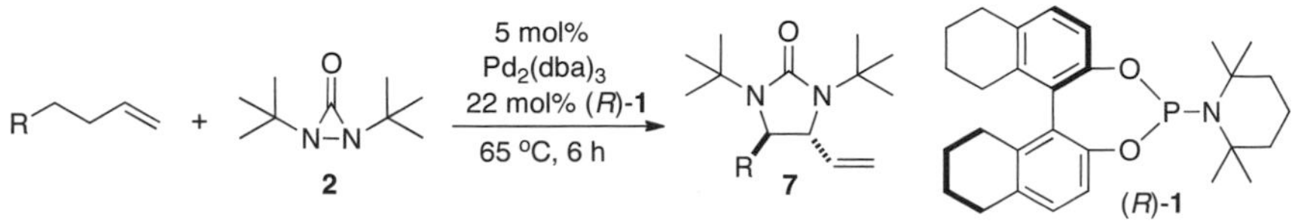

Scheme 3

Table 1. Asymmetric Allylic and Homoallylic C-H Diamination of Terminal Olefins

Entry	Substrate	Product	Yield (%)	ee(%)
1			50	90
2	Ph		67	93
3			51	94
4	n-C$_5$H$_{11}$		81	90
5	n-C$_5$H$_{11}$		69	89
6	Bn–N(Me)		67	92
7	TMSO		70	97:3 (dr)
8	TMSO		66	94:6 (dr)

1. Department of Chemistry, Colorado State University, Fort Collins, CO 80523, USA; yian@lamar.colostate.edu.

2. Department of Chemistry, University of Pittsburgh, Pittsburgh, PA 15260, USA; pwipf@pitt.edu.

3. For leading reviews, see: (a) Lucet, D.; Gall, T. L.; Mioskowski, C. *Angew. Chem., Int. Ed.* **1998**, *37*, 2580-2627; (b) Mortensen, M. S.; O'Doherty, G. A. *Chemtracts: Org. Chem.* **2005**, *18*, 555-561; (c) Kotti, S. R. S. S.; Timmons, C.; Li, G. *Chem. Biol. Drug Des.* **2006**, *67*, 101-114; (d) de Figueiredo, R. M. *Angew. Chem. Int. Ed.* **2009**, *48*, 1190-1193.

4. For leading references on metal-mediated diaminations, see: (a) Gomez Aranda, V.; Barluenga, J.; Aznar, F. *Synthesis* **1974**, 504-505; (b) Chong, A. O.; Oshima, K.; Sharpless, K. B. *J. Am. Chem. Soc.* **1977**, *99*, 3420-3426; (c) Bäckvall, J.-E. *Tetrahedron Lett.* **1978**, 163-166; (d) Barluenga, J.; Alonso-Cires, L.; Asensio, G. *Synthesis* **1979**, 962-964; (e) Becker, P. N.; White, M. A.; Bergman, R. G. *J. Am. Chem. Soc.* **1980**, *102*, 5676-5677; (f) Fristad, W. E.; Brandvold, T. A.; Peterson, J. R.; Thompson, S. R. *J. Org. Chem.* **1985**, *50*, 3647-3649; (g) Muñiz, K.; Nieger, M. *Synlett* **2003**, 211-214; (h) Muñiz, K. *Eur. J. Org. Chem.* **2004**, 2243-2252; (i) Muñiz, K.; Nieger, M. *Chem. Commun.* **2005**, 2729-2731; (j) Zabawa, T. P.; Kasi, D.; Chemler, S. R. *J. Am. Chem. Soc.* **2005**, *127*, 11250-11251; (k) Zabawa, T. P.; Chemler, S. R. *Org. Lett.* **2007**, *9*, 2035-2038.

5. For leading references on metal-catalyzed diaminations, see: (a) Li, G.; Wei, H.-X.; Kim, S. H.; Carducci, M. D. *Angew. Chem. Int. Ed.* **2001**, *40*, 4277-4280; (b) Bar, G. L. J.; Lloyd-Jones, G. C.; Booker-Milburn, K. I. *J. Am. Chem. Soc.* **2005**, *127*, 7308-7309; (c) Streuff, J.; Hövelmann, C. H.; Nieger, M.; Muñiz, K. *J. Am. Chem. Soc.* **2005**, *127*, 14586-14587; (d) Muñiz, K.; Streuff, J.; Hövelmann, C. H.; Núñez, A. *Angew. Chem., Int. Ed.* **2007**, *46*, 7125-7127; (e) Muñiz, K. *J. Am. Chem. Soc.* **2007**, *129*, 14542-14543; (f) Muñiz, K.; Hövelmann, C. H.; Streuff, J. *J. Am. Chem. Soc.* **2008**, *130*, 763-773; (g) Sibbald, P. A.; Michael, F. E. *Org. Lett.* **2009**, *11*, 1147-1149.

6. Greene, F. D.; Stowell, J. C.; Bergmark, W. R. *J. Org. Chem.* **1969**, *34*, 2254-2262.

7. For a leading review on diaziridinones, see: Heine, H. W. In *The Chemistry of Heterocyclic Compounds*; Hassner, A., Ed.; John Wiley & Sons, Inc: New York, 1983; pp 547-628.

8. (a) Du, H.; Zhao, B.; Shi, Y. *J. Am. Chem. Soc.* **2007**, *129*, 762-763; (b) Xu, L.; Du, H.; Shi, Y. *J. Org. Chem.* **2007**, *72*, 7038-7041; (c) Du, H.; Yuan, W.; Zhao, B.; Shi, Y. *J. Am. Chem. Soc.* **2007**, *129*, 7496-7497; (d) Du, H.; Yuan, W.; Zhao, B.; Shi, Y. *J. Am. Chem. Soc.* **2007**, *129*, 11688-11689; (e) Xu, L.; Shi, Y. *J. Org. Chem.* **2008**, *73*, 749-751; (f) Du, H.; Zhao, B.; Shi, Y. *J. Am. Chem. Soc.* **2008**, *130*, 8590-8591; (g) Fu, R.; Zhao, B.; Shi, Y. *J. Org. Chem.* **2009**, *74*, 7577-7580; (h) Zhao, B.; Du, H.; Cui, S.; Shi, Y. *J. Am. Chem. Soc.* **2010**, *132*, 3523-3532.
9. (a) Yuan, W.; Du, H.; Zhao, B.; Shi, Y. *Org. Lett.* **2007**, *9*, 2589-2591; (b) Zhao, B.; Yuan, W.; Du, H.; Shi, Y. *Org. Lett.* **2007**, *9*, 4943-4945; (c) Du, H.; Zhao, B.; Yuan, W.; Shi, Y. *Org. Lett.* **2008**, *10*, 4231-4234.

Appendix
Chemical Abstracts Nomenclature; (Registry Number)

2,2,6,6-Tetramethylpiperidine; (768-66-1)

Phosphorus trichloride (7719-12-2)

(*S*)-H$_8$-BINOL: (1S)-5,5',6,6',7,7',8,8'-octahydro-[1,1'-Binaphthalene]-2,2'-diol; (65355-00-2)

(*S*)-1-(8,9,10,11,12,13,14,15-Octahydro-3,5-dioxa-4-phosphacyclohepta[2,1-a;3,4-a']dinaphthalen-4-yl)-2,2,6,6-tetramethylpiperidine

Pd$_2$(dba)$_3$: Tris(dibenzylideneacetone)dipalladium; (51364-51-3)

Di-*tert*-butyldiaziridinone: 3-Diaziridinone, 1,2-bis(1,1-dimethylethyl)-; (19656-74-7)

(4*S*,5*S*)-1,3-Di-*tert*-butyl-4-phenyl-5-vinylimidazolidin-2-one: (4*R*, 5*R*)-1,3-bis(1,1-dimethylethyl)-4-ethenyl-5-phenyl-2-imidazolidinone; (927902-91-8)

272

Yian Shi was born in Jiangsu, China in 1963. He obtained his B.Sc. degree from Nanjing University in 1983, M.Sc. degree from University of Toronto with Professor Ian W.J. Still in 1987, and Ph.D. degree from Stanford University with Professor Barry M. Trost in 1992. After a postdoctoral study at Harvard Medical School with Professor Christopher Walsh, he joined Colorado State University as assistant professor in 1995 and was promoted to associate professor in 2000 and professor in 2003. His current research interests include the development of new synthetic methods, asymmetric catalysis, and synthesis of natural products.

Baoguo Zhao was born in Hubei, China in 1973. He received his B.Sc. degree from Wuhan University in 1996 and M.Sc. degree from Nanjing University in 2002 under the supervision of Professor Jianhua Xu. After completing his Ph.D. degree under the supervision of Professor Kuiling Ding at Shanghai Institute of Organic Chemistry, Chinese Academy of Science in 2006, he joined the Department of Chemistry at Colorado State University as a postdoctoral fellow with Professor Yian Shi. His current research interests include the development of novel synthetic methodology and asymmetric synthesis.

Haifeng Du was born in Jilin, China in 1974. He received his B.Sc. degree in 1998 and M.Sc. degree in 2001 from Nankai University. He then moved to Shanghai Institute of Organic Chemistry, Chinese Academy of Sciences, and obtained his Ph.D. degree in 2004 under the supervision of Professor Kuiling Ding. In the fall of 2004, he joined the Department of Chemistry at Colorado State University as a postdoctoral fellow with Professor Yian Shi. His research interests include the development of novel synthetic methodology and asymmetric synthesis.

Renzhong Fu was born in Shanxi, China, in 1981. He obtained his B.Sc. degree in Chemistry from Jilin University in 2003. He received his Ph.D. degree in Organic Chemistry in 2008 from Jilin University under the direction of Professor Xu Bai. His doctoral research studies focused on design and synthesis of novel pyrimidine-fused heterocyclic scaffolds. In September 2008, he joined the laboratory of Professor Yian Shi at Colorado State University as a postdoctoral fellow. His current research interest is the development of novel asymmetric catalytic methodology.

Eric Buck was born in 1985 in Fairmont, West Virginia. He graduated from the University of Minnesota with his B.Sc. in 2006. During his freshman year he entered the laboratory of Professor Thomas R. Hoye where he pursued the synthesis of petromyzonamine disulfate analogs with a focus on 5-β petromyzonamine disufate. While attending the U of M he was supported by the David A. and Merece H. Johnson Scholarship. In 2007 he moved to Pittsburgh, Pennsylvania, where he is currently a graduate student under the direction of Professor Peter Wipf. His current research interests include asymmetric synthesis and the synthesis of natural products.

SYNTHESIS OF ENAMIDES FROM KETONES: PREPARATION OF
N-(3,4-DIHYDRONAPHTHALENE-1-YL)ACETAMIDE

Submitted by Hang Zhao,[1] Charles P. Vandenbossche, Stefan G. Koenig, Surendra P. Singh,[1] and Roger P. Bakale.
Checked by Gopal K. Datta and Jonathan A. Ellman.

1. Procedure

> *Caution! Triethylphosphine is extremely pyrophoric and has a strong odor. Care must be taken to avoid exposure to air at all times. The reagent, the reaction and its work-up should be handled in an adequately ventilated fume hood while wearing gloves, safety glasses and laboratory coat.*

A. 3,4-Dihydronaphthalene-1(2H)-one oxime. An oven-dried 250-mL, three-necked, round-bottomed flask equipped with a reflux condenser, nitrogen inlet adapter, glass stopper, and magnetic stir bar (octagonal; 5/16 inch diameter; 1 inch length) is charged with α-tetralone (20.0 g, 137 mmol, 1.0 equiv) (Note 1), methanol (120 mL) (Note 2), hydroxylamine hydrochloride (10.5 g, 150 mmol, 1.1 equiv) (Note 3), and anhydrous sodium acetate (12.3 g, 150 mmol, 1.1 equiv) (Note 4) at room temperature. The flask is flushed with nitrogen, and the nitrogen inlet adapter is then moved to the top of the condenser. A temperature probe is placed into the third neck of the flask. The flask is heated at reflux in an oil bath. After one hour at reflux, the reaction is complete (Note 5). The mixture is cooled to room temperature, and methanol is removed on a rotary evaporator (20–30

°C, 30 mmHg). The resulting slurry is re-dissolved in ethyl acetate (200 mL) (Note 6), and 2 N aqueous NaOH (60 mL) is added. The mixture is transferred to a separatory funnel (500 mL), and the organic phase is separated, washed with DI water (100 mL) and brine (100 mL, 15 wt%), and concentrated at reduced pressure on a rotary evaporator (30–40 °C, ~30 mmHg) to give a brown crystalline solid (21.2 g, 96% yield) (Note 7) that was used in the next step without further purification.

B. *N-(3,4-Dihydronaphthalene-1-yl)acetamide.* An oven-dried 250-mL, three-necked round-bottomed flask equipped with a reflux condenser, thermometer, rubber septum, nitrogen inlet through rubber septum, and magnetic stir bar (octagonal; 5/16 inch diameter; 1 inch length) is charged with crude 3,4-dihydronaphthalen-1(2*H*)-one oxime (20.0 g, 124 mmol, 1.0 equiv) and toluene (freshly distilled and dry, 200 mL) (Note 8). The resulting solution is flushed with nitrogen for 50 min at room temperature. Triethylphosphine (Note 9) in a glass ampule is opened in a glove box, and the liquid triethylphosphine is securely transferred into a Schlenk flask (oven dried and cooled under nitrogen), sealed, and brought inside the fume hood. Triethylphosphine (17.6 g, 22.0 mL, 149.0 mmol, 1.2 equiv) is thereafter transferred into a properly marked volumetric container (oven dried and cooled under nitrogen) via cannula from the Schlenk flask under a nitrogen atmosphere (Figure 1). Triethylphosphine is then added directly to the reaction mixture from the volumetric container via cannula over 1 min (Figure 2). The reaction mixture is stirred at room temperature for 15 min. Acetic anhydride (15.2 g, 14.1 mL, 149.0 mmol, 1.2 equiv) is charged to the reaction using a syringe over 15 min (Note 10), which causes an exothermic reaction and the temperature rises to 39 °C (measured by an inserted thermometer immersed into the reaction mixture) (Figure 3). A yellow slurry solution (Note 11) is formed thereafter. The rubber septum and thermometer are quickly replaced with glass stoppers and the nitrogen inlet is moved to the top of the condenser. The resulting slurry is heated to reflux in an oil bath for 16 h, and reaction progress is monitored by HPLC or TLC (Notes 12 and 13). After complete conversion is achieved (Note 14), the reaction mixture is cooled to room temperature (~25 °C), 6 N aqueous NaOH (60 mL) and *tetra-n*-butylammonium hydroxide in MeOH (1 N, 2.0 mL) (Note 15) are added. The mixture is stirred at room temperature for 2–3 h until complete conversion of imide to enamide is achieved (Note 16). The reaction mixture is diluted with ethyl acetate (500 mL) and transferred to a 1-L separatory funnel, where the organic phase is separated, washed with 1 wt% acetic acid solution in water (100 mL), followed by brine (100 mL, 15 wt%) (Note 17).

Figure 1

Figure 2

Figure 3

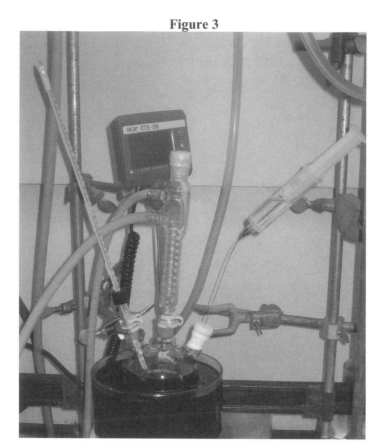

The solution is then concentrated on a rotary evaporator (30–40 °C, ~30 mmHg) to give a slurry (~80 g, ~85 mL) (Note 18). The slurry is then heated in an oil bath to 100 °C (oil bath temperature) to form a clear solution. The toluene solution is cooled to 90 °C and heptane (30 mL) (Note 19) is slowly charged to achieve a cloudy solution. The cloudy solution is stirred at 90 °C for 10 min, until a thin slurry is formed (Note 20). Additional heptane (70 mL) is charged over 5 min at 90 °C. The slurry is stirred at 90 °C for an additional 20 min and then cooled to room temperature over 2 h. After stirring at ambient temperature overnight (Note 21), the solid is collected by filtration on a Büchner funnel, washed with a mixture of 40% toluene and 60% heptanes (v/v, 2 x 40 mL), and dried under vacuum (45 °C at 15 mmHg for 5 h) to give an off-white crystalline solid (20.2 g, 87% yield) (Notes 22 and 23).

2. Notes

1.   α-Tetralone (97% purity) was purchased from Aldrich Chemical Company, Inc., and used as supplied.

2.   Methanol, HPLC grade, was purchased from EMD Chemicals, Inc., and used as supplied.

3.   Hydroxylamine hydrochloride was purchased from Aldrich Chemical Company, Inc., and used as supplied.

4.   Sodium acetate (99% purity) was purchased from Aldrich Chemical Company, Inc., and used as supplied.

5.   The progress of the reaction can be monitored by thin layer chromatography (TLC) on Multiformat (pre-scored to 5 x 20 cm) silica gel 60 F_{254} plates from Merck KGaA, elution with 10:90 (v/v) ethyl acetate/hexanes, and visualization with UV light. R_f values: tetralone, 0.40; oxime, 0.30. Alternatively, the reaction can also be monitored by HPLC. One drop of an aliquot was diluted with 1 mL of eluent and analyzed by: Conditions A (submitters), Waters 2695 HPLC system using a C18 SunFire column 3.5 μm, 4.6 x 150 mm; UV detection at 250 nm; eluent 40:60 of 0.1 vol% aqueous 85% phosphoric acid/acetonitrile, respectively at flow rate of 1 mL/min. Retention times for tetralone and oxime are 3.7 and 3.4 min, respectively: Conditions B (checkers), Hewlett Packard 1100 HPLC system using a Agilent narrow-bore ZORBAX SB-C18 column 5.0 μm, 2.1 x 150 mm; detection at 254 nm, 210 nm, 230 nm, and 280 nm ; eluent 95:5 of H_2O:acetonitrile (0.1 vol% TFA) at a flow rate of 0.4 mL/min. Retention time for tetralone and the oxime are 9.0 and 8.8 min, respectively. After 1 h at reflux, the sample showed >99.0 area% of oxime.

6.   Reagent grade ethyl acetate was purchased from EMD Chemicals, Inc., and used as supplied (submitters).

7.   The crude product displays the following physicochemical properties: Mp 100–102 °C; ^{1}H NMR (400 MHz, CDCl$_3$) δ: 1.84-1.91 (m, 2 H), 2.76 (t, 2 H, J = 6.0 Hz), 2.84 (t, 2 H, J = 6.7 Hz), 7.16 (d, 1 H, J = 7.4 Hz), 7.18–7.28 (m, 2 H), 7.87 (d, 1 H, J = 7.7 Hz); ^{13}C NMR (100 MHz, CDCl$_3$) δ: 21.2, 23.9, 29.8, 124.0, 126.4, 128.6, 129.2, 130.4, 139.8, 155.3; HRMS (ESI) m/z calcd. for $C_{10}H_{12}NO$ ([M+H]$^+$) 162.0919; found 162.0912.

8.   Reagent grade toluene was purchased from EMD Chemicals, Inc., and used as supplied (submitters).

9.   Triethylphosphine (97% purity) was purchased from Cytec Industries, Inc., and used as supplied (submitters), and in a glass ampule,

purchased from Alfa Aesar, stock no: 30177, lot no: G25T018 (checkers). Triphenylphosphine is also effective in this transformation, but removal of the by-product, triphenylphosphine oxide, by flash chromatography, proved difficult at scale. For scales >1 g, we recommend using triethylphosphine which enables easy isolation of the enamide by crystallization from the reaction mixture after work up.

10. Reagent grade acetic anhydride was purchased from Aldrich Chemical Company, Inc., and used as supplied.

11. A slight exotherm was observed with a temperature rise from 23 °C to 39 °C with formation of solid acylated oxime intermediate (Note 12).

12. The reaction can be monitored by HPLC analysis of a sample prepared by diluting one drop of the reaction mixture with 1 mL of eluent.. Retention times of the intermediate, by-product, and product are depicted below (HPLC conditions A: Waters 2695 HPLC system using a C18 SunFire column 3.5 μm, 4.6 x 150 mm; UV detection at 250 nm; eluent 40:60 of 0.1 vol% aqueous 85% phosphoric acid/acetonitrile, respectively at flow rate of 1 mL/minute). By-product imide is formed as a result of acylation of product enamide in presence of acetic anhydride. However, it is hydrolyzed to the enamide with aqueous NaOH in presence of *tetra-n-*butylammonium hydroxide.

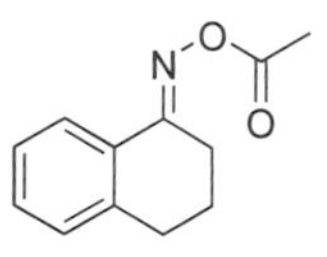

Acylated oxime intermediate	By-product imide	Product
R_t **4.6 min**	R_t **4.3 min**	R_t **2.5 min**

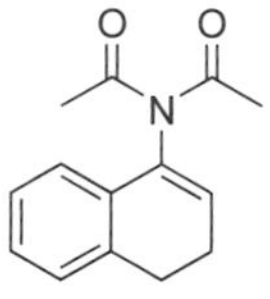

13. Alternatively, the progress of the reaction can also be monitored by thin layer chromatography (TLC) on Silicycle precoated 250-μm silica gel 60 F254 plates, elution with 40:60 ethyl acetate/hexanes, and visualization with UV light. The R_f values for the starting material, intermediate and product are given below with their structures.

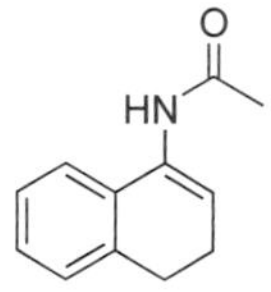

Org. Synth. **2010**, *87*, 275-287

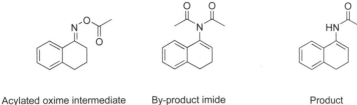

Acylated oxime intermediate	By-product imide	Product
R$_f$ **0.9**	R$_f$ **0.8**	R$_f$ **0.2**

14. The reaction solution was heated at reflux overnight to ensure complete consumption of acylated oxime as determined by TLC (R$_f$ 0.9) (Note 13). By HPLC (conditions A, Note 12), 69.7 area% product, R$_t$ 2.5 min; 21.5 area% imide, R$_t$ 4.3 min .

15. *Tetra-n*-butylammonium hydroxide as a 1 N solution in MeOH was purchased from Aldrich Chemical Company, Inc. and used as supplied.

16. After 2 h, HPLC analysis (conditions A, see Note 12) showed 92.4 area% of enamide (R$_t$ 2.5 min) and no imide (R$_t$ 4.3 min).

17. The color of the organic phase changes from dark brown to light yellow after the acid wash. The pH of the aqueous phase was <7 as determined by pH paper.

18. Toluene can be used to adjust the slurry to ~80 g or 85 mL.

19. Reagent grade heptane was purchased from EMD Chemicals, Inc. and used as supplied.

20. Up to 50 mL of the total 100 mL heptanes may be required to form a cloudy solution; however, in this case only 30 mL of heptanes was added.

21. The reaction mixture was stirred overnight for convenience only. Stirring for 4-6 h at room temperature is typically sufficient.

22. The product displays the following spectroscopic properties: Mp 138–139 °C; ^{1}H NMR (400 MHz, CDCl$_3$, mixture of rotamers) δ: 1.94 (s, 0.7 H), 2.14 (s, 2.3 H), 2.33–2.36 (m, 1.5 H), 2.37–2.41 (m, 0.5 H), 2.75 (t, 1.5 H, *J* = 7.9 Hz), 2.83 (bt, 0.5 H, *J* = 7.9 Hz), 5.95 (m, 0.25 H), 6.39 (t, 0.75 H, *J* = 4.8 Hz), 6.83 (br s, 0.25 H), 7.00 (br s, 0.75 H), 7.12–7.22 (m, 4 H); ^{13}C NMR (100 MHz, CDCl$_3$) δ: 20.3, 22.1, 22.6, 24.1, 27.2, 27.5, 119.6, 120.6, 121.9, 125.9, 126.3, 126.8, 127.4, 127.8, 128.1, 131.5, 136.7, 169.2; Anal. calcd. for C$_{12}$H$_{13}$NO: C, 76.56; H, 7.50; N, 7.44; found: C, 76.30; H, 7.37; N, 7.40.

23. The purity of the product was determined to be >98 area% by HPLC analysis [R$_t$ 2.5 min (conditions A), R$_t$ 7.4 min (conditions B) (Note 5)]. An analytical sample was prepared by dissolving 1 mg of sample in 5

mL of eluent and analyzed on a Waters 2695 HPLC system (conditions A) or Hewlett Packard 1100 HPLC system (conditions B) (Note 5).

Safety and Waste Disposal Information

All hazardous materials should be handled and disposed of in accordance with "Prudent Practices in the Laboratory"; National Academy Press; Washington, DC, 1995.

3. Discussion

Enantiomerically pure amines and their derivatives are valuable building blocks for organic synthesis and are an important class of biologically active compounds. In recent years, significant progress has been made to access chiral amines catalytically by asymmetric hydrogenation of enamides.[2] As a result, there has been an explosive growth in the development of chiral ligands for the catalytic hydrogenation of enamides. Despite the availability of a large number of catalysts for this transformation, the methods for the synthesis of enamides from the corresponding carbonyl compounds are still limited.[3] The procedure reported by Burk et al.[3a] in 1998 is still the most frequently used method for the small scale preparation of enamides from ketones. Due to the difficulty with the scale-up of this procedure, a few alternatives have been reported involving palladium-catalyzed coupling of enol triflates or tosylates, derived from the corresponding ketones, with acetamide.[4] The procedure reported herein incorporates a similar approach as the Burk procedure, but is more scaleable and high-yielding, specifically for the synthesis of tetralone-based enamides.[5]

Various phosphines were found to be effective for this transformation as depicted in Table 1. Due to the highly water-soluble nature of the triethylphosphine oxide by-product, the use of triethylphosphine was preferred, enabling easy isolation of enamide.

Table 1. Effect of Various Phosphines on Enamide Formation[a]

entry	phosphines	reaction time (h)	% yield[b]
1	Ph$_3$P	14	78
2	DPPE[c]	14	89
3	Et$_3$P	16	89
4	(n-Bu)$_3$P	16[d]	84
5	(n-Oct)$_3$P	14	72
6	(Cy)$_3$P	38	78
7	(t-Bu)$_3$P	38	trace
8	(EtO)$_3$P	38	trace

a) Reaction conditions: phosphine (1.2 equiv), Ac$_2$O (1.2 equiv), toluene, reflux. b) Isolated yield of enamide from ketoxime after flash chromatography. c) 0.6 equiv of diphenyl phosphino-ethane (DPPE) was used. d) Reaction was run in o-xylene.

A variety of benzylic ketones and non-benzylic ketones were successfully converted to the corresponding enamides in reasonable yields (Tables 2 and 3). The substituted tetralone-based enamide (entry 4, Table 2) has been successfully scaled up to multi-kilogram scale.

Table 2. Enamides from Benzylic Ketones via Oximes[a]

entry	ketone	oxime (% yield)[b]	enamide reaction time (h)	enamide	enamide (% yield)[c]
1		100	22		74
2	OMe	97	23	OMe	77
3	MeO	96	19	MeO	71
4	Ar Ar = 3,4-dichlorophenyl	99	12	Ar	89
5		100	24		90
6		100	10		76
7		99	10		58
8		99	23		58
9		93	5		78

a) Reaction conditions: (i) NH$_2$OH•HCl (1.2 equiv), NaOAc (1.2 equiv), solvent, reflux; (ii) Et$_3$P (1.2 equiv), Ac$_2$O (1.2 equiv), toluene, reflux. b) Crude yield of ketoxime. c) Unoptimized isolated yield of enamide from ketoxime after column chromatography.

Table 3. Enamides from Non-benzylic Ketones via Oximes[a]

entry	ketone	oxime (% yield)[b]	enamide reaction time (h)	enamide	enamide (% yield)[c]
1		100	22		71
2		99	22		64
3		98	28		54[d,e]
4		100	22		54[d,e]

a) Reaction conditions: (i) NH$_2$OH • HCl (1.2 equiv), NaOAc (1.2 equiv), solvent, reflux; (ii) Et$_3$P (1.2 equiv), Ac$_2$O (1.2 equiv), toluene, reflux. b) Crude yield of ketoxime. c) Unoptimized isolated yield of enamide from ketoxime after chromatography. d) Low yield is due to physical loss during isolation. e) Only the depicted geometric isomer was isolated.

1. Chemical Process Research and Development, Sepracor Inc., 84 Waterford Dr., Marlborough, MA 01752, USA. E-mail: surendra.singh@sepracor.com; hang.zhao@sepracor.com

2. (a) Blaser, H.-U.; Malan, C.; Pugin, B.; Spindler, F.; Steiner, H.; Studer, M. *Adv. Synth. Catal.* **2003**, *345*, 103-151; (b) Van den Berg, M.; Haak, R. M.; Minnaard, A. J.; de Vries, A. H. M.; Vries, J. G.; Feringa, B. L. *Adv. Synth. Catal.* **2002**, *344*, 1003-1007.

3. (a) Burk, M. J.; Casy, G.; Johnson, N. B. *J. Org. Chem.* **1998**, *63*, 6084-6085; (b) Zhu, G.; Casalnuovo, A. L.; Zhang, X. *J. Org. Chem.* **1998**, *63*, 8100-8101; (c) Laso, N. M.; Quiclet-Sire, B.; Zard, S. Z. *Tetrahedron Lett.* **1996**, *37*, 1605-1608; (d) Neugnot, B.; Cintrat, J.-C.;

Rousseau, B. *Tetrahedron* **2004**, *60,* 3575-3579. (e) Brice, J. L.; Meerdink, J. E.; Stahl, S. S. *Org. Lett.* **2004**, *6*, 1845-1848. (f) Harrison, P.; Meek, G. *Tetrahedron Lett.* **2004**, *45*, 9277-9280; (g) Willis, M. C.; Brace, G. N.; Holmes, I. P. *Synth* **2005***, 3229-3234; Guan, Z-H.; Huang, K.; Yu, S.; Zhang, X. *Org. Lett.* **2009**, *11*, 481-483.

4. (a) Wallace, D. J.; Klauber, D. J.; Chen, C.-y.; Volante, R. P. *Org. Lett.* **2003**, *5*, 4749-4752; (b) Klapars, A.; Campos, K. R.; Chen, C.-y.; Volante, R.P. *Org. Lett.* **2005**, *7*, 1185-1188.

5. Zhao, H.; Vandenbossche, C.P.; Koenig, S.G.; Singh, S.P.; Bakale, R.P. *Org. Lett.* **2008**, *10,* 505-507.

Appendix
Chemical Abstract Nomenclature (Registry Number)

3,4-Dihydronaphthalene-1(2*H*)-one oxime; (3349-64-2)

Hydroxylamine hydrochloride: (5470-11-1)

α-Tetralone: 1(2*H*)-Naphthalenone, 3,4-dihydro-; (529-34-0)

N-(3,4-Dihydronaphthalene-1-yl)acetamide: Acetamide, *N*-(3,4-dihydro-1-naphthalenyl)-; (213272-97-0)

Triethylphosphine; (554-70-1)

Acetic anhydride; (108-24-7)

Surendra Singh obtained his Ph.D. in Organic Chemistry in India in 1988. He did postdoctoral work at Technion, Israel with Prof. Ehud Keinan, and at Ohio State University with Prof. Viresh Rawal working in the areas of asymmetric synthesis, radical chemistry and enzymatic reactions. In 1994, he joined Merck & Co in Technical Operations group as Senior Process Scientist. At Merck, he was involved in the development of chemical processes, which spans from laboratory development through production scale and their technical transfer to the manufacturing site. In 2001, he moved to Sepracor where he has been involved in the process research and development of NCEs and improved chemical entities. Currently, he is Director of Chemical Process Research. His research interests include the design of practical, environmentally benign, and economically viable processes for drug substances.

286

Hang Zhao was born in HangZhou, China and received his B.S. degree from Beijing University, China in 1985. In 1999 he completed his doctoral studies under the supervision of Prof. David Mootoo at the City University of New York. He then began postdoctoral research with Prof. Robert Holton at Florida State University working on the reconstruction and SAR studies on Taxol. His research interests include asymmetric synthesis and the development of new synthetic methods.

Charles P. Vandenbossche was born in 1971 in Detroit, MI. He received his B.A. in Chemistry (1994) and M.S. (1996) degrees from Wayne State University (Detroit), working under the supervision of Prof. James H. Rigby. Upon graduation, he relocated to Marlborough, MA, and joined the pharmaceutical company, Sepracor Inc. He is currently a Sr. Research Chemist in the Chemical and Pharmaceutical Sciences Department working as a process development chemist.

Stefan G. Koenig received his undergraduate degree in chemistry from Providence College in 1997. He completed his Ph.D. under Prof. David J. Austin at Yale University by developing a transient-linkage Pauson-Khand cyclization strategy to the marine natural product palau'amine. In 2002, he joined Prof. Andrea T. Vasella at the Swiss Federal Institute of Technology / ETH Zurich to develop a scalable process toward stable analogs of guanofosfocin, an important antifungal. He has been with Sepracor Process Research & Development since 2004, where his contributions have ranged from route scouting to API manufacture.

Gopal K. Datta was born in Calcutta, India. He graduated from Presidency College, the University of Calcutta with a Master of Science in Chemistry and specialized in organic chemistry. In 2008, Gopal K. Datta received his Ph.D. in Medicinal Chemistry from the division of Organic Pharmaceutical Chemistry, Department of Medicinal Chemistry, Faculty of Pharmacy, Uppsala University, Sweden under the supervision of Prof. Mats Larhed and Prof. Anders Hallberg. In 2009, Gopal K. Datta started his postdoctoral work in chemical biology at Prof. Jonathan A. Ellman's laboratory in the Department of Chemistry, University of California at Berkeley.

SAFE AND SCALABLE PREPARATION OF BARLUENGA'S REAGENT
(Bis(pyridine)iodonium(I) tetrafluoroborate)

A. HBF_4 (aq) $\xrightarrow[\text{2. } SiO_2]{\text{1. } Ag_2CO_3}$ $AgBF_4$ on SiO_2

B. $AgBF_4$ on SiO_2 $\xrightarrow[\text{2. } I_2]{\text{1. Pyridine}}$

CH_2Cl_2, 1 h, 23 °C IPy_2BF_4 (1)

Submitted by Justin M. Chalker, Amber L. Thompson, and Benjamin G. Davis.[1]
Checked by Nilesh Zaware and Peter Wipf.[2]

1. Procedure

A. Silver tetrafluoroborate on SiO_2. A 500-mL, one-necked round-bottomed flask is equipped with a Teflon-coated magnetic stir bar (3.9 cm x 1 cm, rectangular). Deionized water (50 mL) is poured into the flask and tetrafluoroboric acid (6.3 mL, 50 mmol, 1.0 equiv) is added by syringe (Note 1). The solution is stirred, open to air, at room temperature (23 °C) for 2 min and silver carbonate (6.89 g, 25.0 mmol, 0.50 equiv) is added in several portions with a spatula over 2 min (Note 2). Carbon dioxide gas evolves. The resulting mixture is stirred rapidly (380 rpm) until all solids are dissolved to give a grey solution (20 min). Silica gel (10.0 g) is poured into the solution and the mixture is stirred for 5 min (Note 3). The stir bar is removed using a magnetic wand, rinsed with deionized water (1 mL) with a glass pipette, and the flask is transferred to a rotary evaporator into a water bath pre-heated to 55 °C. The water in the reaction flask is evaporated (55 °C, 80 mmHg, cooling is effected by circulating chilled water through the condenser) over 35 min to give the $AgBF_4$ on SiO_2 as an off-white/grey solid on the walls of the flask. A spatula is used to scrape all solids to the bottom of the flask, which is then returned to the rotary evaporator and dried with rapid spinning (55 °C, 80 mmHg). After 15 min of drying in this manner, a spatula is used to scrape all solids to the bottom of the flask. A free flowing

Org. Synth. **2010**, *87*, 288-298
Published on the Web 5/13/2010

powder is obtained. The flask is allowed to stand and cool at room temperature for 5 min before the mixture is used directly in the next step.

B. *Bis(pyridine)iodonium(I) tetrafluoroborate* (IPy$_2$BF$_4$, **1**). A Teflon-coated magnetic stir bar (3.9 cm x 1 cm, rectangular) is added to the 500-mL 1-necked flask that contains the AgBF$_4$ on SiO$_2$ from above. Dichloromethane (300 mL) is poured into the flask and the mixture is stirred at room temperature, open to the atmosphere (Note 4). Pyridine (8.1 mL, 100 mmol, 2.0 equiv) is poured into the reaction flask from a 10.0 mL measuring cylinder, and the mixture is stirred at room temperature (Note 5). One minute after the addition of pyridine, iodine flakes (12.69 g, 50.0 mmol, 1.0 equiv) are poured into the reaction flask from a weighing paper (Note 6). A yellow precipitate (AgI) is formed immediately upon addition of the iodine. A glass stopper is used to loosely cap the reaction flask and the mixture is stirred vigorously (380 rpm) for 1 h (Note 7). The solution gradually turns red as the iodine is dissolved. After 1 h of stirring, the solids are removed by filtration through a sintered glass funnel (Note 8) under suction (15 mmHg), and the dark red filtrate is collected directly in a 1-L, 1-necked round-bottomed flask using a 24/40 glass filtration adapter. The reaction flask is rinsed with dichloromethane (100 mL) to ensure complete transfer, and the SiO$_2$/AgI filter cake is then washed with additional dichloromethane (100 mL). The solvent is removed by rotary evaporation (35 °C, 80 mmHg) to give a reddish-yellow solid. This solid is then dissolved in dichloromethane (100 mL) and a Teflon-coated stir bar (5 cm x 1 cm, rectangular) is added. The solution is stirred in an ice bath for 5 min and then diethyl ether (200 mL) is poured into the stirred solution to precipitate the product (Note 9). The mixture is stirred at 0 °C for 10 min to ensure that all IPy$_2$BF$_4$ is precipitated. The product is isolated by filtration under suction (15 mmHg) (Note 10). The stirrer is removed by a magnetic wand, held over the filter funnel and washed with diethyl ether (4 mL) using a glass pipette. The flask is rinsed with diethyl ether (50 mL) to complete the transfer. The off-white powder is dried on the filter with suction for 5 min, transferred to a weighing paper (6" x 6") and dried under vacuum in a desiccator for 4 h (1.2 mmHg) to obtain **1** (13.67 g, 73%) (Notes 11, 12, 13 and 14). This material is used for NMR, IR and Mp analysis. IPy$_2$BF$_4$ can also be recrystallized from dichloromethane. Accordingly, the powdered **1** (13.64 g) is added to a 100-mL, 1-necked round-bottomed flask with a Teflon-coated stir bar (1.5 cm x 1 cm, rectangular). A water bath is preheated to 50 °C and dichloromethane (25 mL) is added to the solid. The suspension is stirred (870 rpm) and lowered into the water bath. Once the dichloromethane is boiling (ca. 1 min), additional dichloromethane is added by glass pipette, with stirring, until all solids are dissolved (48 mL of additional dichloromethane, ca. 73 mL of total solvent). Once a clear solution is obtained, the stirring is stopped and the flask removed from the water bath. Crystals are observed within 3 min. After 30 min, the flask is cooled to 0 °C (ice bath) to facilitate the crystallization. After 2 h on ice, the crystals are isolated by filtration with suction (15 mmHg) (Note 15), collecting the filtrate directly in a 250-mL round-bottomed flask. The suction is removed, the stir bar is held over the filter funnel with a magnetic wand and washed with diethyl ether (2 mL), and the crystals are triturated with diethyl ether (30 mL) on the filter. The crystals are then dried on the filter for 5 min with suction, transferred to a jar and dried under vacuum in a desiccator for 2 h (1.2 mmHg). The first crop of crystals yields 8.34 g (Notes 16 and 17). All remaining material in the crystallization flask and filter is collected by rinsing with dichloromethane (50 mL). The combined filtrates and mother liquor are concentrated by rotary evaporation (35 °C, 80 mm Hg) to give a reddish yellow solid. The recrystallization is repeated twice more to recover the remaining material, using dichloromethane (24 mL) in a 50-mL round-bottomed flask in the second recrystallization (2.46 g of crystals isolated) and dichloromethane (23 mL) in a 25-mL round-bottomed flask in the third and final recrystallization (0.79 g of crystals isolated). The total recovery of **1** from recrystallization is 11.59 g (85% from powdered IPy_2BF_4 and 62% overall).

2. Notes

1. The submitters utilized deionized water from an Elix 10 water purification system equipped with a Progard 1 pretreatment pack (Millipore). The checkers used deionized water from a Barnstead purification system. Tetrafluoroboric acid (8.0 M, 49.5-50.5% w/w) was purchased from Fluka and used as received. The checkers bought fluoroboric acid (50 wt% solution in water) from Acros Organics and used it as received.

2. Silver carbonate (99%) was purchased from Alfa Aesar and used as received. The checkers bought silver carbonate (99.5%) from Alfa Aesar and used it as received.

3.   The submitters purchased SiO_2 (10% <40 μm and 10% >63 μm, Laboratory Reagent Grade for Flash Chromatography) from VWR and used it as received. Checkers purchased SiO_2 (40-63 μm (230-400 mesh) Siliaflash® P60) from Silicycle and used it as received.

4.   Dichloromethane (HPLC grade) was purchased from Fisher Scientific and used as received.

5.   Submitters used pyridine (glass distilled grade) purchased from Rathburn Chemicals Ltd. as received. Checkers used pyridine (GR grade) purchased from EMD Chemicals as received.

6.   Iodine beads (laboratory reagent grade) were purchased from Fisher Scientific and used as received. The checkers bought iodine (flakes, certified ACS) from Fisher Scientific.

7.   The formation of the product, IPy_2BF_4, is difficult to monitor by conventional methods such as TLC. Qualitatively, the formation of AgI (yellow solid) and the dissolution of the iodine beads are indicators of reaction progression. Typically, 1 h of stirring is sufficient for a preparation on the scale described, and prolonged reaction time did not improve the final yield.

8.   The submitters used a Büchner filter funnel (porosity 3 glass frit) with a 250 mL capacity for the filtration. The checkers used a Büchner filter funnel (coarse, porosity 40-60 glass frit) with a 150 mL capacity for the filtration.

9.   The submitters used diethyl ether (glass distilled grade) purchased from Rathburn Chemicals Ltd. as received. The checkers used diethyl ether (GR grade) purchased from EMD Chemicals as received.

10.   The submitters used a Büchner filter funnel (porosity 3 glass frit) with a 250 mL capacity layered with filter paper (QL 100 from Fisher Scientific) for the isolation of IPy_2BF_4. The checkers used a Büchner filter funnel (porosity 10-15 glass frit) with a 150 mL capacity for the isolation of IPy_2BF_4. The checkers did not use a filter paper.

11.   The preparation was run in triplicate on the described scale by the submitters with yields of 13.0 to 13.6 g (70–73%) after precipitation with diethyl ether.

12.   As observed by the submitters, the powdered IPy_2BF_4 is sufficiently pure for most synthetic applications and can be stored for months if protected from light and kept at –20 °C.

13.   Working at 50% scale, the checkers obtained 5.65 g (61%).

14. Spectroscopic data obtained for **1** prior to recrystallization: Mp: 151–173 °C (Lit[3]: 149-151 °C); IR (ATR) 1600, 1453, 1062, 786, 659 cm^{-1}; ^{1}H NMR (300 MHz, CD$_3$CN) δ: 7.64 (dd, J = 7.8, 6.6 Hz, 4 H), 8.26 (tt, J = 7.8, 1.5 Hz, 2 H), 8.79 (dd, J = 6.3, 1.2 Hz, 4 H); ^{13}C NMR (75 MHz, CD$_3$CN) δ: 127.9, 142.3, 149.7; ^{19}F NMR (282 MHz, CD$_3$CN) δ: –151.65, –151.60.

15. The submitters used a Büchner filter funnel (porosity 3 glass frit) with a 250 mL capacity, layered with a filter paper (QL 100 from Fisher scientific) for the isolation of IPy$_2$BF$_4$. The checkers used a Büchner filter funnel (porosity 4-5.5 glass frit) with a 60 mL capacity. Checkers did not use a filter paper.

16. Analytical data obtained for recrystallized **1**: Mp: 151–179 °C (Lit[3]: 149–151 °C); ^{19}F NMR (282 MHz, CD$_3$CN) δ: –151.81, –151.76; Anal. calcd. for C$_{10}$H$_{10}$BF$_4$IN$_2$: C, 32.29; H, 2.71; N, 7.53; found: C, 32.26; H, 2.52; N, 7.56; ^{1}H NMR data matched data listed in Note 14.

17. Submitters prepared single crystals (from the first crop of crystals) suitable for X-ray analysis to unambiguously confirm the identity of the product. Accordingly, 500 mg of the first crop of crystals was dissolved in 3.0 mL of boiling dichloromethane in a straight-sided glass tube (2.0 × 5.0 cm) and left to cool, open to air. After 30 min, the vial was capped, covered in foil, and left to stand overnight at room temperature (23 °C). Colorless plate-like crystals were isolated from the mother liquor and coated in inert perfluoro-polyether oil, and mounted on a hair and frozen in place at 150 K using an Oxford Cryosystems Cryostream N2 open flow cooling device.[4] Diffraction data were collected using a Nonius Kappa-CCD diffractometer and results compared with those reported by Alvarez-Rua et al.[5] Three crystals were tested in all by the collection of a 10° ω-scan which indexed in each case to give the same monoclinic cell (by transformation, similar to that reported previously[6]; see supplementary information, for details). A complete dataset to a maximum resolution of 0.77 Å was recorded for the third crystal. On completion, the crystal was warmed to 250 K at 120 K/h, where the cell was redetermined and found to be a smaller, related C-centered monoclinic cell. A second dataset was recorded for comparison before the crystal was returned to 150 K (at 120 K/h) where the cell was found to have returned to that seen previously at that temperature, demonstrating the occurrence of a reversible structural phase transition. Cell parameters and intensity data for both data sets were processed using the DENZO-SMN package[7] and the structures solved by direct methods using

292

SIR92.[8] Both structures were refined by full-matrix least squares on F^2 using the CRYSTALS suite (Figure 1).[9] Where possible, non-hydrogen atoms were refined with anisotropic displacement parameters and the hydrogen atoms were initially refined with soft restraints, then constrained using a riding model. In the case of the 250 K data, the BF_4 counter-ion was found to be disordered and a full description of the way this was modeled and a brief comparison of the structures is given in the supplementary information, together with the structural data in CIF format. These data have also been deposited with the Cambridge Crystallographic Data Centre as supplementary publications no. CCDC 772585 and CCDC 772586, respectively, and copies can be obtained, free of charge from The Cambridge Crystallographic Data Centre via www. ccdc.cam.ac.uk/data_request/cif.

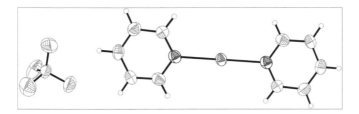

Figure 1.   The molecular structure of bis(pyridine)iodonium(I) tetrafluoroborate, from the crystal structure at 150 K. Only one molecule is shown for clarity and the atomic displacement parameters are drawn at 50% probability.

Safety and Waste Disposal Information

All hazardous materials should be handled and disposed of in accordance with "Prudent Practices in the Laboratory"; National Academy Press; Washington, DC, 1995.

3. Discussion

Bis(pyridine)iodonium(I) tetrafluoroborate (IPy_2BF_4, **1**) – also referred to as Barluenga's Reagent, in recognition of its discoverer[3] – is a mild iodonium source and oxidant. IPy_2BF_4 is a relatively stable solid and is soluble in both organic and aqueous systems. The utility of **1** in organic synthesis has been widely demonstrated and a selection of transformations promoted by IPy_2BF_4 is shown in Scheme 1. Alkenes, alkynes, and arenes are iodinated by **1** under mild conditions (Scheme 1A-D).[3,10,11] This method is mild enough that selective iodination of tyrosine in peptides[12] and proteins[13] has been realized. Barluenga's reagent can oxidize secondary alcohols to ketones under thermal conditions (Scheme 1E). Under photolytic conditions, cycloalkanols are oxidatively cleaved to the corresponding ω-iodocarbonyls when treated with **1**, and 1,2-diols are converted to the respective dicarbonyls (Scheme 1F-G).[14,15] C-H functionalization[16] and heterocycle synthesis[17,18,19] is also enabled by **1** (Scheme 1H-K). Additionally, IPy$_2$BF$_4$ is useful in carbohydrate synthesis (Scheme 1L-M). Thio- and *n*-pentenyl glycosides can be activated for glycosylation or converted to the corresponding glycosyl fluoride with **1**.[20,21] Finally, it is noteworthy that many of the reactions in Scheme 1 involve carbon-carbon bond formations (A, B, H, J, K).

Scheme 1: Barluenga's Reagent (IPy$_2$BF$_4$) in Organic Synthesis

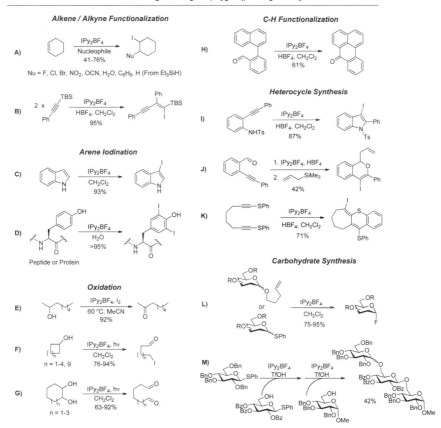

294

Given the synthetic versatility of IPy$_2$BF$_4$, it is useful to have ready access to multi-gram quantities of this reagent. Even though Barluenga's reagent is available from several commercial suppliers, the cost may preclude its use in large-scale synthetic endeavors.[22] Moreover, the preparation of **1** described in the literature uses extremely toxic Hg(II) salts that necessitate a risky and tedious workup.[3,23] The preparation above is adapted procedurally from Barluenga's initial reports, but replaces Hg(II) with Ag(I).[24] The resulting procedure is a much safer and convenient synthesis, with yields comparable to that reported by Barluenga. The protocol reported herein can conveniently supply more than 10 g of IPy$_2$BF$_4$ in less than a day. With a safe, scalable synthesis of IPy$_2$BF$_4$ in hand, we anticipate increased deployment of Barluenga's reagent in synthetic endeavors.

1. Chemistry Research Laboratory, Department of Chemistry, University of Oxford, 12 Mansfield Road, Oxford OX1 3TA, U.K.; ben.davis@chem.ox.ac.uk.

2. Department of Chemistry and Center for Chemical Methodologies & Library Development, University of Pittsburgh, Pittsburgh, Pennsylvania 15260, USA; pwipf@pitt.edu.

3. Barluenga, J.; González, J. M.; Campos, P.J.; Asensio, G. *Angew. Chem. Int. Ed. Engl.* **1985**, *24*, 319-320.

4. Cosier, J.; Glazer, A. M. *J. Appl. Cryst.* **1986**, *19*, 105-107.

5. Álvarez-Rúa, C.; García-Granda, S.; Ballesteros, A.; González-Bobes, F.; González, J. M. *Acta Cryst.* **2002**, *E58*, o1381-o1383.

6. Feast, G. C.; Haestier, J.; Page, L. W.; Robertson, J.; Thompson, A. L.; Watkin, D. J. *Acta Cryst.* **2009**, *C65*, o635-o638.

7. Otwinowski, Z.; Minor, W. *Methods Enzymol.* **1997**, *276*, 307-326.

8. Altomare, A.; Cascarano, G.; Giacovazzo, C.; Guagliardi, A.; Burla, M. C.; Polidori, G.; Camalli, M. *J. Appl. Cryst.* **1994**, *27*, 435.

9. Betteridge, P. W.; Carruthers, J. R.; Cooper, R. I.; Prout, K.; Watkin, D. J. *J. Appl. Cryst.* **2003**, *36*, 1487.

10. Barluenga, J.; Llorente, I.; Alvarez-García, L. J.; González, J. M.; Campos, P. J.; Díaz, M. R.; García-Granda, S. *J. Am. Chem. Soc.* **1997**, *119*, 6933-6934.

11. Barluenga, J. *Pure Appl. Chem.* **1999**, *71*, 431-436.

12. Vilaró, M.; Arsequell, G.; Valencia, G.; Ballesteros, A.; Barluenga, J. *Org. Lett.* **2008**, *10*, 3243-3245.

13. Espuña, G.; Andreu, D.; Barluenga, J.; Pérez, X.; Planas, A.; Arsequell, G.; Valencia, G. *Biochemistry* **2006**, *45*, 5957-5963.
14. Barluenga, J.; González-Bobes, F.; Ananthoju, S. R.; García-Martín, M. A.; González, J. M. *Angew. Chem. Int. Ed.* **2001**, *40*, 3389-3392.
15. Barluenga, J.; González-Bobes, F.; Murguía, M. C.; Ananthoju, S. R.; González, J. M. *Chem. Eur. J.* **2004**, *10*, 4206-4213.
16. Barluenga, J.; Trinicado, M.; Rubio, E.; González, J. M. *Angew. Chem. Int. Ed.* **2006**, *45*, 3140-3143.
17. Barluenga, J.; Trincado, M.; Rubio, E.; González, J. M. *Angew. Chem. Int. Ed.* **2003**, *42*, 2406-2409.
18. Barluenga, J.; Vázquez-Villa, H.; Ballesteros, A.; González, J. M. *J. Am. Chem. Soc.* **2003**, *125*, 9028-9029.
19. Barluenga, J.; Romanelli, G. P.; Alvarez-García, L. J.; Llorente, I.; González, J. M.; García-Rodríguez, E.; García-Granda, S. *Angew. Chem. Int. Ed.* **1998**, *37*, 3136-3139.
20. Huang, K.–T.; Winssinger, N. *Eur. J. Org. Chem.* **2007**, 1887-1890.
21. López, J. C.; Uriel, C.; Guillamón-Martín, A.; Valverde, S.; Gómez, A. M. *Org. Lett.* **2007**, *9*, 2759-2762.
22. Typical prices for IPy$_2$BF$_4$ are $40-60 per gram.
23. Barluenga, J.; Rodríguez, M. A.; Campos, P. J. *J. Org. Chem.* **1990**, *55*, 3104-3106.
24. While silver carbonate ($2-3 per gram) is more expensive than HgO ($0.2-0.3 per gram), this price difference is offset by the low toxicity of Ag(I) relative to Hg(II). Moreover, each gram of Ag$_2$CO$_3$ used in this protocol produces nearly two grams of IPy$_2$BF$_4$.

Appendix
Chemical Abstracts Nomenclature; (Registry Number)

Tetrafluoroboric acid: Borate(1-), tetrafluoro-, hydrogen (1:1); (16872-11-0)
Silver carbonate: Carbonic acid, silver(1+) salt (1:2); (534-16-7)
Pyridine: (110-86-1)
Iodine: (7553-56-2)
Bis(pyridine)iodonium(I) tetrafluoroborate: Iodine(1+), bis(pyridine)-, tetrafluoroborate(1-) (1:1); (15656-28-7)

Benjamin G. Davis completed his B.A. (1993) and D.Phil. (1996) at the University of Oxford. During this time he learned the beauty of carbohydrate chemistry under the supervision of Professor George Fleet. He is now a full Professor at the University of Oxford and Pembroke College. His group's research centers on chemical biology with an emphasis on carbohydrates and therapeutic proteins. In 2008, he was awarded the Wain Medal for Chemical Biology and was the first U.K. recipient of the American Chemical Society's Horace S. Isbell Award. Ben is also a 2009-2010 Novartis Chemistry Lecture Award recipient.

Justin M. Chalker was born in Kansas City, Kansas in 1983. He graduated from the University of Pittsburgh with a B.S. in Chemistry and a B.A. in the History and Philosophy of Science in 2006. Justin carried out undergraduate research in the laboratory of Prof. Theodore Cohen where he investigated the Zn-ene cyclization and its use in total synthesis. As a Rhodes Scholar, Justin is pursuing his Ph.D. at the University of Oxford under the direction of Prof. Benjamin G. Davis where he is exploring novel aqueous chemistry for selective protein modification.

Amber L. Thompson graduated from University of Durham with B.Sc. in chemistry in 2000. She went on to complete her Ph.D. on "Structure-Property Correlations in Novel Spin Crossover Materials" under the supervision of Dr. Andrés Goeta, also at Durham. After a short time working with Prof. Judith Howard, she moved to Oxford, where she runs the Chemical Crystallography Service, teaching chemists to solve crystal structures and working on the wide range of problems they encounter.

Nilesh Zaware obtained his B. Pharm. (2002) from the University of Mumbai, India. He then moved to Duquesne University to pursue his Ph.D. in Medicinal Chemistry under the supervision of Professor Aleem Gangjee, where his research was focused on the synthesis of pyrrolo[2,3-*d*]pyrimidines and pyrimido[4,5-*b*]indoles as inhibitors of multiple receptor tyrosine kinases, folate metabolizing enzymes and tubulin. After completion of his Ph.D. in 2009 he joined the group of Professor Peter Wipf at the Department of Chemistry and Center for Chemical Methodologies & Library Development at the University of Pittsburgh as a postdoctoral research associate and is currently working on the synthesis of structurally and stereochemically diverse tetrahydropyrans.

298

PALLADIUM-CATALYZED ALKYL-ALKYL SUZUKI CROSS-COUPLINGS OF PRIMARY ALKYL BROMIDES AT ROOM TEMPERATURE: (13-CHLOROTRIDECYLOXY)TRIETHYLSILANE

[Silane, [(13-chlorotridecyl)oxy]triethyl-]

Submitted by Sha Lou and Gregory C. Fu.[1]
Checked by Takuya Higo and Tohru Fukuyama.

1. Procedure

A. *1-Bromo-8-chlorooctane* (**1**). An oven-dried, 200-mL, two-necked, round-bottomed flask equipped with an argon inlet and a magnetic stirbar (octagonal, molded pivot ring, 25 mm length and 6 mm diameter) is purged with argon for 5 min and then charged through the open neck with CH_2Cl_2 (50 mL via syringe) (Note 1), imidazole (2.19 g, 32.1 mmol, 1.10 equiv) (Note 2), and dichlorotriphenylphosphorane (10.4 g, 31.2 mmol, 1.07 equiv) (Note 3). The open neck is capped with a rubber septum, and the stirred solution is cooled in an ice bath for 5 min. A solution of 8-bromo-1-octanol (5.0 mL, 6.11 g, 29.2 mmol, 1.00 equiv) (Note 4) in CH_2Cl_2 (10 mL) (Note 1) is added via syringe over 5 min. The reaction mixture is allowed to warm to rt, and the resulting heterogeneous solution (a white precipitate formed) is stirred for 4 h. The progress of the reaction is followed by TLC analysis on SiO_2 (10% EtOAc/hexanes as the eluent; visualization with a $KMnO_4$ stain; the alcohol starting material has an $R_f = 0.2$, and the chloride product has an $R_f = 0.9$) (Note 5). After the alcohol is consumed, the reaction is diluted with pentane (200 mL), and the mixture is filtered through a pad of SiO_2 (7 cm diameter × 6 cm height) in a sintered glass funnel. The SiO_2 is washed with additional pentane (400 mL). The filtrate is concentrated by rotary evaporation (20 mmHg, 30 °C), which furnishes the desired product as a colorless oil (6.23–6.44 g, 94–97 % yield) (Note 6). The product is used in the next step without further purification.

B. *Triethyl(pent-4-enyloxy)silane* (**2**). An oven-dried, 200-mL, two-necked, round-bottomed flask equipped with an argon inlet and a magnetic stirbar (octagonal, molded pivot ring, 25 mm length and 6 mm diameter) is purged with argon for 5 min and then charged through the open neck with *N,N*-dimethylformamide (50 mL via syringe) (Note 7), 4-penten-1-ol (8.93 mL via syringe, 7.50 g, 87.1 mmol, 1.00 equiv) (Note 8), and imidazole (5.93 g, 87.1 mmol, 1.00 equiv) (Note 2). The open neck is capped with a rubber septum. The stirred solution is cooled in an ice bath for 5 min, and then chlorotriethylsilane (14.6 mL, 13.1 g, 87.1 mmol, 1.00 equiv) (Note 9) is added over 4 min via syringe. The reaction mixture is stirred at rt for 24 h. The progress of the reaction is followed by TLC analysis on SiO_2 (20% EtOAc/hexanes as the eluent; visualization with a $KMnO_4$ stain; the alcohol starting material has an $R_f = 0.2$, and the silyl ether product has an $R_f = 0.7$) (Note 5). After the alcohol has been consumed, the reaction mixture is poured into a mixture of pentane (300 mL) and water (60 mL) in a 500-mL separatory funnel. The organic layer is separated and washed with brine (3 × 50 mL). The organic solution is dried over $MgSO_4$ (30 g) and then vacuum filtered through a Büchner funnel containing a bed of celite (1.0 cm height). The filtrate is concentrated by rotary evaporation (20 mmHg, 30 °C), and the residue is transferred to a 50-mL round-bottomed flask equipped with a magnetic stirbar (octagonal, molded pivot ring, 15 mm length and 7 mm diameter) and a short-path distillation head. The residue is distilled under vacuum (bp 77–79 °C at 8 mmHg), which provides the desired silyl ether **2** as a colorless oil (15.2–15.7 g, 87–90% yield) (Note 10).

C. *(5-(9-Borabicyclo[3.3.1]nonan-9-yl)pentyloxy)triethylsilane* (**3**). An oven-dried, 200-mL, round-bottomed flask equipped with an argon inlet and a magnetic stirbar (octagonal, molded pivot ring, 25 mm length and 6 mm

300

diameter) is purged with argon for 10 min. The open neck is capped with a rubber septum, and then a solution of 9-borabicyclo[3.3.1]nonane (9-BBN; 0.50 M in THF; 72 mL, 36 mmol, 1.0 equiv) (Note 11) is added via syringe. Next, triethyl(pent-4-enyloxy)silane (2) (7.21 g, 36.0 mmol, 1.0 equiv) is added via syringe over 3 min to the solution of 9-BBN. The reaction mixture is stirred for 3 h, at which time all of the starting olefin is consumed as determined by TLC analysis (pentane as the eluent; visualization with a KMnO$_4$ stain; the olefin starting material has an R$_f$ = 0.2) (Note 5). This solution is used directly in the next step.

D. *(13-Chlorotridecyloxy)triethylsilane* (4). An oven-dried, 1000-mL, three-necked, round-bottomed flask equipped with a thermometer inlet, a thermometer, a magnetic stir bar (octagonal, molded pivot ring, 40 mm length and 10 mm diameter), and an argon inlet is purged with argon for 10 min. Palladium(II) acetate (270 mg, 1.20 mmol, 0.040 equiv) (Note 12), tricyclohexylphosphine (673 mg, 2.40 mmol, 0.080 equiv) (Note 13), and tripotassium phosphate, monohydrate (K$_3$PO$_4$·H$_2$O; 8.28 g, 36.0 mmol, 1.2 equiv) (Note 14) are added through the open neck of the flask. Then, the open neck is capped with a rubber septum, and the solution of (5-(9-borabicyclo[3.3.1]nonan-9-yl)-pentyloxy)triethylsilane (3) prepared in Step C (36 mmol, 1.2 equiv) is added to the flask via syringe, followed by the addition of 1-bromo-8-chlorooctane (1) (6.83 g, 30.0 mmol, 1.0 equiv). The resulting dark-brown heterogeneous reaction mixture is stirred vigorously at rt for 24 h. The progress of the reaction is followed by TLC analysis on SiO$_2$ (25% CH$_2$Cl$_2$/hexanes as the eluent; visualization with a KMnO$_4$ stain; the alkyl bromide starting material has an R$_f$ = 0.5, and the cross-coupling product has an R$_f$ = 0.4) (Note 5). Next, the mixture is diluted with diethyl ether (200 mL) and filtered through a sintered glass funnel containing SiO$_2$ (7.0 cm diameter × 5.0 cm height). The SiO$_2$ is washed with additional diethyl ether (200 mL), and the combined filtrate is concentrated by rotary evaporation (20 mmHg, 30 °C). The residue is purified by column chromatography on SiO$_2$ (Note 15). The desired cross-coupling product 4 has R$_f$ = 0.7 (TLC analysis on SiO$_2$: 50% CH$_2$Cl$_2$/hexanes as eluent, visualization with KMnO$_4$) (Note 5). The cross-coupling product is obtained as a pale-yellow oil (9.63–10.10 g, 92–96% yield) (Note 16).

2. Notes

1.   Dichloromethane (>99.5%) was purchased from Kanto Chemical Co., Inc. (water content <0.001%) and purified by Glass Contour solvent systems. The submitters purchased dichloromethane (>99.8%) from J.T. Baker (water content <0.02%), which was purified by passage through activated alumina under argon.

2.   Imidazole (99%) was purchased from Alfa Aesar and used as received.

3.   Dichlorotriphenylphosphorane (95%) was purchased from Aldrich and used as received.

4.   8-Bromo-1-octanol (95%) was purchased from Alfa Aesar and used as received.

5.   Analytical thin-layer chromatography was performed using Merck silica gel 60 F254 plates (0.25 mm).

6.   Compound **1** has the following properties: IR (film): 2931, 2855, 1457, 1291, 1245, 913 cm^{-1}; ^{1}H NMR (CDCl$_3$, 400 MHz) δ: 1.31–1.35 (m, 4 H), 1.40–1.47 (m, 4 H), 1.75–1.81 (m, 2 H), 1.82–1.89 (m, 2 H), 3.41 (t, J = 7.0 Hz, 2 H), 3.54 (t, J = 7.0 Hz, 2 H); ^{13}C NMR (CDCl$_3$, 100 MHz) δ: 26.7, 28.0, 28.6, 28.7, 32.5, 32.7, 34.0, 45.1; Anal. Calcd. for C$_8$H$_{17}$BrCl: C, 42.22; H, 7.09; N, 0; found: C, 41.94; H, 6.85; N, 0. The spectral data are in agreement with the reported values.[2] Compound **1** can be purified via column chromatography on SiO$_2$, eluting with pentane (R$_f$ = 0.6, pentane; visualization with KMnO$_4$).

7.   *N,N*-Dimethylformamide (>99.5) was purchased from Wako Pure Chemical Industries, Ltd. and used as received.   The submitters purchased *N,N*-dimethylformamide (ACS reagent grade) from MP Biomedicals, LLC and used it as received.

8.   4-Penten-1-ol (98+%) was purchased from Alfa Aesar and used as received.

9.   Chlorotriethylsilane (98+%) was purchased from Alfa Aesar and used as received.

10.   Compound **2** has the following properties: IR (film): 2947, 2884, 2826, 1653, 1457, 1418, 1015, 743 cm^{-1}; ^{1}H NMR (CDCl$_3$, 400 MHz) δ: 0.60 (q, J = 8.0 Hz, 6 H), 0.96 (t, J = 8.0 Hz, 9 H), 1.60–1.66 (m, 2 H), 2.11 (q, J = 6.8 Hz, 2 H), 3.62 (t, J = 6.4 Hz, 2 H), 4.94–5.05 (m, 2 H), 5.79–5.86 (m, 1 H); ^{13}C NMR (CDCl$_3$, 100 MHz) δ: 4.4, 6.8, 30.1, 32.0, 62.3, 114.5, 138.5; LRMS (DART) *m/z* calcd. for C$_{11}$H$_{25}$OSi ([M+H]$^+$) 201.2; found

201.2. The spectral data are in agreement with the reported values.[2]

11. The solution of 9-BBN (0.50 M in THF) was purchased from Aldrich and used as received.

12. Pd(OAc)₂ (99+%) was purchased from Strem and used as received.

13. PCy₃ (97%) was purchased from Strem and used as received.

14. Tripotassium phosphate, monohydrate (≥94%) was purchased from Fluka and used as received. When anhydrous tripotassium phosphate is employed, essentially no cross-coupling is observed.

15. Column chromatography was performed on Kanto Chemical Silica Gel 60N (wet packed in hexanes; 7 cm diameter × 13 cm height; 200 g), eluting with a gradient of CH_2Cl_2 in hexanes (500 mL of hexanes, 500 mL of 5% CH_2Cl_2/hexanes, 1.0 L of 10% CH_2Cl_2/hexanes, 1.0 L of 15% CH_2Cl_2/hexanes; 100-mL fractions). All of the fractions (8-26) containing the desired product were combined and concentrated by rotary evaporation (20 mmHg, 30 °C).

16. Compound **4** has the following properties: IR (film): 2926, 2875, 2854, 1459, 1414, 1384, 1238, 1098, 1007, 913, 735 cm⁻¹; ¹H NMR (CDCl₃, 400 MHz) δ: 0.60 (q, J = 8.0 Hz, 6 H), 0.96 (t, J = 8.0 Hz, 9 H), 1.26–1.28 (m, 16 H), 1.39–1.45 (m, 2 H), 1.50–1.56 (m, 2 H), 1.73–1.80 (m, 2 H), 3.53 (t, J = 6.8 Hz, 2 H), 3.59 (t, J = 6.4 Hz, 2 H); ¹³C NMR (CDCl₃, 100 MHz) δ: 4.4, 6.8, 25.8, 26.9, 28.9, 29.47, 29.54, 29.59, 29.60, 29.62, 32.6, 32.9, 45.2, 63.0; LRMS (DART) m/z calcd. for $C_{19}H_{42}ClOSi$ ([M+H]⁺) 349.3, found 349.3; Anal. calcd. for $C_{11}H_{42}OSi$: C, 65.93; H, 12.07; N, 0; found: C, 65.99; H, 11.80; N, 0. The spectral data are in agreement with the reported values.[2]

Safety and Waste Disposal Information

All hazardous materials should be handled and disposed of in accordance with "Prudent Practices in the Laboratory"; National Academy Press; Washington, DC, 1995.

3. Discussion

The palladium-catalyzed coupling of organometallic compounds with aryl and vinyl halides is a now-classic method for carbon–carbon bond formation.[3] In contrast, until recently, the corresponding reactions of *alkyl* halides were relatively uncommon.[3] Slow oxidative addition and facile β-hydride elimination have been suggested as two of the possible culprits for this comparative lack of success (Scheme 1). With respect to the Suzuki reaction, prior to 2001 only one somewhat general method had been described for achieving cross-couplings of unactivated, β-hydrogen containing alkyl electrophiles, specifically, a Pd(PPh₃)₄-catalyzed process for coupling alkyl iodides.[5]

Scheme 1. A Generalized Mechanism for Palladium-Catalyzed Cross-Coupling of an Alkyl Electrophile.

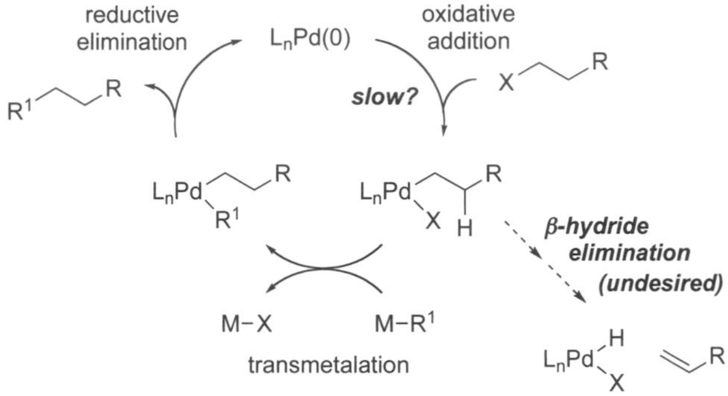

We have determined that, through the appropriate choice of ligand, palladium-catalyzed Suzuki cross-couplings can be accomplished with an array of unactivated alkyl bromides, chlorides, and tosylates that bear β hydrogens.[2,6,7] Specifically, bulky electron-rich trialkylphosphines furnish particularly active catalysts. For example, in the case of alkyl bromides, Pd(OAc)₂/PCy₃ achieves the desired coupling under mild conditions (room temperature; Table 1). The process is compatible with a broad spectrum of functional groups, including amines, alkenes, esters, alkynes, ethers, and nitriles. Furthermore, an alkyl bromide can be cross-coupled selectively in the presence of an alkyl chloride (equation D). Not only alkylboranes (entries 1-6), but also vinylboranes (entry 7), serve as suitable coupling partners.

Org. Syn. **2010**, *87*, 299-309

Table 1. Pd/PCy$_3$-Catalyzed Suzuki Cross-Couplings of Unactivated Alkyl Bromides at Room Temperature.

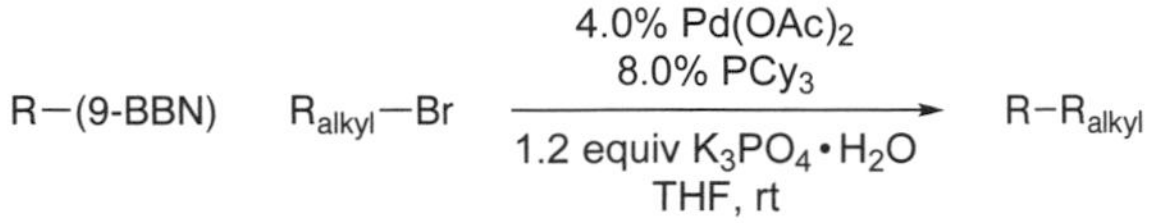

$$R-(9\text{-}BBN) \quad R_{alkyl}-Br \quad \xrightarrow[\substack{1.2 \text{ equiv } K_3PO_4 \cdot H_2O \\ THF, \text{ rt}}]{\substack{4.0\% \text{ Pd(OAc)}_2 \\ 8.0\% \text{ PCy}_3}} \quad R-R_{alkyl}$$

entry	R—(9-BBN)	R$_{alkyl}$—Br	yield (%)
1	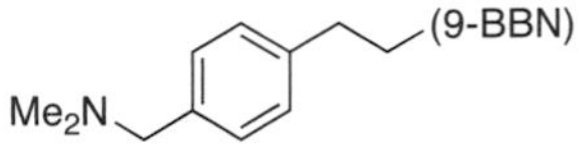 Me$_2$N—⟨benzyl⟩—CH$_2$CH$_2$(9-BBN)	n-Dodec—Br	78
2	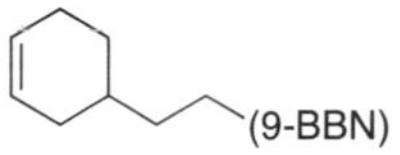 ⟨cyclohexenyl⟩—CH$_2$CH$_2$(9-BBN)	n-Dodec—Br	85
3	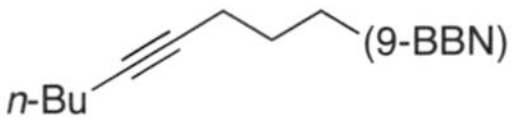 n-Bu—C≡C—CH$_2$CH$_2$CH$_2$(9-BBN)	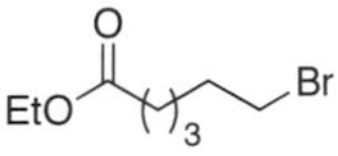 EtO(C=O)(CH$_2$)$_3$—Br	58
4	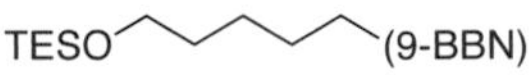 TESO—(CH$_2$)$_5$(9-BBN)	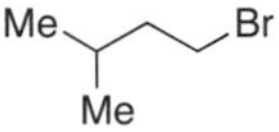 Me$_2$CH—CH$_2$CH$_2$—Br	72
5	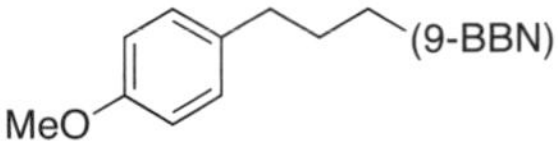 MeO—⟨benzyl⟩—CH$_2$CH$_2$CH$_2$(9-BBN)	n-Hex—Br	80
6	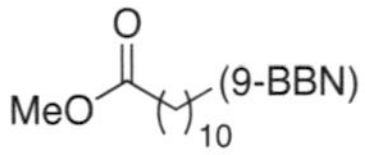 MeO(C=O)(CH$_2$)$_{10}$(9-BBN)	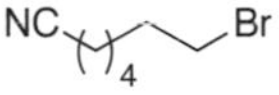 NC—(CH$_2$)$_4$—Br	81
7	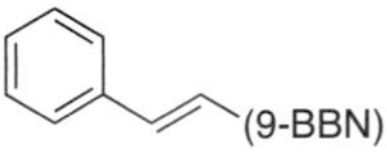 ⟨Ph⟩—CH=CH—CH$_2$(9-BBN)	n-Dodec—Br	66

This method for Suzuki coupling of alkyl bromides has been employed by others, e.g., by Phillips to achieve late-stage fragment couplings in natural-product total synthesis (Scheme 2).[8]

Scheme 2. Applications in Total Synthesis of Pd/PCy$_3$-Catalyzed Alkyl–Alkyl Suzuki Reactions: Fragment Couplings.

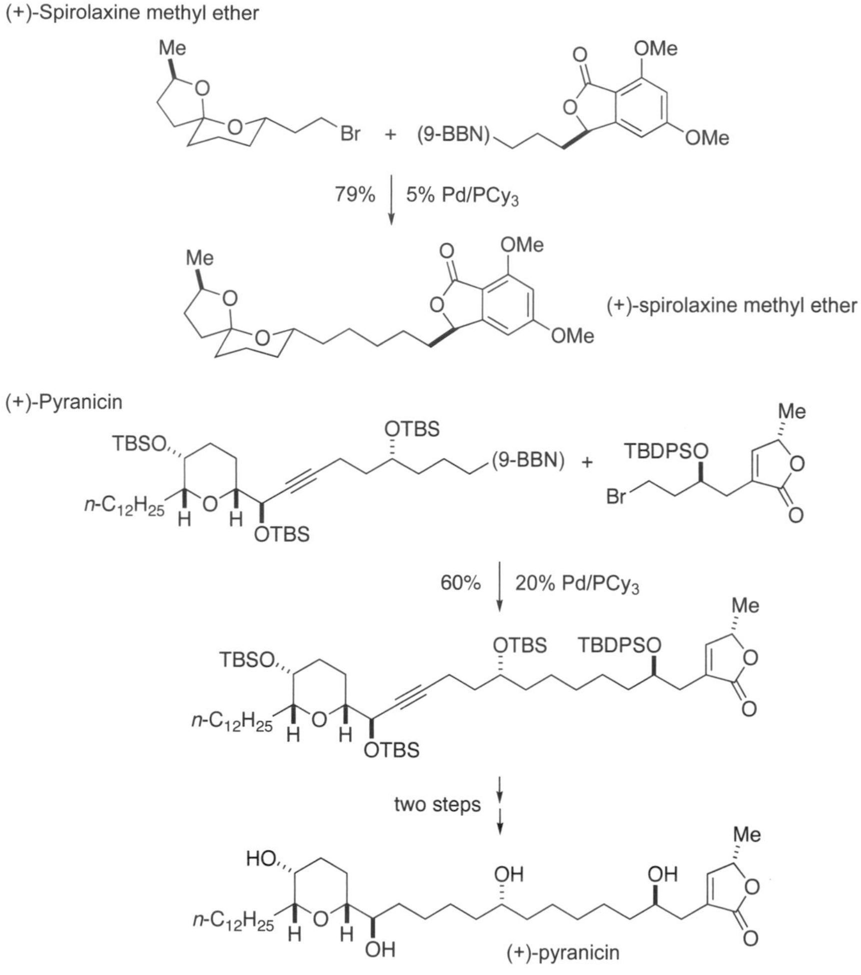

(+)-Spirolaxine methyl ether

(+)-spirolaxine methyl ether

(+)-Pyranicin

(+)-pyranicin

1. Department of Chemistry, Room 18-290, Massachusetts Institute of Technology, Cambridge, Massachusetts 02139; E-mail: gcf@mit.edu.
2. Netherton, M. R.; Dai, C.; Neuschütz, K.; Fu, G. C. *J. Am. Chem. Soc.* **2001**, *123*, 10099–10100.
3. For some leading references, see: (a) *Metal-Catalyzed Cross-Coupling*

306

Org. Syn. **2010**, *87*, 299-309

Reactions; de Meijere, A., Diederich, F., Eds.; Wiley–VCH: New York, 2004. (b) *Cross-Coupling Reactions: A Practical Guide*; Miyaura, N., Ed.; *Topics in Current Chemistry* Series 219; Springer-Verlag: New York, 2002. (c) *Handbook of Organopalladium Chemistry for Organic Synthesis*; Negishi, E.-i., Ed.; Wiley Interscience: New York, 2002.

4. For overviews, see: (a) Frisch, A. C.; Beller, M. *Angew. Chem., Int. Ed.* **2005**, *44*, 674–688. (b) Netherton, M. R.; Fu, G. C. *Topics in Organometallic Chemistry: Palladium in Organic Synthesis*; Tsuji, J., Ed.; Springer: New York, 2005; pp 85–108.

5. Ishiyama, T.; Abe, S.; Miyaura, N.; Suzuki, A. *Chem. Lett.* **1992**, 691–694.

6. (a) Alkyl chlorides and trialkylboranes: Kirchhoff, J. H.; Dai, C.; Fu, G. C. *Angew. Chem., Int. Ed.* **2002**, *41*, 1945–1947. (b) Alkyl tosylates and trialkylboranes: Netherton, M. R.; Fu, G. C. *Angew. Chem., Int. Ed.* **2002**, *41*, 3910–3912. (c) Alkyl bromides and boronic acids: Kirchhoff, J. H.; Netherton, M. R.; Hills, I. D.; Fu, G. C. *J. Am. Chem. Soc.* **2002**, *124*, 13662–13663.

7. For mechanistic studies, see: (a) References 2 and 6. (b) Hills, I. D.; Netherton, M. R.; Fu, G. C. *Angew. Chem., Int. Ed.* **2003**, *42*, 5749–5752.

8. (a) Keaton, K. A.; Phillips, A. J. *Org. Lett.* **2007**, *9*, 2717–2719. (b) Griggs, N. D.; Phillips, A. J. *Org. Lett.* **2008**, *10*, 4955–4957.

Appendix
Chemical Abstracts Nomenclature; (Registry Number)

8-Bromo-1-octanol: 1-Octanol, 8-bromo-; (50816-19-8)

Dichlorotriphenylphosphorane: Phosphorane, dichlorotriphenyl-; (2526-64-9)

Imidazole: 1H-Imidazole; (288-32-4)

1-Bromo-8-chlorooctane: Octane, 1-bromo-8-chloro-; (28598-82-5)

4-Penten-1-ol; (821-09-0)

Chlorotriethylsilane: Silane, chlorotriethyl-; (994-30-9)

Triethyl(pent-4-enyloxy)silane: Silane, triethyl(4-penten-1-yloxy)-; (374755-00-7)

9-BBN: 9-Borabicyclo[3.3.1]nonane; (280-64-8)

(5-(9-Borabicyclo[3.3.1]nonan-9-yl)pentyloxy)triethylsilane:
9-Borabicyclo[3.3.1]nonane, 9-[5-[(triethylsilyl)oxy]pentyl]-;
(157123-09-6)
Palladium(II) acetate: Acetic acid, palladium(2+) salt (2:1); (3375-31-3)
Tricyclohexylphosphine: Phosphine, tricyclohexyl-; (2622-14-2)
Tripotassium phosphate, monohydrate: Phosphoric acid, tripotassium salt,
monohydrate (8CI,9CI); (27176-10-9)
(13-Chlorotridecyloxy)triethylsilane: Silane,
[(13-chlorotridecyl)oxy]triethyl-; (374754-99-1)

Prof. Greg Fu was born in Galion, Ohio, in 1963. He received a B.S. degree in 1985 from MIT, where he worked in the laboratory of Prof. K. Barry Sharpless. After earning a Ph.D. from Harvard in 1991 under the guidance of Prof. David A. Evans, he spent two years as a postdoctoral fellow with Prof. Robert H. Grubbs at Caltech. In 1993, he returned to MIT, where he is currently the Firmenich Professor of Chemistry. Prof. Fu received the Springer Award in Organometallic Chemistry in 2001, the Corey Award of the American Chemical Society in 2004, and the Mukaiyama Award of the Society of Synthetic Organic Chemistry of Japan in 2006. He is a fellow of the Royal Society of Chemistry and of the American Academy of Arts and Sciences. Prof. Fu serves as an associate editor of the *Journal of the American Chemical Society*. His current research interests include metal-catalyzed coupling reactions, chiral-ligand design, and enantioselective nucleophilic catalysis.

Sha Lou was born in He Bei, China, in 1979. He received a B.S. in Chemistry from Beijing University of Chemical Technology in 2002, where he conducted undergraduate research on fullerene functionalizations with Professor Shenyi Yu. He obtained a Ph.D. degree in January 2008 from Boston University under the direction of Professor Scott E. Schaus. His graduate research focused on transition metal- and organic molecule-catalyzed asymmetric carbon–carbon bond-forming reactions and synthesis. In 2008, he joined the group of Professor Greg Fu at MIT as a postdoctoral fellow. His current research involves the development of transition metal-catalyzed enantioselective cross-coupling reactions.

Takuya Higo was born in Saga, Japan in 1986. He received his B.S. in 2009 from the University of Tokyo. In the same year, he began his graduate studies at the Graduate School of Pharmaceutical Sciences, the University of Tokyo, under the guidance of Professor Tohru Fukuyama. His research interests are in the area of the total synthesis of natural products.

SYNTHESIS OF CHIRAL PYRIDINE BIS(OXAZOLINE) LIGANDS FOR NICKEL-CATALYZED ASYMMETRIC NEGISHI CROSS-COUPLINGS OF SECONDARY ALLYLIC CHLORIDES WITH ALKYLZINCS: 2,6-BIS[(4R)-4,5-DIHYDRO-4-(2-PHENYLETHYL)-2-OXAZOLYL]-PYRIDINE

[Pyridine, 2,6-bis[(4R)-4,5-dihydro-4-(2-phenylethyl)-2-oxazolyl]-]

Submitted by Sha Lou and Gregory C. Fu.[1]
Checked by Gavin Chit Tsui, David I. Chai, and Mark Lautens.

1. Procedure

A. *(R)-2-Amino-4-phenylbutan-1-ol* (**1**). An oven-dried, 500-mL, two-necked, round-bottomed flask equipped with a condenser fitted with an argon inlet and a magnetic stir bar (31.8 x 15.9 mm, egg-shaped) is purged with argon for 10 min. Anhydrous THF (200 mL) (Note 1) is added into the flask by syringe through the open neck under a positive pressure of argon. Lithium aluminum hydride (4.56 g, 120 mmol, 1.50 equiv) (Note 2) is added over 5 min through the open neck under a positive pressure of argon. The solution is stirred at rt for 5 min, and then D-homophenylalanine ethyl ester hydrochloride (19.5 g, 80.0 mmol, 1.00 equiv) (Note 3) is added in portions (500 mg per portion) over 30 min through the open neck under a positive pressure of argon, during which time the solution gently refluxes. Upon the completion of the addition, the open neck is capped with a glass stopper, and the heterogeneous gray reaction mixture is heated at reflux in an oil bath for 24 h. Next, the mixture is allowed to cool to rt, THF (100 mL) is added by

310

syringe, and the solution is cooled to 0 °C in an ice bath. Water (10 mL) is added dropwise by syringe through the condenser over 30 min (Note 4), and then a solution of NaOH (2.5 M, 20 mL) is added in one portion. The resulting mixture is heated at reflux for 20 min until the color of the precipitate changes from gray to white. The warm solution (approx 60–70 °C) is filtered through a Büchner funnel that contains a bed of celite (1.0 cm height), and the precipitate is washed with warm THF (100 mL). The filtrate is concentrated under reduced pressure (20 mmHg) to remove the THF and most of the water. The residue is dried azeotropically with toluene (two 50-mL portions) by rotary evaporation (20 mmHg) and then under vacuum (10 mmHg) for 6 h, which yields the desired β-amino alcohol 1 as an off-white solid (12.7–13.1 g, 96–99% yield) (Note 5) that is used in the next step without further purification.

B. *2,6-Bis[(4R)-4,5-dihydro-4-(2-phenylethyl)-2-oxazolyl]pyridine* (**2**). (Note 6) An oven-dried, 250-mL, two-necked, round-bottomed flask equipped with a condenser fitted with an argon inlet and a magnetic stirbar (31.8 x 15.9 mm, egg-shaped) is purged with argon for 10 min. 2,6-Pyridinedicarbonitrile (2.58 g, 19.4 mmol, 1.00 equiv) (Note 7), anhydrous toluene (140 mL; added by syringe) (Note 8), and zinc trifluoromethanesulfonate (368 mg, 0.99 mmol, 0.051 equiv) (Note 9) are added through the open neck under a positive pressure of argon, the open neck is capped with a glass stopper, and the mixture is stirred at rt for 5 min. (*R*)-2-Amino-4-phenylbutan-1-ol (**1**) (7.64 g, 43.5 mmol, 2.24 equiv) is added through the open neck under a positive pressure of argon, the open neck is capped with a glass stopper, and the reaction mixture is heated at reflux in an oil bath for 24 h. Next, the reaction mixture is allowed to cool to rt and diluted with ethyl acetate (300 mL). The solution is washed with a saturated aqueous solution of NaHCO$_3$ (200 mL) and brine (200 mL), and then it is dried over anhydrous MgSO$_4$ (20 g) and filtered through a Büchner funnel that contains a bed of celite (1.0 cm height). The filtrate is concentrated under reduced pressure on a rotary evaporator (20 mmHg), and the residue is purified by column chromatography on SiO$_2$ (5 cm diameter × 20 cm height), eluting with ethyl acetate:hexanes:Et$_3$N (1:1:0.02) (Note 10), which affords the desired pyridine derivative **2** (6.18 g, 75% yield, >99% ee) as a white solid (Notes 11 and 12).

2. Notes

1.   THF was distilled from Na/benzophenone ketyl. Submitters used THF (99+%) that was purchased from J.T. Baker (water content: 24 ppm) and purified by passage through activated alumina under argon.

2.   Lithium aluminum hydride (97%) was purchased from Alfa Aesar and used as received.

3.   D-Homophenylalanine ethyl ester hydrochloride (≥98%) was purchased from Fluka and used as received.

4.   Quenching excess lithium aluminum hydride with water is a highly exothermic process that produces H_2. Dropwise addition of water is recommended, and care should be taken to efficiently cool and stir the reaction mixture. The quenching should be conducted in an efficient fume hood in order to safely vent the H_2 gas.

5.   The purity of compound **1** (94%) was determined by HPLC analysis (t_r = 1.65 min) with an Agilent 1100 Series HPLC system equipped with an Eclipse XDB-C18 column (length 150 mm, I.D. 4.6 mm, particle size 5 μm), using 98% MeOH/(0.2% AcOH in water) for 6 min, with a flow rate of 0.8 mL/min. Compound **1** can be purified via column chromatography on SiO_2, eluting with MeOH:CH_2Cl_2:Et_3N (5:95:2): R_f = 0.4 (MeOH:CH_2Cl_2; 1:9). Compound **1** has the following physical properties: Mp 38-40 °C; $[\alpha]^{27}_D$ −1.2 (c 1.0, $CHCl_3$); IR (film) 3349, 2924, 1601, 1497, 1454, 1367, 1057, 748, 698 cm^{-1}; ^{1}H NMR ($CDCl_3$, 400 MHz) δ: 1.54-1.60 (m, 1 H), 1.71-1.77 (m, 1 H), 1.92 (br s, 3 H), 2.61-2.66 (m, 1 H), 2.70-2.76 (m, 1 H), 2.82-2.86 (m, 1 H), 3.31 (dd, J = 10.5, 7.5 Hz, 1 H), 3.58 (dd, J = 10.5, 3.5 Hz, 1 H), 7.16-7.18 (m, 3 H), 7.26-7.30 (m, 2 H); ^{13}C NMR ($CDCl_3$, 100 MHz) δ: 32.4, 36.0, 52.3, 66.5, 125.9, 128.2, 128.4, 141.7; HRMS (ESI) m/z calcd. for $C_{10}H_{16}NO$ ($[M+H]^+$) 166.1226; found 166.1233.

6.   2,6-Bis[(4S)-4,5-dihydro-4-(2-phenylethyl)-2-oxazolyl]pyridine (97%) is available from Aldrich.

7.   2,6-Pyridinedicarbonitrile (97%) was purchased from Aldrich and used as received.

8.   Toluene was distilled from sodium. Submitters used toluene (99+%) that was purchased from J.T. Baker (water content: <0.01%) and purified by passage through activated alumina under argon.

9.   Zinc trifluoromethanesulfonate (98%) was purchased from Strem and used as received.

10.   Column chromatography was performed on Silicycle 60 Å silica

312

gel. Submitters used Sorbent Technologies 60 Å silica gel.

11. On half-scale runs using 1.29 g of the starting 2,6-pyridinedicarbonitrile (97%), 2.83–2.88 g of the product was obtained (69–70% yield). On full-scale runs, submitters obtained 6.89–7.06 g of the product (81–83% yield). By conducting the reaction with anhydrous zinc chloride as the catalyst (10%) and chlorobenzene as the solvent at 120 °C for 24 h, the pybox ligand can be generated in comparable yield.[2]

12. Compound **2** has the following properties: R_f = 0.13 (ethyl acetate:hexanes:Et$_3$N, 1:1:0.02); Mp 87–89 °C; $[\alpha]^{26}_D$ +163 (c 1.15, CHCl$_3$); IR (film) 3437, 2924, 1643, 1574, 1497, 1454, 1358, 1246, 1161, 1107, 1072, 976, 910, 748, 702 cm^{-1}; ^{1}H NMR (CDCl$_3$, 400 MHz) δ: 1.90-1.96 (m, 2 H), 2.05-2.11 (m, 2 H), 2.73-2.89 (m, 4 H), 4.16 (t, J = 8.5 Hz, 2 H), 4.33-4.39 (m, 2 H), 4.59 (t, J = 8.5 Hz, 2 H), 7.20-7.31 (m, 10 H), 7.89 (t, J = 7.9 Hz, 1 H), 8.21 (d, J = 7.9 Hz, 2 H); ^{13}C NMR (CDCl$_3$, 100 MHz) δ: 32.3, 37.5, 66.4, 73.1, 125.7, 125.9, 128.4 (2), 137.2, 141.4, 146.8, 162.3; HRMS (EI) m/z calcd. for C$_{27}$H$_{27}$N$_3$O$_2$ ([M]$^+$) 425.2103; found 425.2111. The enantiomeric excess of **2** was determined on an Agilent 1100 Series HPLC system: Chiralcel OD-H column (length 250 mm, I.D. 4.6 mm); solvent system: 20% ^{i}PrOH, 1.5 mL/min; retention time: 53.9 min.

Safety and Waste Disposal Information

All hazardous materials should be handled and disposed of in accordance with "Prudent Practices in the Laboratory"; National Academy Press; Washington, DC, 1995.

3. Discussion

Chiral pyridine-2,6-bisoxazolines (pybox) form tridentate complexes with a variety of metals, and they have proved to be highly effective ligands in asymmetric catalysis.[3] We have reported several examples of enantioselective Negishi cross-couplings of secondary alkyl halides with organozinc reagents that are catalyzed by Ni/pybox complexes.[4,5] When α-bromo amides[4a] and benzylic bromides[4b] are employed as electrophiles, commercially available i-Pr-Pybox is the ligand of choice. However, for Negishi couplings of allylic chlorides, CH$_2$CH$_2$Ph-pybox is the optimal ligand from the standpoints of enantioselectivity and yield.[4c] Herein, we describe a procedure for the synthesis of CH$_2$CH$_2$Ph-pybox.

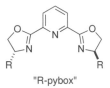

"R-pybox"

The procedure is based on a general route developed by Pires and coworkers.[6] (R)- and (S)-2-Amino-4-phenylbutan-1-ol are readily available by reduction of homophenylalanine or an ester derivative.

1. Department of Chemistry, Room 18-290, Massachusetts Institute of Technology, Cambridge, Massachusetts 02139; E-mail: gcf@mit.edu.
2. Witte, H; Seeliger, W. *Angew. Chem., Int. Ed.* **1972**, 11, 287–288. See also: Chelucci, G.; Deriu, S.; Pinna, G. A.; Saba, A.; Valenti, R. *Tetrahedron: Asymmetry* **1999**, *10*, 3803–3809.
3. For leading references to the use of oxazolines as ligands in asymmetric catalysis, see: (a) Hargaden, G. C.; Guiry, P. J. *Chem. Rev.* **2009**, *109*, 2505–2550. (b) McManus, H. A.; Guiry, P. J. *Chem. Rev.* **2004**, *104*, 4151–4202. (c) Desimoni, G.; Faita, G.; Quadrelli, P. *Chem. Rev.* **2003**, *108*, 3119–3154.
4. (a) Fischer, C.; Fu, G. C. *J. Am. Chem. Soc.* **2005**, *127*, 4594–4595. (b) Arp, F. O.; Fu, G. C. *J. Am. Chem. Soc.* **2005**, *127*, 10482–10483. (c) Son, S.; Fu, G. C. *J. Am. Chem. Soc.* **2008**, *130*, 2756–2757. (d) Smith, S. W.; Fu, G. C. *J. Am. Chem. Soc.* **2008**, *130*, 12645–12647. (e) Lundin, P. M.; Esquivias, J.; Fu, G. C. *Angew. Chem., Int. Ed.* **2009**, *48*, 154–156.
5. For an overview, see: Rudolph, A.; Lautens, M. *Angew. Chem., Int. Ed.* **2009**, *48*, 2656–2670.
6. Cornejo, A.; Fraile, J. M.; García, J. I.; Gil, M. J.; Martínez-Merino, V.; Mayoral, J. A.; Pires, E.; Villalba, I. *Synlett* **2005**, 2321–2324.

Appendix
Chemical Abstracts Nomenclature; (Registry Number)

L-Homophenylalanine: Benzenebutanoic acid, α-amino-, (αS)-; (943-73-7)

D-Homophenylalanine ethyl ester hydrochloride: Benzenebutanoic acid, α-amino-, ethyl ester, hydrochloride (1:1), (αR)-; (90940-54-8)

Lithium aluminum hydride: Aluminate(1-), tetrahydro-, lithium (1:1), (T-4)-; (16853-85-3)

(S)-2-Amino-4-phenylbutan-1-ol: Benzenebutanol, β-amino-, (βS)-; (27038-09-1)

(R)-2-Amino-4-phenylbutan-1-ol: Benzenebutanol, β-amino-, (βR)-; (761373-40-4)

2,6-Pyridinedicarbonitrile; (2893-33-6)

Zinc trifluoromethanesulfonate: Methanesulfonic acid, 1,1,1-trifluoro-, zinc salt (2:1); (54010-75-2)

2,6-Bis[(4R)-4,5-dihydro-4-(2-phenylethyl)-2-oxazolyl]pyridine: Pyridine, 2,6-bis[(4R)-4,5-dihydro-4-(2-phenylethyl)-2-oxazolyl]-

Prof. Greg Fu was born in Galion, Ohio, in 1963. He received a B.S. degree in 1985 from MIT, where he worked in the laboratory of Prof. K. Barry Sharpless. After earning a Ph.D. from Harvard in 1991 under the guidance of Prof. David A. Evans, he spent two years as a postdoctoral fellow with Prof. Robert H. Grubbs at Caltech. In 1993, he returned to MIT, where he is currently the Firmenich Professor of Chemistry. Prof. Fu received the Springer Award in Organometallic Chemistry in 2001, the Corey Award of the American Chemical Society in 2004, and the Mukaiyama Award of the Society of Synthetic Organic Chemistry of Japan in 2006. He is a fellow of the Royal Society of Chemistry and of the American Academy of Arts and Sciences. Prof. Fu serves as an associate editor of the *Journal of the American Chemical Society*. His current research interests include metal-catalyzed coupling reactions, chiral-ligand design, and enantioselective nucleophilic catalysis.

Sha Lou was born in He Bei, China, in 1979. He received a B.S. in Chemistry from Beijing University of Chemical Technology in 2002, where he conducted undergraduate research on fullerene functionalizations with Professor Shenyi Yu. He obtained a Ph.D. degree in January 2008 from Boston University under the direction of Professor Scott E. Schaus. His graduate research focused on transition metal- and organic molecule-catalyzed asymmetric carbon–carbon bond-forming reactions and synthesis. In 2008, he joined the group of Professor Greg Fu at MIT as a postdoctoral fellow. His current research involves the development of transition metal-catalyzed enantioselective cross-coupling reactions.

Gavin Chit Tsui was born in China and raised in Hong Kong. He obtained his Hon. B. Sc. (2004) and M. Sc. (2006, supervisor: Prof. William Tam) degrees from the University of Guelph, Ontario, Canada. He then joined pharmaceutical companies (NPS Pharmaceuticals, Inc. and Merck Frosst Canada) as a research chemist in the medicinal chemistry division. Currently he is pursuing a Ph. D. degree at the University of Toronto under the supervision of Prof. Mark Lautens to investigate novel rhodium-catalyzed reactions.

316

NICKEL-CATALYZED ASYMMETRIC NEGISHI CROSS-COUPLINGS OF RACEMIC SECONDARY ALLYLIC CHLORIDES WITH ALKYLZINCS: (*S,E*)-ETHYL 6-(1,3-DIOXOLAN-2-YL)-4-METHYLHEX-2-ENOATE

[2-Hexenoic acid, 6-(1,3-dioxolan-2-yl)-4-methyl-, ethyl ester, (2*E*,4*S*)-]

Submitted by Sha Lou and Gregory C. Fu.[1]

Checked by David I. Chai, Gavin Chit Tsui, and Mark Lautens.

1. Procedure

A. *Ethyl (E)-4-hydroxy-2-pentenoate* (**1**). An oven-dried, 1000-mL, two-necked, round-bottomed flask equipped with a rubber septum, an argon inlet, and a magnetic stir bar [41.3 x 19.0 mm, egg-shaped] is purged with argon and charged with Et$_2$O (500 mL) (Note 1) and ethyl

(*E*)-4-oxo-2-butenoate (24.1 mL, 25.6 g, 200 mmol, 1.0 equiv) (Note 2) by syringe. The solution is cooled to –78 °C in a dry ice/acetone bath, and then MeMgBr (3.0 M in Et$_2$O; 70.0 mL, 210 mmol, 1.05 equiv) (Note 3) is added dropwise via syringe pump (2.0 mL/min) over 35 min (the slow addition can also be conducted via a 100-mL pressure-equalizing addition funnel fitted with a septum). After the addition is complete, the reaction mixture is stirred for an additional 1 h at –78 °C. Additional MeMgBr (3.0 M in Et$_2$O; 60.0 mL, 180 mmol, 0.90 equiv) (Note 3) is added dropwise via syringe pump (2.0 mL/min) over 35 min, and the reaction mixture is stirred for an additional 1 h at –78 °C. During the course of the reaction, the precipitation of white solids leads to a heterogeneous reaction mixture. The progress of the reaction can be followed by TLC analysis on SiO$_2$ (60% Et$_2$O/pentane as the eluent; visualization with a UV lamp or with a KMnO$_4$ stain; the aldehyde starting material has an R$_f$ = 0.8 and the alcohol product has an R$_f$ = 0.4) (Note 4). Then, the dry ice/acetone bath is removed, and the reaction mixture is carefully poured over 5 min into a rapidly stirred, 0 °C saturated aqueous solution of NH$_4$Cl (300 mL) in a 2-L Erlenmeyer flask equipped with a magnetic stir bar. The two-necked flask is washed with a saturated aqueous solution of NH$_4$Cl (100 mL) and then Et$_2$O (100 mL). The aqueous solution and the organic solution are combined and transferred to a 2-L separatory funnel. The aqueous phase is separated and extracted with Et$_2$O (200 mL × 3). The organic layers are combined, dried over MgSO$_4$ (30 g), and filtered through a Büchner funnel containing a bed of celite (1.0 cm height). The filtrate is concentrated by rotary evaporation (30 mmHg). The residue is loaded onto a wet-packed (hexanes) column of SiO$_2$ (8 cm diameter × 22 cm height; 250 g) and eluted with a gradient of Et$_2$O in hexanes (500 mL of hexanes; 1.0 L of 15% Et$_2$O/hexanes, 1.0 L of 20% Et$_2$O in hexanes, 1.0 L of 30% Et$_2$O/hexanes, 1.0–2.0 L of 50% Et$_2$O/hexanes; fraction size: 100 mL) (Note 5) (TLC analysis on SiO$_2$: 60% Et$_2$O/pentane as eluent, visualization with a UV lamp and a KMnO$_4$ stain, R$_f$ = 0.4) (Note 4). The desired allylic alcohol **1** (22.3 g, 77% yield) is obtained as a yellow oil (Note 6).

   B. *Ethyl (E)-4-chloropent-2-enoate* (**2**). An oven-dried, 500-mL, two-necked, round-bottomed flask equipped with a rubber septum, an argon inlet, and a magnetic stir bar [31.8 x 15.9 mm, egg-shaped] is purged with argon and charged with CH$_2$Cl$_2$ using a 50-mL syringe (200 mL) (Note 7) and ethyl (*E*)-4-hydroxy-2-pentenoate (**1**) (14.4 g, 100 mmol, 1.00 equiv). The solution is cooled to 0 °C in an ice bath, and

318   *Org. Synth.* **2010**, *87*, 317-329

dichlorotriphenylphosphorane (40.0 g, 120 mmol, 1.20 equiv) (Note 8) is added, followed by imidazole (8.17 g, 120 mmol, 1.20 equiv) (Note 9). The solution is allowed to warm to rt, and then it is stirred for 6 h, during which time a white precipitate forms. The progress of the reaction can be followed by TLC analysis on SiO_2 (50% Et_2O/pentane as the eluent; visualization with a UV lamp or with a $KMnO_4$ stain; the alcohol starting material has an R_f = 0.2, and the chloride product has an R_f = 0.8) (Note 4). Pentane (150 mL) and ether (150 mL) are poured into the reaction flask slowly over 3 min, and the white precipitate is removed by filtration through SiO_2 (8 cm diameter × 10 cm height). The SiO_2 is washed with pentane/ether (1/1; 400 mL). The combined organic layers are concentrated by rotary evaporation (30 mmHg), and the residue is transferred to a 50-mL round-bottomed flask equipped with a magnetic stir bar. The product is distilled under reduced pressure (16–20 mmHg, 75–80 °C) through a short-path distillation head to afford the desired allylic chloride (**2**) (13.2 g, 82% yield) as a colorless oil (Note 10).

C. *2-[2-(1,3-Dioxolan-2-yl)ethyl]zinc bromide* (**3**). An oven-dried, 200-mL, two-necked, round-bottomed flask equipped with a rubber septum and a magnetic stir bar [15.9 x 9.5 mm, octagon-type] is purged with argon for 5 min and charged with zinc powder (9.80 g, 150 mmol, 1.50 equiv) (Note 11). The zinc is heated under vacuum (0.5 mmHg) at 70 °C in an oil bath for 30 min. Anhydrous *N,N*-dimethylacetamide (DMA; 100 mL) (Note 12) is added using a 50-mL syringe, and I_2 (634 mg, 2.50 mmol, 0.0250 equiv) (Note 13) is added in one portion via the side neck of the round-bottomed flask. The red-gray solution is stirred at 70 °C until the red color fades (~5 min). 2-(1,3-Dioxolan-2-yl)ethyl bromide (12.0 mL, 100 mmol, 1.00 equiv) (Note 14) is added via syringe over 3 min, and then the reaction mixture is stirred at 70 °C for 12 h. Next, the mixture is allowed to cool to rt, and the unreacted zinc powder is removed by filtration under argon (see Figure 1). The clear yellow solution is employed in the next step without further purification. ^{1}H NMR spectroscopy with a D_2O insert is used to determine that the concentration of the alkylzinc solution is 0.697 M (Note 15).

Figure 1.

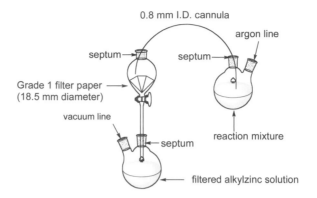

D. *(S,E)-Ethyl 6-(1,3-dioxolan-2-yl)-4-methylhex-2-enoate* (**4**). An oven-dried, 1000-mL, three-necked, round-bottomed flask equipped with a thermometer inlet and thermometer, an argon inlet, a rubber septum, and a magnetic stir bar [41.3 x 19.0 mm, egg-shaped] is purged with argon for 20 min. Under a positive pressure of argon, anhydrous NaCl (10.4 g, 177 mmol, 4.00 equiv) (finely ground and dried at 120 °C under vacuum overnight) (Note 16 and Note 17) is added, followed by NiCl$_2$·glyme (485 mg, 2.22 mmol, 0.0500 equiv) (Note 18) and (*R*)-CH$_2$CH$_2$Ph-Pybox (1.04 g, 2.44 mmol, 0.0550 equiv) (Note 19 and Note 20). Anhydrous DMF (222 mL) (Note 21) and anhydrous DMA (155 mL) (Note 22) are then added via 50-mL syringe, and the flask is capped with a rubber septum. This solution of the catalyst is stirred at 0 °C for 20 min in an isopropanol bath cooled by a chiller system (Note 23). Ethyl (*E*)-4-chloropent-2-enoate (**2**) (7.20 g, 44.3 mmol, 1.00 equiv) is added via 50-mL syringe, and then a solution of 2-[2-(1,3-dioxolan-2-yl)ethyl]zinc bromide **3** (75.7 mL, 0.70 M in DMA, 53 mmol, 1.2 equiv) is added via another 50-mL syringe over 10 min. Throughout the addition of the alkylzinc reagent, the temperature of the solution is maintained at 0 °C. The reaction mixture is stirred for 24 h at 0 °C. The progress of the reaction can be followed by GC analysis (Note 24). After the reaction is complete, water (400 mL) is added over 20 min to quench the excess organozinc reagent, during which time the temperature of the reaction mixture is kept below 25 °C. After the addition of water is complete, the solution is stirred at rt for an additional 10 min, and then it is extracted with Et$_2$O (200 mL × 3). The combined organic layers are washed with water (200 mL × 2), dried over anhydrous Na$_2$SO$_4$ (30 g), filtered through a Büchner funnel containing a bed of celite (1.0 cm height), and

320

concentrated by rotary evaporation (30 mmHg). The residue is loaded onto a wet-packed (hexanes) column of SiO_2 (4 cm diameter × 30 cm height; 350 g) and eluted with a gradient of Et_2O in hexanes (500 mL of hexanes, 2.0 L of 5% Et_2O/hexanes, 2.0 L of 10% Et_2O/hexanes, 2.0 – 4.0 L of 15% Et_2O/hexanes) (Note 5) (TLC analysis on SiO_2: 50% Et_2O/pentane as eluent, visualization with a $KMnO_4$ stain, R_f = 0.5) (Note 4). The desired cross-coupling product **4** (9.0 g, 89% yield, 97% ee) is obtained as a colorless oil (Note 24).

2. Notes

1. Et_2O (99+%) was purchased from Caledon and purified by passage through activated alumina under nitrogen.

2. Ethyl (*E*)-4-oxo-2-butenoate (96%) was purchased from Alfa Aesar and used as received.

3. MeMgBr (3.0 M in Et_2O) was purchased from Aldrich and used as received.

4. Analytical thin-layer chromatography was performed using EMD 0.25 mm silica gel 60-F plates.

5. Column chromatography was performed on Silicycle 60 Å silica gel.

6. On half-scale, compound **1** was isolated in 80% yield (11.6 g); Compound **1** has the following physical properties: ^{1}H NMR (CDCl$_3$, 400 MHz) δ: 1.30 (t, *J* = 7.2 Hz, 3 H), 1.34 (d, *J* = 6.8 Hz, 3 H), 2.03 (br s, 1 H), 4.20 (q, *J* = 6.8 Hz, 2 H), 4.46–4.49 (m, 1 H), 6.00 (dd, *J* = 15.6, 1.6 Hz, 1 H), 6.95 (dd, *J* = 15.6, 4.8 Hz, 1 H); ^{13}C NMR (CDCl$_3$, 100 MHz) δ: 14.2, 22.7, 60.5, 67.2, 119.6, 151.0, 166.7. The spectral data are in agreement with the reported values.[2] The purity (95%) was determined by GC analysis (t_r = 2.61 min; Agilent Technologies 7890A Series GC system equipped with a DB-1 column (length 30 m, I.D. 0.25 mm, film 0.25 μm); carrier gas: H_2 (2 mL/min); gradient temperature: hold at 100 °C for 1 min, 100 → 280 °C at 40 °C/min, hold at 280 °C for 2 min).

7. Dichloromethane (>99.8%) was purchased from ACP Chemicals and purified by passage through activated alumina under nitrogen.

8. Dichlorotriphenylphosphorane (95%) was purchased from Aldrich and used as received.

9. Imidazole (99%) was purchased from Alfa Aesar and used as received.

10. On half-scale, compound **2** was isolated in 85% yield (13.7 g); compound **2** has the following physical properties: IR (neat) 2982, 2932, 2905, 1724, 1658, 1446, 1369, 1342, 1273, 1226, 1180, 1033, 1015, 976 cm^{-1}; ^{1}H NMR (CDCl$_3$, 400 MHz) δ: 1.30 (t, J = 7.2 Hz, 3 H), 1.64 (d, J = 6.4 Hz, 3 H), 4.22 (q, J = 6.8 Hz, 2 H), 4.63 (d of pentets, J = 6.8, 1.2 Hz, 1 H), 6.00 (dd, J = 15.6, 1.2 Hz, 1 H), 6.93 (dd, J = 15.2, 6.8 Hz, 1 H); ^{13}C NMR (CDCl$_3$, 100 MHz) δ: 14.2, 24.3, 54.8, 60.8, 121.8, 147.0, 165.9; GCMS m/z calcd. for C$_7$H$_{11}$ClO$_2$ (M$^+$) 162.0448; found 162.0447. The purity (96%) was determined by GC analysis (t$_r$ = 2.51 min), using the same conditions as for compound **1**.

11. Zinc powder (99.9%, ~100 mesh) was purchased from Alfa Aesar and used as received.

12. Anhydrous DMA (99.8%) purchased from Aldrich (water content: 14 ppm) wasused as received.

13. I$_2$ (≥99%, chips) was purchased from Aldrich and used as received.

14. 2-(1,3-Dioxolan-2-yl)ethyl bromide (95%) was purchased from Alfa Aesar and used as received.

15. For the NMR experiment, a ^{1}H NMR spectrum of the organozinc solution with a D$_2$O insert (prepared in a melting point capillary, and sealed with flame) was taken to determine the concentration of the solution. The resonances for DMA appear at δ 3.02, 2.84, and 2.02. The resonance for the CH$_2$Zn protons of RZnBr appears at δ 0.00. The resonance for the terminal Me group of the reduction product appears at δ 0.87. The molar ratio of DMA : RZnBr : RH (X : 1 : Y) was determined by No-D NMR to be 15.12 : 1 : 0.20. The density (d) of the solution was measured to be 1.10 g/mL. The final concentration of alkylzinc was calculated to be 0.697 M according to the equation: [RZnBr] = d × 1000/(X × MW$_{DMA}$ + MW$_{RZnBr}$ + Y × MW$_{RH}$); For half-scale, X = 16.457, Y = 0.317, d = 1.17 g/mL, and the concentration of alkylzinc was 0.68 M.

16. NaCl (99.999%) was purchased from Aldrich, ground to a fine powder and dried under vacuum at 150 °C overnight before use. The use of anhydrous NaCl beads (99.99% from Aldrich) or NaCl (99.5% from Mallinckrodt Chemicals) powder that was oven-dried in the air furnished lower yields (30% and 74%, respectively).

17. The presence of NaCl has a substantial impact on the rate of the cross-coupling, but little impact on the ee. The role of the NaCl may be to increase the ionic strength of the reaction mixture (the use of more polar

solvents is generally advantageous) or to activate the organozinc reagent.[3]

18.   NiCl$_2$·glyme was purchased from Strem and used as received.

19.   (R)- or (S)-CH$_2$CH$_2$Ph-Pybox can be prepared via the reaction of 2-amino-4-phenylbutan-1-ol with 2,6-pyridinedicarbonitrile catalyzed by 5% zinc triflate in toluene at reflux.[4]

20.   (S)-CH$_2$CH$_2$Ph-Pybox (97%) may be purchased from Aldrich and used as received.

21.   Anhydrous DMF (≥99.5%, over molecular sieves) was purchased from Aldrich (water content: <0.005%) and used as received.

22.   When anhydrous DMA (99.8%) purchased from Aldrich (water content: 14 ppm) was used as received, the cross-coupling product was obtained in good yield (>90%). However, the use of DMA with a higher water content (e.g, 77 ppm) led to a lower yield of product (~75%).

23.   When the reaction was conducted at –10 °C on a 10-g scale, the cross-coupling proceeded more slowly, resulting in a 52% yield after 24 h.

24.   For half-scale, compound **4** was isolated in 90% yield (4.54 g); The product was generated with >20:1 E:Z selectivity and >20:1 regioselectivity. Compound **4** has the following physical properties: $[\alpha]^{26}_D$ +23.6 (c 1.67, CHCl$_3$); IR (neat) 2958, 2931, 2878, 1716, 1651, 1458, 1408, 1369, 1265, 1203, 1180, 1142, 1037, 988, 945, 887, 864 cm^{-1}; ^{1}H NMR (CDCl$_3$, 400 MHz) δ: 1.07 (d, J = 6.8 Hz, 3 H), 1.29 (d, J = 7.2 Hz, 3 H), 1.48–1.54 (m, 2 H), 1.61–1.68 (m, 2 H), 2.34 (septet, J = 7.2 Hz, 1 H), 3.81–3.89 (m, 2 H), 3.92–4.00 (m, 2 H), 4.18 (q, J = 7.2 Hz, 2 H), 4.84 (br t, J = 4.4 Hz, 1 H), 5.75 (d, J = 16.0 Hz, 1 H), 6.84 (dd, J = 16.0, 8.0 Hz, 1 H); ^{13}C NMR (CDCl$_3$, 100 MHz) δ: 14.3, 19.4, 30.0, 31.5, 36.4, 60.2, 64.9, 104.4, 120.2, 153.8, 166.8; GCMS m/z calcd. for C$_{12}$H$_{19}$O$_4$ (M$^+$) 227.1283; found 227.1286. The ee was determined by HPLC analysis with an Agilent 1100 Series HPLC system equipped with a CHIRALPAK AD-H column (length 250 mm, I.D. 4.6 mm; hexanes/isopropanol 99:1, 1.0 mL/min) with t$_r$ (minor) = 12.9 min, t$_r$ (major) = 13.3 min. The purity (94%) was determined by GC analysis (t$_r$ = 4.35 min), using the same conditions as for compound **1**.

Safety and Waste Disposal Information

All hazardous materials should be handled and disposed of in accordance with "Prudent Practices in the Laboratory"; National Academy Press; Washington, DC, 1995.

3. Discussion

The transition metal-catalyzed enantioselective coupling of allylic electrophiles with carbon nucleophiles has been the focus of intense investigation.[5] Salient examples include palladium-catalyzed couplings with enolates, nickel-catalyzed couplings with Grignard reagents, and copper-catalyzed couplings with Grignard and diorganozinc reagents.[6] Despite impressive progress, the development of methods that have broader scope with respect to the nucleophile, as well as improved functional-group compatibility, persist as important challenges.

This work describes a nickel-based catalyst for the asymmetric cross-coupling of racemic secondary allylic chlorides with readily available alkylzinc halides (Eq 1).[7,8,9] The process is stereoconvergent: both enantiomers of the substrate are transformed into the same enantiomer of the product with high stereoselectivity.

In the case of "symmetrical" allylic halides (e.g., entries 1-4 of Table 1), regioselectivity in the carbon–carbon bond formation is not an issue. As the steric demand of the substituents on the termini of the allyl fragment increases, the ee decreases (entries 1-3). A bulky 1,2,3-trisubstituted allylic chloride undergoes cross-coupling with excellent enantioselectivity,

Org. Synth. **2010**, *87*, 317-329

although with modest yield (entry 4). An unactivated alkyl chloride is compatible with the reaction conditions (entry 2).

Table 1. Enantioselective, Stereoconvergent Negishi Cross-Couplings of Allylic Chlorides (for the reaction conditions, see Eq 1).

entry	allylic chloride	R–ZnBr	ee (%)	yield (%)
1	Me⟍⟋⟍(Cl)⟍Me	(1,3-dioxolan-2-yl)—ZnBr	90	93
2	n-Pr⟍⟋⟍(Cl)⟍n-Pr	Cl⟍(⟍)₄ZnBr	79	81
3	i-Pr⟍⟋⟍(Cl)⟍i-Pr	TBSO⟍⟍⟍ZnBr	69	57
4	Me⟍⟋(Me)⟍(Cl)⟍Me	MeO-C₆H₄-CH(Me)(CH₂)₃ZnBr	98	54
5[a]	n-Bu⟍⟍(Cl)⟍Me	Ph⟍⟍⟍ZnBr	83	97
6	i-Pr⟍⟍(Cl)⟍Me	(1,3-dioxolan-2-yl)⟍ZnBr	84	95
7	t-Bu⟍⟍(Cl)⟍Me	MeO₂C⟍⟍⟍ZnBr	81	85
8	EtO₂C⟍⟍(Cl)⟍Me	Me₂CH-CH=CH-CH₂ZnBr	96	86
9	Me(MeO)NOC⟍⟍(Cl)⟍Me	TBSO⟍⟍⟍ZnBr	93	91
10	(EtO)₂OP⟍⟍(Cl)⟍Me	n-Hex–ZnBr	90	63

[a] Regioselectivity: 1.9:1; ee of the minor regioisomer: 88%.

In the case of unsymmetrical allylic chlorides, if the two substituents on the termini of the allyl group have similar steric and electronic properties, then the regioselectivity is modest (e.g., 1.9:1 for *n*-Bu vs. Me; entry 5 of Table 1). On the other hand, good regioselectivity and enantioselectivity are observed when there is substantial steric (entries 6 and 7; >20:1

regioselectivity) or electronic (entries 8-10; >20:1 regioselectivity) differentiation. Carbon–carbon bond formation occurs preferentially at the less-hindered site (entries 5-7) or at the site that preserves conjugation (entries 8 and 9).

Ethyl (*E*)-4-chloropent-2-enoate (**2**) and 2-[2-(1,3-dioxolan-2-yl)ethyl]zinc bromide (**3**) were selected as the cross-coupling partners because the product serves as an intermediate in the formal total synthesis of fluvirucinine A₁ (Scheme 1).[7,10] The Negishi reaction proceeded with a strong preference for carbon–carbon bond formation at the γ position of the electrophile, providing compound **4** in excellent yield (94-96%), regioselectivity (>20:1), and ee (94-95%). Reduction of **4**, followed by bromination, furnished **5**, which was converted into the organozinc reagent and subjected to a second asymmetric Negishi cross-coupling with a racemic allylic chloride to afford intermediate **6** in good yield (82%), regioselectivity (>20:1), and stereoselectivity (>98% ee; 15:1 dr). This coupling product can be transformed into aldehyde **7**, which was an intermediate in Suh's total synthesis of fluvirucinine A₁.[10]

Scheme 1. Formal Total Synthesis of Fluvirucinine A₁ via Two Catalytic Enantioselective Negishi Cross-Couplings of Allylic Chlorides ((*R*)-CH₂CH₂Ph-Pybox was used; for the reaction conditions, see Eq 1).

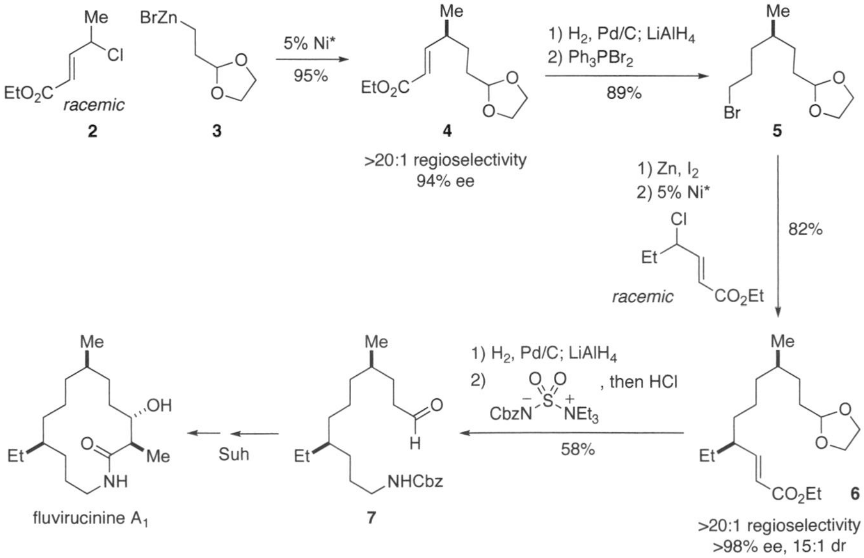

1. Department of Chemistry, Room 18-290, Massachusetts Institute of Technology, Cambridge, Massachusetts 02139; E-mail: gcf@mit.edu.
2. Rodríguez, S.; Kneeteman, M.; Izquierdo, J.; López, I.; González, F. V.; Peris, G. *Tetrahedron* **2006**, *62*, 11112–11123.
3. For a review of halide effects in transition-metal catalysis, see: Fagnou, K.; Lautens, M. *Angew. Chem., Int. Ed.* **2002**, *41*, 26–47.
4. Lou, S.; Fu, G. C. *Org. Synth.,* **2010**, *87*, 310-316.
5. For reviews, see: (a) Pfaltz, A.; Lautens, M. In *Comprehensive Asymmetric Catalysis*; Jacobsen, E. N., Pfaltz, A., Yamamoto, H., Eds.; Springer: New York, 1999; Vol. 2, Chapter 24. (b) Trost, B. M.; Van Vranken, D. L. *Chem. Rev.* **1996**, *96*, 395–422. See also: Kar, A.; Argade, N. P. *Synthesis* **2005**, 2995–3022.
6. For additional examples and for reviews, see: (a) Consiglio, G.; Morandini, F.; Piccolo, O. *J. Chem. Soc., Chem. Commun.* **1983**, 112–114. (b) Gomez-Bengoa, E.; Heron, N. M.; Didiuk, M. T.; Luchaco, C. A.; Hoveyda, A. H. *J. Am. Chem. Soc.* **1998**, *120*, 7649–7650. (c) Chung, K.-G.; Miyake, Y.; Uemura, S. *J. Chem. Soc., Perkin Trans. 1* **2000**, 15–18. (d) Chen, H.; Deng, M.-Z. *J. Organomet. Chem.* **2000**, *603,* 189–193. (e) Yorimitsu, H.; Oshima, K. *Angew. Chem., Int. Ed.* **2005**, *44*, 4435–4439. (f) Alexakis, A.; Malan, C.; Lea, L.; Tissot-Croset, K.; Polet, D.; Falciola, C. *Chimia* **2006**, *60*, 124–130. (g) Novak, A.; Fryatt, R.; Woodward, S. *C. R. Chim.* **2007**, *10*, 206–212.
7. Son, S.; Fu, G. C. *J. Am. Chem. Soc.* **2008**, *130*, 2756–2757.
8. For a convenient method for the preparation of alkylzinc halides from alkyl halides, see: Huo, S. *Org. Lett.* **2003**, *5*, 423–425.
9. For leading references on cross-couplings of alkyl electrophiles, see: (a) Rudolph, A.; Lautens, M. *Angew. Chem., Int. Ed.* **2009**, *48*, 2656–2670. (b) Frisch, A. C.; Beller, M. *Angew. Chem., Int. Ed.* **2005**, *44*, 674–688. (c) Netherton, M. R.; Fu, G. C. In *Topics in Organometallic Chemistry: Palladium in Organic Synthesis*; Tsuji, J., Ed.; Springer: New York, 2005; pp 85–108. (d) Netherton, M. R.; Fu, G. C. *Adv. Synth. Catal.* **2004**, *346*, 1525–1532.
10. For the total synthesis of fluvirucinine A_1, see: Suh, Y.-G.; Kim, S.-A.; Jung, J.-K.; Shin, D.-Y.; Min, K.-H.; Koo, B.-A.; Kim, H.-S. *Angew. Chem., Int. Ed.* **1999**, *38*, 3545–3547.

Appendix
Chemical Abstracts Nomenclature; (Registry Number)

Ethyl (*E*)-4-oxo-2-butenoate: 2-Butenoic acid, 4-oxo-, ethyl ester, (2*E*)-; (2960-66-9)

Methylmagnesium bromide: Magnesium, bromomethyl-; (75-16-1)

Ethyl (*E*)-4-hydroxy-2-pentenoate: 2-Pentenoic acid, 4-hydroxy-, ethyl ester, (2*E*)-; (10150-92-2)

Dichlorotriphenylphosphorane: Phosphorane, dichlorotriphenyl-; (2526-64-9)

Ethyl (*E*)-4-chloropent-2-enoate: 2-Pentenoic acid, 4-chloro-, ethyl ester, (2*E*)-; (872455-00-0)

2-(1,3-Dioxolan-2-yl)ethyl bromide: 1,3-Dioxolane, 2-(2-bromoethyl)-; (18742-02-4)

2-[2-(1,3-Dioxolan-2-yl)ethyl]zinc bromide: Zinc, bromo[2-(1,3-dioxolan-2-yl)ethyl]-; (864501-59-7)

Nickel(II) chloride, dimethoxyethane adduct (NiCl$_2$·glyme): Nickel, dichloro[1,2-di(methoxy-κO)ethane]-; (29046-78-4)

2,6-Bis[(*S*)-4,5-dihydro-4-(2-phenylethyl)-2-oxazolyl]pyridine: Pyridine, 2,6-bis[(4*S*)-4,5-dihydro-4-(2-phenylethyl)-2-oxazolyl]-; (1012042-02-2)

(*S,E*)-Ethyl 6-(1,3-dioxolan-2-yl)-4-methylhex-2-enoate: 2-Hexenoic acid, 6-(1,3-dioxolan-2-yl)-4-methyl-, ethyl ester, (2*E*,4*S*)-; (1012041-93-8)

 Prof. Greg Fu was born in Galion, Ohio, in 1963. He received a B.S. degree in 1985 from MIT, where he worked in the laboratory of Prof. K. Barry Sharpless. After earning a Ph.D. from Harvard in 1991 under the guidance of Prof. David A. Evans, he spent two years as a postdoctoral fellow with Prof. Robert H. Grubbs at Caltech. In 1993, he returned to MIT, where he is currently the Firmenich Professor of Chemistry. Prof. Fu received the Springer Award in Organometallic Chemistry in 2001, the Corey Award of the American Chemical Society in 2004, and the Mukaiyama Award of the Society of Synthetic Organic Chemistry of Japan in 2006. He is a fellow of the Royal Society of Chemistry and of the American Academy of Arts and Sciences. Prof. Fu serves as an associate editor of the *Journal of the American Chemical Society*. His current research interests include metal-catalyzed coupling reactions and enantioselective nucleophilic catalysis.

Sha Lou was born in He Bei, China, in 1979. He received a B.S. in Chemistry from Beijing University of Chemical Technology in 2002, where he conducted undergraduate research on fullerene functionalizations with Professor Shenyi Yu. He obtained a Ph.D. degree in January 2008 from Boston University under the direction of Professor Scott E. Schaus. His graduate research focused on transition metal- and organic molecule-catalyzed asymmetric carbon–carbon bond-forming reactions and synthesis. In 2008, he joined the group of Professor Greg Fu at MIT as a postdoctoral fellow. His current research involves the development of transition metal-catalyzed enantioselective cross-coupling reactions.

David I. Chai was born in Korea in 1981. He obtained a B.S. in Chemistry in 2006 from the University of British Columbia while working in the laboratory of Professor Marco Ciufolini. He is currently a chemistry graduate student with Professor Mark Lautens, working on palladium-catalyzed carbo- and hetero-cycle synthesis.

Gavin Chit Tsui was born in China and raised in Hong Kong. He obtained his Hon. B. Sc. (2004) and M. Sc. (2006, supervisor: Prof. William Tam) degrees from the University of Guelph, Ontario, Canada. He then joined pharmaceutical companies (NPS Pharmaceuticals, Inc. and Merck Frosst Canada) as a research chemist in the medicinal chemistry division. Currently he is pursuing a Ph. D. degree at the University of Toronto under the supervision of Prof. Mark Lautens to investigate novel rhodium-catalyzed reactions.

NICKEL-CATALYZED ENANTIOSELECTIVE NEGISHI CROSS-COUPLINGS OF RACEMIC SECONDARY α-BROMO AMIDES WITH ALKYLZINC REAGENTS: (S)-N-BENZYL-7-CYANO-2-ETHYL-N-PHENYLHEPTANAMIDE

[Heptanamide, 7-cyano-2-ethyl-N-phenyl-N-(phenylmethyl)-, (2S)-]

Submitted by Sha Lou and Gregory C. Fu.[1]
Checked by Christopher S. Bryan and Mark Lautens.

1. Procedure

A. *(5-Cyanopentyl)zinc(II) bromide* (**1**). An oven-dried, 200-mL pear-shaped Schlenk flask equipped with a magnetic stirbar (egg shaped, 25.4 × 12.7 mm) and an argon line connected to the standard taper outer joint is purged with argon for 5 min. Zinc powder (9.80 g, 150 mmol, 1.50 equiv) (Note 1) is added through the open neck, and then the flask is capped with a rubber septum and heated in an oil bath under high vacuum (0.5 mmHg) at 70 °C for 30 min. Then, the flask is refilled with argon, and anhydrous 1,3-dimethyl-2-imidazolidinone (DMI; 100 mL) (Note 2) is added via syringe. Iodine (I_2) (634 mg, 2.50 mmol, 0.0250 equiv) (Note 3) is added in one portion through the neck. The neck is re-capped with a rubber septum, and the reaction mixture is stirred at 70 °C in an oil bath until the red color fades (~5 min). 6-Bromohexanenitrile (13.2 mL, 100 mmol, 1.00 equiv) (Note 4) is added via syringe over 4 min, and the reaction mixture is stirred at 70 °C for 12 h. Then, the oil bath is removed, and the mixture is allowed to cool at rt for 1 h without stirring. During this time, the unreacted zinc powder settles at the bottom of the flask. The flask is equipped with a

330

fritted filter tube of medium porosity capped with an oven-dried, 2-necked 250-mL round-bottom flask, and the supernatant solution is filtered under argon by inverting the set-up (Note 5). The resulting clear yellow solution is employed in the next step without further purification. ^{1}H NMR spectroscopy is used to determine that the concentration of the alkylzinc solution is 0.72 M (Note 6). This organozinc solution can be stored under argon at 0–4 °C for up to 3 weeks without deterioration.

B. *(S)-N-Benzyl-7-cyano-2-ethyl-N-phenylheptanamide* (**2**). An oven-dried, 1000-mL, three-necked, round-bottomed flask equipped with a thermometer inlet, thermometer, magnetic stirbar (egg shaped, 41.3 × 19.0 mm), and argon inlet is purged with argon for 20 min. NiCl$_2$·glyme (536 mg, 2.44 mmol, 0.0697 equiv) (Note 7), (*R*)-(*i*-Pr)-Pybox (960 mg, 3.19 mmol, 0.0910 equiv) (Note 8), and *N*-benzyl-2-bromo-*N*-phenylbutanamide (11.6 g, 35.0 mmol, 1.00 equiv) (Note 9) are then added through the open neck under a positive pressure of argon. The open neck is capped with a rubber septum, and DMI (72 mL) (Note 2) and THF (17.5 mL) (Note 10) are each added via syringe. The resulting orange solution is stirred at rt for 10 min, and then the organozinc reagent (0.72 M in DMI; 63.2 mL, 45.5 mmol, 1.30 equiv) is added via syringe over 6 min, maintaining the internal temperature below 30 °C. The resulting dark-brown reaction mixture is stirred at rt (temperature of the solution: 23 °C) for 24 h. The progress of the reaction can be monitored by ^{1}H NMR analysis of an aliquot of the reaction mixture. After the α-bromo amide starting material is completely consumed (determined by observing the disappearance of a triplet at δ 3.95), the excess organozinc reagent is quenched by adding ethanol (15 mL). The brown reaction mixture is diluted with Et$_2$O (500 mL), and the resulting solution is transferred to a 1-L separatory funnel and washed with H$_2$O (300 mL × 3). The organic layer is dried over anhydrous Na$_2$SO$_4$ (30 g) and filtered through a Büchner funnel containing a bed of celite (1.0 cm height). The filtrate is concentrated by rotary evaporation (40 °C, 20 mmHg), and the resulting orange oil is purified by column chromatography on SiO$_2$ (wet packed in hexanes; 8 cm diameter × 30 cm height; 350 g; eluting with a gradient of EtOAc in hexanes (500 mL of 10% EtOAc/hexanes, 1.5 L of 15% EtOAc/hexanes, 1.5 L of 20% EtOAc/hexanes, 1.0 L of 30% EtOAc/hexanes, 1.0 L of 40% EtOAc/ hexanes; 100-mL fractions) (Note 11). The cross-coupling product **2** has an R_f = 0.5 (TLC analysis on SiO$_2$ (30% EtOAc/hexanes, visualization with a UV lamp) (Note 12). The desired product is obtained as a white solid (10.4–10.8 g, 85–89% yield, 91% ee) (Note 13).

2. Notes

1. Zinc powder (99.9%, ~100 mesh) was purchased from Alfa Aesar and used as received.

2. 1,3-Dimethyl-2-imidazolidinone (≥ 99.5%, over molecular sieves, water content: ≤0.04%) was purchased from Aldrich and used as received.

3. I_2 (≥99%, chips) was purchased from Aldrich and used as received.

4. 6-Bromohexanenitrile (95%) was purchased from Aldrich and used as received.

5. The submitters described an alternative filtration procedure as depicted in Figure 1.

Figure 1.

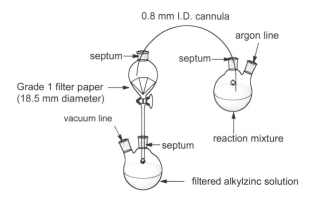

6. For the NMR experiment, a sealed tube of D_2O was placed in an NMR tube containing the alkylzinc solution, and a standard 1D ^1H NMR spectrum was obtained using D_2O as a reference. The resonances for DMI appear at δ 3.31 and 2.74. The resonance for the CH_2Zn protons of RZnBr appears at δ 0.13. The resonance for the terminal Me group of the reduction product appears at δ 0.94. The molar ratio of DMI : RZnBr : RH (X : 1 : Y) was determined by No-D NMR to be 11.7 : 1 : 0.088. The density (d) of the solution was measured to be 1.14 g/mL. The final concentration of alkylzinc was calculated to be 0.72 M according to the equation: $[RZnBr] = d \times 1000/(X \times MW_{DMI} + MW_{RZnBr} + Y \times MW_{RH})$.

The submitters measured the concentration using No-D NMR spectroscopy. For the No-D NMR experiment, a ^1H NMR spectrum of a blank $CDCl_3$ sample was taken to lock the sample and determine the

Org. Synth. **2010**, *87*, 330-338

reference. Then, an unlocked No-D NMR spectrum of the alkylzinc sample in DMI was taken (pw = 1.5). The resonances for DMI appear at δ 4.08 and 3.51. The resonance for the CH_2Zn protons of RZnBr appears at δ 0.89. The resonance for the terminal Me group of the reduction product appears at δ 1.71. The molar ratio of DMI : RZnBr : RH (X : 1 : Y) was determined by No-D NMR to be 11.2 : 1 : 0.13. The density (d) of the solution was measured to be 1.20 g/mL. The final concentration of alkylzinc was calculated to be 0.78 M according to the equation: [RZnBr] = d × 1000/(X × $MW_{DMI} + MW_{RZnBr} + Y × MW_{RH}$).

7. NiCl$_2$·glyme was purchased from Strem and used as received.

8. (R)-(i-Pr)-Pybox (99%) was purchased from Aldrich and used as received.

9. N-Benzyl-2-bromo-N-phenylbutanamide (97%) may be purchased from Aldrich.

10. THF (99+%) was purchased from J.T. Baker (water content: 24 ppm) and purified by passage through an activated alumina column under argon.

11. Column chromatography was performed on SiO_2 (0.040-0.063 mm, pore diameter 6 nm purchased from Silicycle). The submitters report separation with smaller volumes of eluent (500 mL of 10% EtOAc/hexanes, 1.0 L of 15% EtOAc/hexanes, 1.0 L of 20% EtOAc/hexanes, 1.0 L of 30% EtOAc/hexanes, 500 mL of 40% EtOAc/hexanes) using an equal amount of Sorbent Technologies 60 Å silica gel.

12. Analytical thin-layer chromatography was performed using EMD 0.25 mm silica gel 60-F plates.

13. Compound 2 has the following physical properties: Mp: 67–69 °C; $[α]^{26}_D$ −30.5 (c 1.05, CHCl$_3$); ^{1}H NMR (CDCl$_3$, 500 MHz) δ: 0.80 (t, J = 7.4 Hz, 3 H), 1.11–1.28 (m, 2 H), 1.30–1.42 (m, 4 H), 1.56–1.69 (m, 4 H), 2.13–2.20 (m, 1 H), 2.27 (t, J = 7.1 Hz, 2 H), 4.86 (d, J = 14.0 Hz, 1 H), 4.95 (d, J = 14.0 Hz, 1 H), 6.90–6.93 (m, 2 H), 7.18–7.20 (m, 2 H), 7.22–7.28 (m, 3 H), 7.30–7.35 (m, 3 H); ^{13}C NMR (CDCl$_3$, 100 MHz) δ: 12.3, 17.2, 25.3, 26.4, 26.8, 28.9, 32.7, 43.7, 53.2, 120.0, 127.6, 128.1, 128.5, 129.2, 129.6, 138.0, 142.5, 175.9; IR (film): 3061, 2943, 2861, 2239, 1640, 1594, 1497, 1453, 1408, 1262, 1246, 1211, 1079, 1014, 782, 738, 701 cm^{-1}; MS (ESI) m/z (rel. intensity) 387.2 (4), 372.2 (8), 371.2 (30), 350.2 (22), 349.2 (100); HRMS (ESI) calcd for $C_{23}H_{29}N_2O$ ([M+H]$^+$) 349.2274, found, 349.2285; Anal. calcd. for $C_{23}H_{28}N_2O$: C, 79.27; H, 8.10; N, 8.04; found: C, 79.23; H, 8.40; N, 8.10. The spectral data are in agreement with reported values.[2] The ee was determined by HPLC analysis with an Agilent 1100 Series HPLC

system equipped with a CHIRALPAK AD-H column (length 250 mm, I.D. 4.6 mm) (hexanes/isopropanol 93:7, 1.0 mL/min) with t_r (major) = 20.2 min, t_r (minor) = 21.7 min. The peak areas were measured at λ = 210 nm. The ee can also be determined by supercritical fluid chromatography (SFC) analysis on a Berger SFC MiniGram system: CHIRALPAK AD-H column (length 250 mm, I.D. 4.6 mm); solvent system: 15% MeOH, 3.0 mL/min; retention times: t_r (major) = 2.39 min, t_r (minor) = 2.57 min. The purity of compound **2** (98%) was determined by HPLC analysis (t_r = 6.30 min) with an Agilent 1100 Series HPLC system equipped with an Eclipse XDB-C18 column (length 150 mm, I.D. 4.6 mm, particle size 5 µm), using 98% MeOH/(0.2% AcOH in water) for 5 min, with a flow rate of 0.8 mL/min. Peak areas were measured at λ = 210 nm.

Safety and Waste Disposal Information

All hazardous materials should be handled and disposed of in accordance with "Prudent Practices in the Laboratory"; National Academy Press; Washington, DC, 1995.

3. Discussion

Substantial advances have recently been described in the development of catalysts that cross-couple alkyl electrophiles.[3] For couplings of unsymmetrical secondary electrophiles, there is the potential to control the stereochemistry at the carbon that bears the leaving group.[4] This stereochemical aspect adds an important dimension to carbon–carbon bond-forming reactions of alkyl electrophiles. In 2003, we reported that Ni(cod)$_2$/(s-Bu)-Pybox catalyzes Negishi reactions of secondary alkyl bromides and iodides.[5] Our observation that a chiral Pybox is the ligand of choice raised the obvious question of whether it might be possible to develop a highly enantioselective alkyl–alkyl cross-coupling.

We have determined that this objective can be achieved with certain electrophiles, including α-bromo amides.[6,7] Thus, NiCl$_2$·glyme/(i-Pr)-Pybox catalyzes the cross-coupling of a racemic mixture of an α-bromo amide with an array of alkylzinc reagents[8] in good ee and yield (Table 1). The catalyst tolerates a variety of functional groups, such as olefins, ethers, imides, and nitriles.

334

Table 1. Enantioselective, Stereoconvergent Negishi Cross-Couplings of α-Bromo Amides with Alkylzinc Reagents.

entry	R	R^1-ZnX	ee (%)	yield (%)
1	Et	Hex−ZnBr	96	90
2^a	Et	(cyclohexylmethyl)−ZnBr	92	58
3	n-Bu	Ph(CH$_2$CH$_2$CH$_2$)−ZnBr	96	79
4	i-Bu	MeZnI	87	78
5	Et	Me$_2$C=CH(CH$_2$CH$_2$)−ZnBr	95	78
6	Et	PhCH$_2$O(CH$_2$)$_4$−ZnBr	96	77
7	Et	phthalimido-(CH$_2$)$_4$−ZnBr	96	51
8	Et	NC(CH$_2$)$_5$−ZnBr	93	70

^{a} The coupling was conducted at room temperature.

In this stereoconvergent process, both enantiomers of the racemic substrate are transformed into the same enantiomer of the product with good stereoselectivity. There is no evidence for kinetic resolution of the starting material during the catalytic asymmetric Negishi reaction. The cross-coupling product can be converted into other useful families of compounds, such as primary alcohols (reduction with LiAlH$_4$).

1. Department of Chemistry, Room 18-290, Massachusetts Institute of Technology, Cambridge, Massachusetts 02139; E-mail: gcf@mit.edu.

2. Fischer, C.; Fu, G. C. *J. Am. Chem. Soc.* **2005**, *127*, 4594–4595.

3. For leading references, see: (a) Rudolph, A.; Lautens, M. *Angew. Chem. Int. Ed.* **2009**, *48*, 2656–2670. (b) Frisch, A. C.; Beller, M. *Angew. Chem. Int. Ed.* **2005**, *44*, 674–688. (c) Netherton, M. R.; Fu, G. C. In *Topics in Organometallic Chemistry: Palladium in Organic Synthesis*; Tsuji, J., Ed.; Springer: New York, 2005; pp 85–108. (d) Netherton, M. R.; Fu, G. C. *Adv. Synth. Catal.* **2004**, *346*, 1525–1532.

4. For some examples of other types of metal-catalyzed enantioselective cross-coupling processes, see: (a) Hayashi, T. In *Comprehensive Asymmetric Catalysis*; Jacobsen, E. N., Pfaltz, A., Yamamoto, H., Eds.; Springer: New York, 1999; Chapter 25. (b) Yin, J.; Buchwald, S. L. *J. Am. Chem. Soc.* **2000**, *122*, 12051–12052. (c) Shimada, T.; Cho, Y.-H.; Hayashi, T. *J. Am. Chem. Soc.* **2002**, *124*, 13396–13397. (d) Hamada, T.; Chieffi, A.; Åhman, J.; Buchwald, S. L. *J. Am. Chem. Soc.* **2002**, *124*, 1261–1268. (e) Willis, M. C.; Powell, L. H. W.; Claverie, C. K.; Watson, S. J. *Angew. Chem. Int. Ed.* **2004**, *43*, 1249–1251.

5. Zhou, J.; Fu, G. C. *J. Am. Chem. Soc.* **2003**, *125*, 14726–14727.

6. (a) α-bromo amides (Negishi): Fischer, C.; Fu, G. C. *J. Am. Chem. Soc.* **2005**, *127*, 4594–4595. (b) 1-haloindanes (Negishi): Arp, F. O.; Fu, G. C. *J. Am. Chem. Soc.* **2005**, *127*, 10482–10483. (c) allylic chlorides (Negishi): Son, S.; Fu, G. C. *J. Am. Chem. Soc.* **2008**, *130*, 2756–2757. (d) α-bromo esters (Hiyama): Dai, X.; Strotman, N. A.; Fu, G. C. *J. Am. Chem. Soc.* **2008**, *130*, 3302–3303. (e) homobenzylic bromides (Suzuki): Saito, B.; Fu, G. C. *J. Am. Chem. Soc.* **2008**, *130*, 6694–6695. (f) propargylic halides (Negishi): Smith, S. W.; Fu, G. C. *J. Am. Chem. Soc.* **2008**, *130*, 12645–12647. (g) α-bromo ketones (Negishi): Lundin, P. M.; Esquivias, J.; Fu, G. C. *Angew. Chem. Int. Ed.* **2009**, *48*, 154–156.

7. For the work of others, see: Caeiro, J.; Sestelo, J. P.; Sarandeses, L. A. *Chem. Eur. J.* **2008**, *14*, 741–746.

8. Huo, S. *Org. Lett.* **2003**, *5*, 423–425. Preliminary efforts to employ commercially alkylzinc halides (Aldrich) were not successful.

Appendix
Chemical Abstracts Nomenclature; (Registry Number)

6-Bromohexanenitrile: Hexanenitrile, 6-bromo-; (6621-59-6)

Zinc; (7440-66-6)

N-Benzyl-2-bromo-*N*-phenylbutanamide: Butanamide,
2-bromo-*N*-phenyl-*N*-(phenylmethyl)-; (851073-30-8)

Nickel(II) chloride, dimethoxyethane adduct (NiCl₂·glyme): Nickel,
dichloro[1,2-di(methoxy-κO)ethane]-; (29046-78-4)

(*R*)-*i*-Pr-Pybox: Pyridine, 2,6-bis[(4*R*)-4,5-dihydro-4-(1-methylethyl)-2-
oxazolyl]-; (131864-67-0)

(*S*)-*N*-Benzyl-7-cyano-2-ethyl-*N*-phenylheptanamide: Heptanamide,
7-cyano-2-ethyl-*N*-phenyl-*N*-(phenylmethyl)-, (2*S*)-; (851073-44-4)

Prof. Greg Fu was born in Galion, Ohio, in 1963. He received a B.S. degree in 1985 from MIT, where he worked in the laboratory of Prof. K. Barry Sharpless. After earning a Ph.D. from Harvard in 1991 under the guidance of Prof. David A. Evans, he spent two years as a postdoctoral fellow with Prof. Robert H. Grubbs at Caltech. In 1993, he returned to MIT, where he is currently the Firmenich Professor of Chemistry. Prof. Fu received the Springer Award in Organometallic Chemistry in 2001, the Corey Award of the American Chemical Society in 2004, and the Mukaiyama Award of the Society of Synthetic Organic Chemistry of Japan in 2006. He is a fellow of the Royal Society of Chemistry and of the American Academy of Arts and Sciences. Prof. Fu serves as an associate editor of the *Journal of the American Chemical Society*. His current research interests include metal-catalyzed coupling reactions, chiral-ligand design, and enantioselective nucleophilic catalysis.

Sha Lou was born in He Bei, China, in 1979. He received a B.S. in Chemistry from Beijing University of Chemical Technology in 2002, where he conducted undergraduate research on fullerene functionalizations with Professor Shenyi Yu. He obtained a Ph.D. degree in January 2008 from Boston University under the direction of Professor Scott E. Schaus. His graduate research focused on transition metal- and organic molecule-catalyzed asymmetric carbon–carbon bond-forming reactions and synthesis. In 2008, he joined the group of Professor Greg Fu at MIT as a postdoctoral fellow. His current research involves the development of transition metal-catalyzed enantioselective cross-coupling reactions.

Christopher Bryan was born in Winnipeg, Canada in 1982. He received his B.Sc. degree with distinction in 2005 from the University of Victoria, where he worked in the laboratory of Dr. Scott McIndoe. While an undergraduate, he worked as a Co-op student in the medicinal chemistry department at Boehringer-Ingelheim Pharmaceuticals in Laval, PQ. He is currently pursuing his Ph. D. at the University of Toronto under the supervision of Professor Mark Lautens. His research is focused on the synthesis of heterocycles via metal-catalyzed tandem processes.

ONE-POT SYNTHESIS OF 5*H*-INDAZOLO-[3,2-*b*]BENZO[*d*]-1,3-OXAZINE: TWO EFFICIENT PREPARATIVE METHODS

A.

B.

Submitted by Danielle M. Solano,[1] Jeffrey D. Butler,[1] Makhluf J. Haddadin,[2] and Mark J. Kurth.[1*]
Checked by Kyle D. Baucom and Margaret M. Faul.

1. Procedure

> *Caution: This procedure should be carried out in an efficient fume hood due to flammable gas evolution during the reaction.*

A. 5H-Indazolo-[3,2-b]benzo[d]-1,3-oxazine. To a 500-mL round-bottomed flask equipped with a Teflon-coated magnetic stir bar (5 cm x 2 cm egg-shaped) is added 2-nitrobenzaldehyde (**1**) (4.08 g, 27.0 mmol, 1.0 equiv) and 2-aminobenzyl alcohol (**2**) (3.49 g, 28.3 mmol, 1.05 equiv) (Note 1). The solids are dissolved in methanol (70 mL), and glacial acetic acid (7.73 mL, 0.135 mol) is added to give a clear, yellow-orange solution (Note 2). 2-Picoline borane (3.50 g, 28.3 mmol, 1.05 equiv) (Note 3) is added in one portion, resulting in gas evolution and a mild exotherm (5 °C). The reaction flask is sealed with a septum, vented with a needle, and stirred for 3 h at room temperature at which point the reaction is darker orange in color. Reductive amination to (2-(2-nitrobenzylamino)phenyl)methanol is complete as indicated by the disappearance of 2-nitrobenzaldehyde by TLC analysis (Note 4). Additional methanol (200 mL) (Note 2) is added followed by a solution of potassium hydroxide (22.7 g, 0.405 mol) in water (30 mL)

(Note 5). The septum is removed, the flask is fitted with a water-cooled reflux condenser, and the solution is heated, with stirring, to reflux (85 °C oil bath temperature, internal reaction temperature 72 °C) for 1.5 h to affect the bis-heterocyclization reaction (Note 6). The flask is removed from the oil bath, the solution is cooled to room temperature, and concentrated by rotary evaporation under reduced pressure (45–50 °C, 200–230 mmHg) to remove methanol, at which point a tan precipitate forms. The mixture is cooled to room temperature, and 1 M HCl (300 mL) is added (Note 7). The mixture is transferred to a 1-L separatory funnel, and ethyl acetate (200 mL) is added, which dissolves the precipitate. The organic layer is separated, and the aqueous layer is further extracted with ethyl acetate (2 x 100 mL). The combined organic extracts are washed with 1 M HCl (1 x 200 mL), 1 M K_2CO_3 (2 x 200 mL), and brine (1 x 200 mL). The organic extracts are dried over anhydrous sodium sulfate (40 g), filtered, and the solid residue is rinsed with ethyl acetate (2 x 50 mL). The combined extracts are concentrated by rotary evaporation under reduced pressure (45–50 °C, 200–230 mmHg). Purification by flash column chromatography on SiO_2 (Note 8) affords 5.29–5.38 g of 5*H*-indazolo[3,2-*b*]benzo[*d*]-1,3-oxazine (**3**) as a white solid in 88.2–89.7% yield (Note 9).

> *Caution: This procedure should be carried out in an efficient fume hood. Appropriate precautions should be taken to avoid inhalation or direct contact with 2-nitrobenzyl bromide, a known lachrymator.*

    B. 5H-Indazolo-[3,2-b]benzo[d]-1,3-oxazine. To a 500-mL round-bottomed flask equipped with a Teflon-coated magnetic stir bar (5 cm x 2 cm egg-shaped) is added 2-nitrobenzyl bromide (**4**) (6.00 g, 27.8 mmol, 1.0 equiv) and 2-aminobenzyl alcohol (**2**) (4.11 g, 33.4 mmol, 1.2 equiv) (Note 10). The solids are dissolved in methanol (120 mL), and diisopropylethyl amine (DIEA, 9.5 mL, 55.6 mmol, 2.0 equiv) is added (Note 11). The reaction flask is fitted with a water-cooled reflux condenser, and the solution is heated to reflux (85 °C oil bath temperature, internal reaction temperature 73 °C) for 4 h. At this point, the displacement reaction is complete as indicated by the disappearance of 2-nitrobenzyl bromide by TLC analysis (Note 12). The solution is cooled to room temperature, then potassium hydroxide pellets (23.4 g, 0.417 mol) (Note 13) are added followed by water (12 mL). Upon addition, the solution turns orange in color and an exotherm (20 °C) is seen. The reaction vessel is again fitted with a water-cooled reflux

340

condenser and heated to reflux to affect the bis-heterocyclization reaction (85 °C oil bath temperature, internal reaction temperature 72 °C). After stirring at this temperature for 1.5 h (Note 14), the flask is removed from the oil bath, the solution is cooled to room temperature, and concentrated by rotary evaporation under reduced pressure (45–50 °C, 200–230 mmHg) to remove methanol, during which time a heterogeneous emulsion is produced. The mixture is cooled in an ice/water bath, 1 M HCl (450 mL) is added (Note 15), and a white precipitate is observed. The mixture (room temperature) is transferred to a 1-L separatory funnel. Ethyl acetate is added (200 mL) to dissolve the precipitate. The organic layer is separated, and the aqueous layer is further extracted with ethyl acetate (2 x 100 mL). The combined organic extracts are washed with brine (1 x 200 mL), dried over anhydrous sodium sulfate (40 g), filtered, and the solid residue is rinsed with ethyl acetate (50 mL). The combined extracts are concentrated by rotary evaporation under reduced pressure (45–50 °C, 200–230 mmHg). Purification by flash column chromatography on SiO_2 (Note 16) affords 4.87–5.10 g of 5*H*-indazolo[3,2-*b*]benzo[*d*]-1,3-oxazine (**3**) as a pale-yellow solid in 78.8–82.9% yield (Note 17).

2. Notes

1. 2-Nitrobenzaldehyde (98%) and 2-aminobenzyl alcohol (98%) were obtained from Sigma-Aldrich and used as received.

2. Methanol from Sigma Aldrich (99%) and glacial acetic acid from JT Backer were obtained and used as received.

3. 2-Picoline borane (synonyms: borane-2-picoline complex, pic·BH$_3$) was obtained from Sigma Aldrich (95%) and used as received.

4. Reaction progress can be monitored by TLC analysis (Silica gel 60 F_{254} 0.5 mm, Merck) using 30% ethyl acetate in hexane as the eluent with UV visualization (R_f 2-aminobenzyl alcohol = 0.10, R_f 2-nitrobenzaldehyde = 0.43, R_f (2-(2-nitrobenzylamino)phenyl)methanol = 0.26).

5. Potassium hydroxide pellets were obtained from JT Baker.

6. Reaction progress can be monitored by TLC analysis (Silica gel 60 F_{254} 0.5 mm, Merk) using 30% ethyl acetate in hexane as the eluent with UV visualization (R_f 2-aminobenzyl alcohol = 0.10, R_f (2-(2-nitrobenzylamino)phenyl)methanol = 0.26, R_f product = 0.45).

7. The pH of the reaction was ~6 as measured with pH paper.

8.	Flash chromatography is performed using a 3.4 cm diameter column containing 120 g of SiO$_2$ (0.035-0.070, 60 Å, Acros Organics). The column is packed by adding 10% ethyl acetate in hexane to the dry silica and eluting several times (until no cracks or imperfections are visible). The crude material is prepared for loading by dissolving it in ethyl acetate, adding 10 g of Celite® 503 (J.T. Baker), and then concentrating to dryness by rotary evaporation (45–50 °C, 200–230 mmHg). The celite-impregnated material is dry-loaded onto the column. The system is eluted with 15% ethyl acetate in hexane and 50 mL fractions are collected. Fractions are examined by TLC (30% ethyl acetate in hexane as the eluent, product R$_f$ = 0.45, UV visualization). All fractions containing the desired product are combined and concentrated by rotary evaporation under reduced pressure (45–50 °C, 200–230 mmHg), and then under high vacuum at 25 mmHg.

9.	Analytical data for 5H-indazolo[3,2-b]benzo[d]-1,3-oxazine (3): Mp 100–102 °C; IR (neat) v_{max} 1631, 1530, 1503, 1460, 1157, 1095, 737 cm^{-1}; ^{1}H NMR (400 MHz, DMSO-d_6) δ: 5.63 (s, 2 H), 6.93 (dd, J = 8.2, 6.8 Hz, 1 H), 7.26 (ddd, J = 8.9, 6.6, 1.0 Hz, 1 H), 7.39 (td, J = 7.4, 1.0 Hz, 1 H), 7.48 (t, J = 8.4 Hz, 2 H), 7.52 – 7.58 (m, 2 H), 7.88 (d, J = 8.0 Hz, 1 H); ^{13}C NMR (125 MHz, DMSO-d_6) δ: 67.9, 105.9, 115.1, 116.9, 119.1, 120.0, 121.8, 125.5, 127.0, 127.9, 129.6, 132.8, 143.4, 148.0; ESI MS m/z calcd. for C$_{14}$H$_{11}$N$_2$O ([M+H]$^+$) 223.1, found: 223.1. Purity was determined to be >99% by HPLC integration. HPLC specifications are as follows: electrospray (+) ionization, mass range 150 - 1500 Da, 20 V cone voltage, and Xterra® MS C$_{18}$ column (2.1 mm x 50 mm x 3.5 μm). The product is stable in the solid form under refrigeration for months; however minor decomposition has been observed when stored in solution at room temperature (i.e. TLC standards) over several days.

10.	2-Nitrobenzyl bromide was obtained from Alfa-Aesar (98+%) and used as received. 2-Aminobenzyl alcohol was obtained from Sigma-Aldrich (98%) and used as received.

11.	Methanol was obtained from Sigma Aldrich (99%) and used as received. Diisopropylethyl amine (DIEA) was obtained from EMD (98%) and used as received.

12.	Reaction progress can be monitored by TLC analysis (Silica gel 60 F$_{254}$ 0.5 mm, Merck) using 30% ethyl acetate in hexane as the eluent with UV visualization (R$_f$ 2-aminobenzyl alcohol = 0.10, R$_f$ (2-(2-nitrobenzylamino)phenyl)methanol = 0.26, R$_f$ 2-nitrobenzyl bromide = 0.58).

342

13. Potassium hydroxide pellets were obtained from JT Baker.

14. Reaction progress can be monitored by TLC analysis (Silica gel 60 F_{254} 0.5 mm, Merck) using 30% ethyl acetate in hexane as the eluent with UV visualization (R_f 2-aminobenzyl alcohol = 0.10, R_f (2-(2-nitrobenzylamino)phenyl)methanol = 0.26, R_f product = 0.45).

15. The pH of the reaction was ~1 measured with pH paper.

16. Flash chromatography is performed using a 3.4 cm diameter column containing 120 g of SiO_2 (0.035-0.070, 60 Å, Acros Organics). The column is packed by adding 500 mL of 10% ethyl acetate in hexane to the dry silica and eluting several times (until no cracks or imperfections are visible). The crude material is prepared for loading by dissolving it in ethyl acetate, adding 10 g of Celite® 503 (J.T. Baker), and then concentrating to dryness by rotary evaporation (45–50 °C, 200–230 mmHg). The celite-impregnated material is dry-loaded onto the column. The system is eluted with 15% ethyl acetate in hexane and 50 mL fractions are collected. Fractions are examined by TLC (30% ethyl acetate in hexane as the eluent, product R_f = 0.45, UV visualization). All fractions containing the desired product are combined and concentrated by rotary evaporation under reduced pressure (45–50 °C, 200–230 mmHg), and then under high vacuum at 25 mmHg.

17. Analytical data for 5H-indazolo[3,2-b]benzo[d]-1,3-oxazine (3): Mp 100–102 °C; IR (neat) ν_{max} 1631, 1530, 1503, 1460, 1157, 1095, 737 cm^{-1}; ^{1}H NMR (400 MHz, DMSO-d_6) δ: 5.62 (s, 2 H), 6.93 (dd, J = 8.2, 6.8 Hz, 1 H), 7.25 (ddd, J = 8.9, 6.5, 0.9 Hz, 1 H), 7.38 (td, J = 7.4, 1.0 Hz, 1 H), 7.48 (t, J = 8.4 Hz, 2 H), 7.51 - 7.58 (m, 2 H), 7.88 (d, J = 7.8 Hz, 1 H); ^{13}C NMR (125 MHz, DMSO-d_6) δ: 67.9, 105.9, 115.1, 116.9, 119.2, 120.0, 121.8, 125.5, 127.0, 127.9, 129.6, 132.8, 143.4, 148.0; ESI MS m/z calcd. for $C_{14}H_{11}N_2O$ ([M+H]$^+$) 223.1, found 223.1. Purity was determined to be >99% by HPLC integration. HPLC specifications are as follows: electrospray (+) ionization, mass range 150 - 1500 Da, 20 V cone voltage, and Xterra® MS C_{18} column (2.1 mm x 50 mm x 3.5 μm). The product is stable in the solid form under refrigeration for months, however minor decomposition has been observed when stored in solution at room temperature (i.e. TLC standards) over several days.

Safety and Waste Disposal Information

All hazardous materials should be handled and disposed of in accordance with "Prudent Practices in the Laboratory"; National Academy Press; Washington, DC, 1995.

3. Discussion

The indazolobenzoxazine ring system (**3**) is a heterocycle comprised of 2*H*-indazole (**5**) and dihydrobenzo-1,3-oxazine (**6**) substructures (Fig. 1). Indazoles[3] and benzoxazines[4] are well represented in the chemical literature, and compounds containing these heterocycles display interesting biological activities.[5] Until our recent report,[6] the parent indazolobenzoxazine ring system (**3**) and derivatives thereof were unknown and thus good candidates for discovery-based synthesis.

Figure 1. Heterocycles of interest: indazolobenzoxazine **3**, 2*H*-indazole (**5**), dihydrobenzo-1,3-oxazine **6**, and the 3-alkoxy-2*H*-indazole **7**.

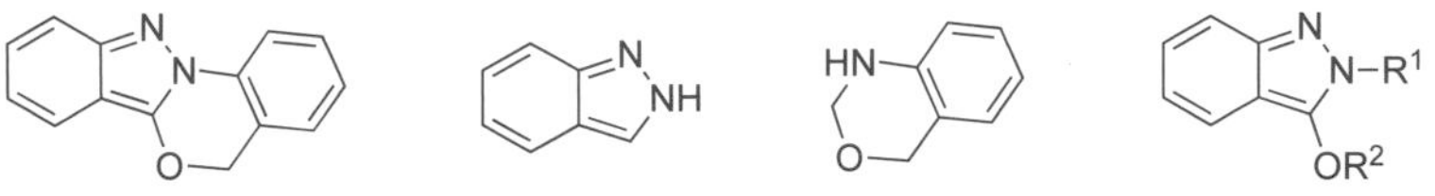

It has been demonstrated that 3-alkoxy-2*H*-indazoles (**7**) can be obtained from *o*-nitrobenzylamines via an *N*,*N*-bond-forming heterocyclization reaction mediated by potassium hydroxide in alcoholic solvent.[7] Further, we recently reported that an appropriately designed intramolecular variant of this reaction forms two fused heterocycles in a single step, providing an effective entry into the interesting indazolobenzoxazines and expanding the scope of this intriguing heterocyclization.[6] In this report, indazolobenzoxazine **3** was prepared stepwise by first performing a reductive amination of the imine resulting from the reaction of 2-nitrobenzaldehyde (**1**) with 2-aminobenzyl alcohol (**2**) to yield *o*-nitrobenzylamine **8** (Scheme 1).[6] Bis-heterocyclization of **8** to **3**, presumably via a nitroso imine intermediate, was achieved under basic conditions using aqueous KOH in isopropanol or methanol. It is of note that intramolecular cyclization (**8** → **3**) is preferred to the potentially competing intermolecular possibility (e.g., **8** → **7** where R^2 is dependant on the solvent employed).[7]

This stepwise approach, Scheme 1, delivered **3** in a 61% overall yield, demonstrating this to be a viable route for the preparation of *5H*-indazolo[3,2-*b*]benzo[*d*]-1,3-oxazines.

Scheme 1. Our initial stepwise route to the parent indazolobenzoxazine **3**.[6]

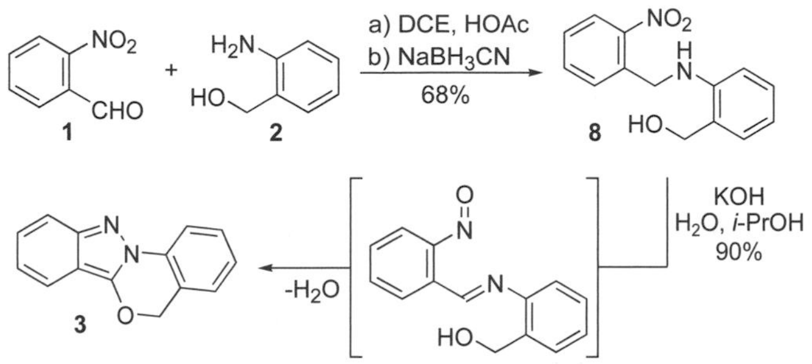

We then focused on optimization of our reported conditions to improve the efficiency and yield of this reaction. It was first determined that *o*-nitrobenzylamine **8** did not require purification; the crude material could be used in the subsequent bis-heterocyclization with no loss in yield. Further, it was established that switching the reductive amination solvent from 1,2-dicholoroethane (DCE) to methanol reduced reaction time without diminishing the yield. As the intramolecular bis-heterocyclization proceeded well in methanol, it seemed prudent to attempt both reactions in one-pot. Further modifications included a slight increase in the equivalents of 2-aminobenzyl alcohol (**2**) and addition of a potassium carbonate washing step to the workup. Finally, it was determined that the reaction time of the bis-heterocyclization could be significantly decreased by raising the reaction temperature to reflux. By integrating all of these modifications, indazolobenzoxazine **3** can be synthesized from **1** and **2** in one-pot, with a total reaction time of less than 6 h.

Sodium cyanoborohydride, the reducing agent of our initial approach, is highly toxic and residual cyanide is often detected in both the product and in generated waste.[8] We therefore proceeded to explore alternative reducing agents. Sodium borohydride successfully yielded indazolobenzoxazine **3**, but in very low yield and purity. 2-Picoline borane (pic·BH₃) has been reported as an excellent alternative to sodium cyanoborohydride and has been used to perform reductive amination reactions in methanol.[8] To our delight, this reagent was perfectly compatible with our one-pot method. Using pic·BH₃, indazolobenzoxazine **3** is obtained in 88.2-89.7% yield, with a reaction time of less than 5 h (**Method A**) and, more importantly, cyanide contamination is completely avoided.

Concurrently, we pursued an alternate route to *o*-nitrobenzylamine **8**. Considering that **8** can also be prepared via *N*-alkylation, a one-pot protocol to synthesize indazolobenzoxazine **3** from 2-nitrobenzyl bromide (**4**) and 2-aminobenzyl alcohol (**2**) was explored. Though this nucleophilic displacement proceeds in methanol, reaction times are quite long (more than 24 h at room temperature). By heating to reflux, the reaction time was decreased to 4 h and the overall yield was increased. Subsequent addition of potassium hydroxide and further heating affords indazolobenzoxazine **3** in 78.8–82.9% yield (**Method B**). This method, although slightly lower in yield, significantly increases the scope of our methodology by extending the list of possible starting materials to include *o*-nitrobenzyl bromides. Further, properly modified benzyl alcohols (-OMs, -OTs, -OTf) are also potential starting points for the synthesis of indazolobenzoxazines.

In summary, we have developed two efficient, one-pot, scalable methods to prepare the 5*H*-indazolo-[3,2-*b*]benzo[*d*]-1,3-oxazine system, each from commercially available materials. Either method gives pure product in excellent yield (with **Method A** affording yields approaching 90%) and will allow access to a variety of novel indazolobenzoxazines.

1. Department of Chemistry, University of California, One Shields Avenue, Davis, CA 95616. E-mail: mjkurth@ucdavis.edu; Fax: (530) 752-8995; Tel: (530) 554-2145. Authors Solano and Butler contributed equally to this work.
2. Department of Chemistry, American University of Beirut, Beirut, Lebanon.
3. A review on indazoles: Stadlbauer, W. *Science of Synthesis* **2002**, *12*, 227-324.
4. Selected examples of benzoxazines: (a) Basheer, A.; Rappoport, Z. *J. Org. Chem.* **2006**, *71*, 9743-9750. (b) Spagnol, G.; Rajca, A.; Rajca, S. *J. Org. Chem.* **2007**, *72*, 1867-1869.
5. Selected examples of benzoxazine bioactivity: (a) Clark, R. D.; Caroon, J. M.; Kluge, A. F.; Repke, D. B.; Roszkowski, A. P.; Strosberg, A. M.; Baker, S.; Bitter, S. M.; Okada, M. D. *J. Med. Chem.* **1983**, *26*, 657-661. (b) Goodman, K. B.; Cui, H.; Dowdell, S. E.; Gaitanopoulos, D. E.; Ivy, R. L.; Sehon, C. A.; Stavenger, R. A.; Wang, G. Z.; Viet, A. Q.;

Xu, W.; Ye, G.; Semus, S. F.; Evans, C.; Fries, H. E.; Jolivette, L. J.; Kirkpatrick, R. B.; Dul, E.; Khandekar, S. S.; Yi, T.; Jung, D. K.; Wright, L. L.; Smith, G. K.; Behm, D. J.; Bentley, R.; Doe, C. P.; Hu, E.; Lee, D. *J. Med. Chem.* **2007**, *50*, 6-9. Selected examples of indazole bioactivity: (c) Huang, L.-J.; Shih, M.-L.; Chen, H.-S.; Pan, S.-L.; Teng, C.-M.; Lee, F.-Y.; Kuo, S.-C. *Bioorg. Med. Chem.* **2006**, *14*, 528-536. (d) Matsuoka, H.; Ohi, N.; Mihara, M.; Suzuki, H.; Miyamoto, K.; Maruyama, N.; Tsuji, K.; Kato, N.; Akimoto, T.; et al. *J. Med. Chem.* **1997**, *40*, 105-111.

6. Butler, J. D.; Solano, D. M.; Robins, L. I.; Haddadin, M. J.; Kurth, M. J. *J. Org. Chem.* **2008**, *73*, 234-240.

7. (a) Mills, A. D.; Maloney, P.; Hassanein, E.; Haddadin, M. J.; Kurth, M. J. *J. Comb. Chem.* **2007**, *9*, 171-177. (b) Mills, A. D.; Nazer, M. Z.; Haddadin, M. J.; Kurth, M. J. *J. Org. Chem.* **2006**, *71*, 2687-2689.

8. Sato, S.; Sakamoto, T.; Miyazawa, E.; Kikugawa, Y. *Tetrahedron* **2004**, *60*, 7899-7906.

Appendix
Chemical Abstracts Nomenclature; (Registry Number)

2-Nitrobenzaldehyde; (552-89-6)

2-Aminobenzyl alcohol: Benzenemethanol, 2-amino-; (5344-90-1)

2-Picoline borane: Boron, trihydro(2-methylpyridine)-, (T-4)-; (3999-38-0)

5*H*-Indazolo-[3,2-b]benzo[d]-1,3-oxazine: 5*H*-Indazolo[2,3-a][3,1]benzoxazine; (342-86-9)

2-Nitrobenzyl bromide: Benzene, 1-(bromomethyl)-2-nitro-; (3958-60-9)

Diisopropylethyl amine: 2-Propanamine, *N*-ethyl-*N*-(1-methylethyl)-; (7087-68-5)

Mark J. Kurth received his B.A. in chemistry from the University of Northern Iowa in 1975 where he did undergraduate research with Professor James G. MacMillan. He obtained his Ph.D. from the University of Minnesota in the laboratory of Professor Thomas R. Hoye (1980) and subsequently did a one year postdoctoral study with Professor Wolfgang Oppolzer at the Université de Genève. He has been a faculty member in the Department of Chemistry at the University of California, Davis since 1981. His research interests focus on heterocyclic chemistry with emphasis on compounds of biological importance.

Danielle M. Solano received her B.S. in Biochemistry from the California Polytechnic State University, San Luis Obispo in 2004. She obtained her M.A. in Chemistry from Boston University in 2006, and is currently pursuing her Ph.D. at the University of California, Davis under the supervision of Professor Mark J. Kurth. Her research is focused on heterocyclic chemistry and includes methodology development, library synthesis, insecticidal applications, and the development of α4β1 integrin antagonists towards the treatment of lymphoma.

Jeffrey D. Butler, originally from Oklahoma, received his B.S. degree in 2006 from California State University, Bakersfield where he worked in the laboratory of Professor Carl Kemnitz investigating the reactive intermediates of ozonolysis proposed by Rudolf Criegee. He is currently pursuing his Ph.D. at the University of California, Davis under the supervision of Professor Mark J. Kurth. His research is focused on the synthesis of novel heterocyclic libraries, asymmetric catalysis, and xenobiotics associated with primary bilary cirrocis (PBC).

Makhluf J. Haddadin received his B.Sc. and M.Sc. (Professor C.H. Issidorides) from the American University of Beirut, Lebanon; and his Ph.D. (Professor A. Hassner) from the University of Colorado, Boulder. He spent two years of post-doctoral work at Harvard University (Professor L.F. Fieser) and is currently Professor of Chemistry at the American University of Beirut. His research interests lie in synthetic heterocyclic chemistry.

Kyle D. Baucom obtained his B.S. degree in Chemistry from University of California at Santa Barbara in 2006 where he worked as an undergrad under the supervision of Jeffery Bode. In the fall of 2006 he began working at Amgen Inc. as a Process Chemist in the Chemical Process Research and Development group. He is currently a Senior Research Associate for Amgen.

PREPARATION OF CHIRAL AND ACHIRAL TRIAZOLIUM SALTS: CARBENE PRECURSORS WITH DEMONSTRATED SYNTHETIC UTILITY

A.

B.

C.

Submitted by Harit U. Vora, Stephen P. Lathrop, Nathan T. Reynolds, Mark S. Kerr, Javier Read de Alaniz, and Tomislav Rovis.[1]
Checked by Spandan Chennamadhavuni and Huw M. L. Davies.

1. Procedure

A. 4,4a,9,9a-Tetrahydro-1-oxa-4-aza-fluoren-3-one (1). A 2.0 L, single-necked, round-bottom flask fitted with a magnetic stir bar (1 ¼ inch x 5/8 inch egg shaped), and a rubber septum with a needle inlet is flame dried and then cooled under an argon atmosphere. The septum is removed and the flask is charged with sodium hydride (3.48 g, 60% in mineral oil, 87.15 mmol, 1.3 equiv) (Note 1) via a powder funnel, followed by *n*-pentane

350

(300 mL) (Note 2). The septum is then replaced and the heterogeneous mixture is subjected to an argon atmosphere by means of a needle in the septum followed by agitation for 20 min. Agitation is stopped and the sodium hydride is allowed to settle out of solution for 30 min. *n*-Pentane is removed via cannula by increasing the positive pressure of argon in the flask. Residual mineral oil present in sodium hydride is removed by washing with an additional 100 mL of pentane as above. The combined pentane solutions containing residual sodium hydride are quenched with isopropanol (Note 3). The septum is briefly removed to add anhydrous tetrahydrofuran (1.0 L) (Note 4) by means of an addition funnel followed by replacement of the septum and refilling the reaction flask with argon. The reaction flask is placed in a −10 °C ice-brine bath (Note 5) with agitation and allowed to cool for 30 min. The septum is then replaced with a powder funnel and (1*R*,2*S*)-(-)-*cis*-1-amino-2-indanol (5 g, 33.52 mmol) (Note 6) is added followed by refilling of the reaction flask with argon. Within 10 min, the reaction mixture turns heterogeneous and the color changes to light purple. The septum is then replaced with a powder funnel and a further batch of (1*R*,2*S*)-(-)-*cis*-1-amino-2-indanol (5 g, 33.52 mmol) (Note 6) is added followed by refilling of the reaction flask with argon. The flask is then fitted with a reflux condenser and the stirred reaction mixture is heated in a 70 °C oil bath for 40 min under an argon atmosphere. The solution is then placed into a −10 °C ice-brine bath and allowed to cool for 1 h while stirring. Ethyl chloroacetate (8.38 g, 7.32 mL, 68.37 mmol, 1.02 equiv) (Note 7) is added via syringe over 5 min. The reaction flask is removed from the ice-brine bath and the mixture is stirred for 30 min at room temperature. The solution is then placed in an oil bath heated to 70 °C and stirred for 1 h under argon. After cooling to room temperature, brine (30 mL) is slowly added to the reaction flask to quench the reaction, and then THF is removed under reduced pressure using a rotarory evaporator. The homogeneous deep purple reaction mixture is then poured into a 2.0-L separatory funnel and brine (200 mL) is added. The reaction mixture is extracted using ethyl acetate (2 x 200 mL) (Note 8). The combined organic layers are then dried using anhydrous MgSO$_4$ (50 g) (Note 9) followed by vacuum filtration through a coarse fritted funnel. The resultant purple solution is transferred into a 1.0-L round-bottomed flask in approx. 500 mL portions and concentrated *in vacuo* to dryness. To the 1.0-L round bottom flask containing the crude light brown solid is added a stir bar and hexanes (300 mL) (Note 10). The sides of the round-bottomed flask are scraped with a metal spatula to ensure that all of

the crude solid material rests at the bottom of the flask. The flask is fitted with a reflux condenser left open to the atmosphere and the heterogeneous mixture is stirred vigorously in an oil bath heated to 70 °C for a period of 2 h. After cooling to room temperature, vacuum filtration affords an off-white solid which is transferred into a 100-mL round-bottomed flask by means of a powder funnel and dried under vacuum (2 mmHg) in an oil bath heated to 70 °C for 1 h, affording 10.21–10.95 g (80–86%) of **1** (Note 11).

B. *2-Pentafluorophenyl-6,10b-dihydro-4H,5aH-5-oxa-3,10c-diaza-2-azonia-cyclopenta[c]fluorene; tetrafluoroborate (2)*. To a flame-dried 1.0-L single-necked, round-bottomed flask with magnetic stir bar (1 ¼ inch x 5/8 inch egg shaped) is added morpholinone **1** (10.00 g, 52.85 mmol, 1.0 equiv) by means of a powder funnel. The flask is then evacuated and back-filled with argon. Methylene chloride (300 mL) (Note 12) and trimethyloxonium tetrafluoroborate (7.82 g, 52.85 mmol, 1.0 equiv) (Note 13) are then added via powder funnel and the flask is fitted with a septum and a needle connected to an argon line. The heterogeneous mixture is stirred at room temperature until the reaction is homogeneous (Note 14). Pentafluorophenylhydrazine (10.47 g, 52.85 mmol, 1.0 equiv) (Note 15) is added in a single portion by means of a powder funnel followed by replacement of the powder funnel with a septum and a needle connected to an argon line, and then the reaction mixture is stirred for 4 h (Note 16). The magnetic stir bar (Note 17) is removed followed by removal of the solvent *in vacuo*. The 1.0-L flask is then placed in an oil bath heated to 100 °C for 1 h under vacuum (2 mm Hg). A magnetic stir bar, chlorobenzene (300 mL) (Note 18) and triethyl orthoformate (19.58 g, 21.97 mL, 132.13 mmol, 2.5 equiv) (Note 19) are then added using a syringe, and the flask is fitted with a reflux condenser, placed in an oil bath heated to 130 °C and stirred for 24 h open to the atmosphere. Triethyl orthoformate (19.58 g, 21.97 mL, 132.13 mmol, 2.5 equiv) (Note 18) is then added via syringe followed by continued agitation for 24 h. A third portion of triethyl orthoformate (19.58 g, 21.97 mL, 132.13 mmol, 2.5 equiv) (Note 19) is added via syringe followed by continued agitation for 24 h. After removal of the reaction flask from the oil bath and cooling to room temperature, the solution is added to a 1.0-L round-bottomed flask containing toluene (300 mL) (Note 20) that is agitated with a magnetic stir bar. The reaction flask is then rinsed with toluene (50 mL) (Note 20) followed by addition of the heterogeneous mixture to the 1-L flask containing the crude product. The slurry is stirred for 10 min followed by vacuum filtration. The filtrate is rinsed with toluene

352

(200 mL) (Note 20) and hexane (200 mL) (Note 10). The solid is then transferred to a 125-mL Erlenmeyer flask containing a stir bar by means of a powder funnel, triturated with ethyl acetate (20 mL) (Note 8) and methanol (5 mL) (Note 21) and stirred vigorously for 30 min. The slurry is then filtered through a medium frit funnel and the filter cake is washed with cold ethyl acetate (15 mL) (Note 8) via a glass pipette to yield **2** as an off-white solid. The off-white solid is transferred to a 100-mL round-bottomed flask by means of a powder funnel, placed in an oil bath heated to 100 °C and subjected to vacuum (2 mmHg) for 1 h, affording 15.06–15.79 g (61–64%) of **2** (Note 22).

   C. 2-Pentafluorophenyl-6,7-dihydro-5H-pyrrolo[2,1-c][1,2,4]triazol-2-ium; tetrafluoroborate (3). Into a flame-dried 1.0-L flask equipped with a magnetic stir bar is added 2-pyrrolidinone (5.00 g, 58.75 mmol, 1.0 equiv) (Note 23). The flask is then evacuated and back-filled with argon. Methylene chloride (300 mL) (Note 12) and trimethyloxonium tetrafluoroborate (8.69 g, 58.75 mmol, 1.0 equiv) (Note 13) are added by means of a powder funnel and the flask is then fitted with a septum and a needle connected to an argon line. The heterogeneous mixture is stirred at room temperature until the solution is homogeneous (Note 14). Pentafluorophenylhydrazine (11.64 g, 58.75 mmol, 1.0 equiv) (Note 15) is added in a single portion by means of a powder funnel followed by replacement of the powder funnel with a septum and a needle connected to an argon line. The reaction mixture is stirred for 4 h (Note 16). The magnetic stir bar is removed (Note 17) followed by removal of the solvent in vacuo. The 1.0-L flask is then placed in an oil bath heated to 100 °C for 1 h under vacuum (2 mmHg). A magnetic stirring bar, chlorobenzene (300 mL) (Note 18) and triethyl orthoformate (17.41 g, 19.54 mL, 117.5 mmol, 2.0 equiv) (Note 19) are then added via syringe and the flask is fitted with a reflux condenser, placed in an oil bath heated to 130 °C and stirred for 24 h open to the atmosphere. Triethyl orthoformate (17.41 g, 19.54 mL, 117.5 mmol, 2.0 equiv) (Note 19) is then added via syringe followed by continued agitation for 24 h. After removal of the reaction vessel from the oil bath and cooling to room temperature, the reaction mixture is added to a 1.0-L round-bottomed flask containing toluene (300 mL) (Note 20) that is agitated with a magnetic stir bar for 1 h. The reaction flask is then rinsed with toluene (50 mL) (Note 20) followed by addition of the heterogeneous mixture to the 1-L round-bottomed flask containing the crude product. The slurry is stirred for 10 min followed by vacuum filtration. The filtrate is rinsed with toluene (200 mL) (Note 20) and hexanes (200 mL) (Note 10). The crude brown solid is then transferred to a 125-mL Erlenmeyer flask containing a magnetic stir bar by means of a powder funnel. The solid is triturated with ethyl acetate (20 mL) (Note 8) and methanol (3 mL) (Note 20) and stirred vigorously for 30 min. The heterogeneous mixture is then filtered through a medium fritted funnel under vacuum (120 mmHg). The filter cake is then washed with cold ethyl acetate (15 mL) via a glass pipette to yield **3** (15.91–16.13 g) as an off-white powder. The off-white solid is transferred to a 100-mL round-bottomed flask by means of a powder funnel, placed in an oil bath heated to 100 °C and subjected to vacuum (2 mmHg) for 1 h, affording 15.91–16.13 g (74–76%) of **3** (Note 24).

2. Notes

1. Sodium hydride (60% dispersion in mineral oil) was purchased from Aldrich. It was determined that removal of the mineral oil in the reaction resulted in higher yields of the desired product.

2. Pentane (99.0%) was purchased from Fisher Scientific and used as received.

3. Isopropanol was purchased from Aldrich Chemical Co., and used without further purification.

4. THF (HPLC grade, H_2O = 0.003%) was purchased from Fisher Scientific Company and purified by pressure filtration under argon through activated alumina. The checkers found that the reaction gave similar results with freshly distilled THF over sodium benzophenone ketyl.

5. It was observed that cooling the reaction with an ice-brine bath resulted in an increase in chemoselectivity for the *N*-acylation product to give a higher yield of the desired regioisomer. It is crucial to maintain the reaction temperature between –10 °C and –15 °C in order to obtain a single regioisomer of the product. The checkers monitored the internal temperature of the reaction mixture using a digital thermometer.

6. (1*R*,2*S*)-(+)-*cis*-1-Amino-2-indanol (99%) was purchased from Aldrich, and used without further purification.

7. Ethyl chloroacetate (99%) was purchased from Aldrich and used without further purification.

8. Ethyl acetate (99.9%) was purchased from Fischer Scientific and used without further purification.

354

9. Anhydrous magnesium sulfate was purchased from Fischer Scientific.

10. Hexanes (HPLC grade, H_2O = 0.0005%) was purchased from Aldrich and used as received.

11. The spectral data for 4,4a,9,9a-tetrahydro-1-oxa-4-aza-fluoren-3-one (**1**) are as follows: R_f = 0.18 (EtOAc); Mp 180–183 °C (dec.); $[\alpha]_D^{23}$ - 17.3 (*c* 1.08, MeOH); IR (neat) 3179, 3050, 2910, 1681, 1646, 1484, 1417 cm^{-1}; ^1H NMR (400 MHz, acetone-d_6) δ: 2.97 (d, J = 16.8 Hz, 1 H), 3.23 (dd, J = 16.6, 4.9 Hz, 1 H), 3.89 (d, J = 16.2 Hz, 1 H), 4.05 (d, J = 16.2 Hz, 1 H), 4.57 (t, J = 4.6 Hz, 1 H), 4.82 (t, J = 4.0 Hz, 1 H), 7.23-7.29 (m, 3 H), 7.46-7.49 (m, 1 H), 8.21 (br s, 1 H); ^{13}C NMR (101 MHz, acetone-d_6) δ: 38.2, 59.4, 67.1, 77.1, 124.7, 125.7, 127.6, 128.4, 140.8, 143.1, 168.5; HRMS (APCI+) *m/z* calcd. for $C_{11}H_{12}NO_2$ (M^+) 190.0863; found 190.0862; Anal. calcd. for $C_{11}H_{11}NO_2$: C, 69.83; H, 5.86; N, 7.40; found: C, 69.68; H, 5.83; N, 7.33.

12. Methylene chloride (99.9%) was purchased from Fischer Scientific and purified by pressure filtration under argon through activated alumina. A Karl-Fisher Titrator was used to determine the water content in methylene chloride (<20 ppm/mL).

13. Trimethyloxonium tetrafluoroborate (95%) was purchased from Aldrich Chemical Co. The bottles were stored in the refrigerator at –15 °C and a freshly opened bottle was used for each reaction. Direct contact of with skin trimethyloxonium tetrafluoroborate must be avoided because of its caustic nature and alkylating properties. It is absolutely critical to add only one equivalent of trimethyloxonium tetrafluoroborate. Excess of this reagent will result in lower yields and the product may not precipitate out. The checker also found that it is essential to add trimethyloxonium tetrafluoroborate under an argon atmosphere because of its highly hygroscopic nature.

14. Upon complete dissolution of the trimethyloxonium tetrafluoroborate an aliquot is removed from the reaction vessel and concentrated in vacuo. The oil is then dissolved in acetone-d_6 and ^1H NMR is used to verify the consumption of morpholinone. It usually takes 1–2 h for the reaction mixture to turn completely homogenous.

15. Pentafluorophenylhydrazine (97%) was purchased from Aldrich Chemical Co. and used without further purification.

16. Disappearance of activated amidate (typically ~4 h) could also be observed by measuring the ^1H NMR of an aliquot.

17. The magnetic stir bar was removed, because otherwise it will stick firmly to the bottom of the flask and will not stir during the next sequence of the procedure.

18. Chlorobenzene (99%) was purchased from Aldrich Chemical Co. and used without further purification.

19. Triethylorthoformate (98%) was purchased from Acros Organics, and used without further purification.

20. Toluene (99.9%) reagent grade was purchased from Fisher Scientific and used without further purification.

21. Methanol (99.9%) was purchased from Fischer Scientific and used without further purification.

22. The spectral data for 2-pentafluorophenyl-6,10b-dihydro-4H,5aH-5-oxa-3,10c-diaza-2-azoniacyclopenta[c]fluorine, tetrafluoroborate (**2**) are as follows: R_f = 0.22 (3:1; CH$_2$Cl$_2$:acetone); Mp 223–226 °C; $[\alpha]_D^{23}$ -130.80 (c 1.28, MeCN); IR(neat) 3147, 3106, 3028, 2967, 1595, 1530, 1517, 1487, 1461, 1056, 1046, 998 cm^{-1}; ^{1}H NMR (400 MHz, acetone-d$_6$) δ: 3.28 (d, J = 17.2 Hz, 1 H), 3.55 (dd, J = 17.1, 4.9 Hz, 1 H), 5.19 (t, J = 4.5 Hz, 1 H), 5.25 (d, J = 16.4 Hz, 1 H), 5.39 (d, J = 16.4 Hz, 1 H), 6.33 (d, J = 4.0 Hz, 1 H), 7.34 (t, J = 7.3 Hz, 1 H), 7.43 (q, J = 7.4 Hz, 2 H), 7.63 (d, J = 7.6 Hz, 1 H), 11.09 (br s, 1 H); ^{13}C NMR (101 MHz, acetone-d$_6$) δ: 37.9, 60.8, 63.5, 78.2, 125.2, 126.4, 128.1, 130.4, 136.2, 141.7, 147.1, 152.5; HRMS (APCI+) m/z calcd. for C$_{18}$H$_{11}$N$_3$OF$_5$ (M$^+$) 380.0817; found 380.0816; Anal. calcd. for C$_{18}$H$_{11}$BF$_9$N$_3$O: C, 46.28; H, 2.37; N, 9.00; found: C, 45.95; H, 2.23; N, 9.03. Based on the observed analytical results we propose the salt to be between anhydrous and a half-hydrate. Anal. calcd. for C$_{18}$H$_{11}$BF$_9$N$_3$O•1/2H$_2$O: C, 45.41; H, 2.54; N, 8.83. The checker's efforts to get an elemental analysis of an anhydrous sample were unsuccessful.

23. 2-Pyrrolidinone (≥ 99%) was purchased from Aldrich Chemical Co, and used without further purification.

24. The spectral data for 2-pentafluorophenyl-6,7-dihydro-5H-pyrrolo[2,1-c][1,2,4]triazol-2-ium; tetrafluoroborate (**3**) are as follows: R_f = 0.13 (3:1; CH$_2$Cl$_2$:acetone); Mp 242–245 °C; IR (neat) 3145, 3097, 2983, 1655, 1604, 1524, 1499, 1028, 994, 875 cm^{-1}; ^{1}H NMR (400 MHz, acetone-d$_6$) δ: 3.00 (ddd, J = 15.0, 7.7, 7.7 Hz, 2 H), 3.42 (t, J = 8.0 Hz, 2 H), 4.76 (t, J = 8.0 Hz, 2 H), 10.19 (s, 1 H); ^{13}C NMR (101 MHz, acetone-d$_6$) δ: 22.6, 27.6, 49.5, 137.8, 140.4, 142.9, 143.3, 144.5, 145.6, 145.8, 165.8; HRMS (APCI+) m/z calcd. for C$_{11}$H$_7$N$_3$F$_5$ (M$^+$) 276.0555; found 276.0556; Anal.

calcd. for $C_{11}H_7BF_9N_3$: C, 36.40; H, 1.94; N, 11.58; found: C, 36.45; H, 1.76; N, 11.54.

Safety and Waste Disposal Information

All hazardous materials should be handled and disposed of in accordance with "Prudent Practices in the Laboratory"; National Academy Press; Washington, DC, 1998.

3. Discussion

In recent years, *N*-heterocyclic carbenes have attracted significant attention as catalysts in facilitating umpolung reactivity.[2] Ukai's utilization of a thiazolium salt as a carbene precursor in the benzoin reaction[3] and Breslow's elucidation of the mechanism[4] of this reaction displayed the potential use of these nucleophilic carbene precatalysts. These early reports led to the subsequent development of carbene precatalysts to facilitate a multitude of Umpolung reactions. In 1996, Enders and coworkers demonstrated the asymmetric benzoin reaction with a triazolium salt as the carbene precursor in moderate yield and enantioselectivity.[5] Bicyclic triazolium carbene precursors were next studied to see if greater asymmetric induction could be obtained in various Umpolung reactions.

The aminoindanol-derived carbene scaffold introduced by us in 2002 has proven to be one of the most general for these carbene precursors.[6] In subsequent work, we noted that the electronic nature of the N-aryl substituent has a profound effect on reactivity and control of enantioselectivity in the Stetter reaction. This led to the development of triazolium precatalyst **2**. Application of triazolium **2** in the asymmetric Stetter reaction generates quaternary stereocenters in high enantiomeric excess and very good chemical yields.[7] The umpolung reactivity with triazolium precatalyst **2** can also be extended towards the desymmetrization of cyclohexadienones to yield hydrobenzofuranones in excellent enantioselectivities and chemical yields.[8] Extension of this work to vinylphosphine oxides and vinylphosphonates as the Michael acceptor can be achieved to generate keto phosphonates and keto phospine oxides in excellent enantioselectivities and yields.[9] The application of precatalyst **2** towards the rapid assembly of the tetracyclic core of natural product FD-838 was accomplished.[10] The asymmetric synthesis of α-chloro esters via an acyl anion redox reaction was also accomplished with precatalyst **2** to yield the respective enantioenriched aryl esters.[11]

1. Department of Chemistry, Colorado State University, Fort Collins, CO, 80523, USA.
2. (a) Enders, D.; Niemeier, O; Hensler, A. *Chem. Rev.* **2007**, 107, 5606–5655. (b) Moore, J. L.; Rovis, T. *Top. Curr. Chem.* **2009**, 290, 77-144.
3. Ukai, T.; Tanaka, R.; Dokawa, T. *J. Pharm. Soc. Jpn.* **1943**, 63, 296-300.
4. Breslow, R. *J. Am. Chem. Soc.* **1958**, 80, 3719-3726.
5. Enders, D.; Breuer, K.; Teles, J. H. *Helv. Chim. Acta* **1996**, 79, 1217-1221.
6. Kerr, M. S.; Read de Alaniz, J.; Rovis, T. *J. Am. Chem. Soc.* **2002**, 124, 10298-10299.
7. (a) Kerr, M. S.; Rovis, T.; *J. Am. Chem. Soc.* **2004**, 126, 8876-8877. (b) Moore, J. L.; Kerr, M. S., Rovis, T. *Tetrahedron* **2006**, 128, 2552-2553. (c) Read de Alaniz, J.; Kerr, M. S.; Moore, J. L.; Rovis, T. *J. Org. Chem.* **2008**, 73, 2033-2040. (d) Read de Alaniz, J.; Rovis, T. *Syn. Lett.* **2009**, 1189-1207.
8. (a) Liu, Q.; Rovis, T. *J. Am. Chem. Soc.* **2006**, 128, 2552-2553. (b) Liu, Q.; Rovis, T. *Org. Proc. Res. Dev.* **2007**, 11, 598-604.
9. Cullen, S. C.; Rovis, T. *Org. Lett.* **2008**, 10, 3141-3144.
10. Orellana, A.; Rovis, T. *Chem. Comm.* **2008**, 730-732.
11. Reynolds N. T.; Rovis, T. *J. Am. Chem. Soc.* **2005**, 127, 16406-16407.

Appendix
Chemical Abstracts Nomenclature; (Registry Number)

4,4a,9,9a-Tetrahydro-1-oxa-4-azafluoren-3-one: Indeno[2,1-*b*]-1,4-oxazin-3(2*H*)-one, 4,4a,9,9a-tetrahydro-, (4a*R*,9a*S*)-; (862095-79-2)

(1*R*, 2*S*)-(+)-*cis*-1-Amino-2-indanol; (136030-00-7)

Ethyl chloroacetate; (105-39-5)

2-Pentafluorophenyl-6,10b-dihydro-4*H*,5a*H*-5-oxa-3,10c-diaza-2-azoniacyclopenta[*c*]fluorene; tetrafluoroborate; (740816-14-2)

Trimethyloxonium tetrafluoroborate; (420-37-1)

Pentafluorophenylhydrazine; (828-73-9)

358

Triethyl orthoformate: Ethane, 1,1',1"-[methylidynetris(oxy)]tris-; (122-51-0)

2-Pentafluorophenyl-6,7-dihydro-5*H*-pyrrolo[2,1-*c*][1,2,4]triazol-2-ium; tetrafluoroborate; (862095-91-8)

2-Pyrrolidinone; (616-45-5)

Tomislav Rovis was born in Zagreb in the former Yugoslavia but was largely raised in Southern Ontario, Canada. Following his undergraduate studies at the University of Toronto, he earned his Ph.D. degree at the same institution in 1998 under the direction of Professor Mark Lautens. From 1998-2000, he was an NSERC postdoctoral fellow at Harvard University with Professor David A. Evans. In 2000, he began his independent career at Colorado State University and was promoted in 2005 to Associate Professor and in 2008 to Professor. He currently holds the John K. Stille Chair in Chemistry.

Harit U. Vora was born in 1979 in Calcutta, India and raised in Tobyhanna, PA. He received his undergraduate education at the University of Pittsburgh where he carried out research under the guidance of Professor Paul E. Floreancig. He then obtained a position with the Medicinal Chemistry division at Roche in Palo Alto, CA. He is now pursuing his graduate studies at Colorado State University where his research under the guidance of Professor Tomislav Rovis involves the development of new synthetic methods involving *N*-heterocyclic carbenes.

Stephen P. Lathrop was born in Indianapolis, Indiana in 1982. He received his BS degree in Chemistry in 2005 from Indiana University. Upon graduation, he began his doctoral work at Colorado State University with Tomislav Rovis. His graduate research has focused on the intramolecular Stetter reaction and its applications towards natural product synthesis.

Nathan T. Reynolds was born in 1977 in Ithaca, New York. After graduating from the State University of New York at Binghamton, he obtained his PhD in 2006 at Colorado State University under the direction of Professor Tomislav Rovis, where he worked on the preparation and synthetic applications of N-heterocyclic carbenes. He is currently a Senior Research Scientist at AMRI.

Mark S. Kerr was born in Duluth, Minnesota in 1978. He received his Bachelor of Arts in Chemistry from Coe College in Cedar Rapids, IA in 2000. That year he began his doctoral studies at Colorado State University under Professor Tomislav Rovis. His graduate work focused on the development of a family of chiral nucleophilic carbenes as catalysts for the intramolecular Stetter reaction. Upon completion of his PhD. in 2006, he joined the laboratory of professor David W. C. MacMillan at the Merck Center for Catalysis at Princeton University, working on the implementation of organocascade catalysis in natural product synthesis. He is currently a Scientist at Eli Lilly.

360

Javier Read de Alaniz was born and raised in Las Vegas, New Mexico. He received his B.S. degree from Fort Lewis College in 1999 where he conducted undergraduate research under the direction of Professor William R. Bartlett. His Ph.D. degree was obtained under the direction of Tomislav Rovis at Colorado State University in 2006. During that time he was the recipient of an NIH Ruth L. Kirschstein minority pre-doctoral fellowship. From 2006-2009, he was a University of California President's Postdoctoral Fellow with Larry E. Overman at the University of California, Irvine. He is currently an Assistant Professor of Chemistry at University of California, Santa Barbara.

Spandan Chennamadhavuni received his BS degree in chemistry (2000) from Kakatiya University (India), and earned his MS degree in organic chemistry (2002) from Osmania University (India). He worked as research assistant for Jon C. Antilla at University of Mississippi for three years where he carried out methodological studies on Bronsted acid catalytic amidation of imines, amidation of enones and developed enantio and diastereoselective versions of imine amidation. He moved to SUNY Buffalo to pursue doctoral studies under the supervision of Professor Huw M. L. Davies. He moved to Emory University along with rest of the Davies group to continue his PhD studies. His current project includes synthesis of various pharmaceutical agents utilizing rhodium carbenoid chemistry.

SYNTHESIS OF A *N*-MESITYL SUBSTITUTED AMINOINDANOL-DERIVED TRIAZOLIUM SALT
[(5a*S*,10b*R*)-5a,10b-Dihydro-2-(2,4,6-trimethylphenyl)-4*H*,6*H*-indeno[2,1-b]-1,2,4-triazolo[4,3-d]-1,4-oxazinium chloride]

Submitted by Justin R. Struble and Jeffrey W. Bode.[1]
Checked by Yajing Lian and Huw M. L. Davies.

1. Procedure

A. 2-Mesitylhydrazinium chloride (1). A 3-necked, 1-L round-bottomed flask is equipped with a Teflon-coated overhead stirrer in the center port (Note 1) and placed in a –10 °C acetone bath (Note 2). The flask is charged

Orgn. Synth. **2010**, *87*, 362-376
Published on the Web 6/15/2010

with conc. HCl (26.5 mL) and water (55 mL) (Notes 3 and 4). Another port of the reaction vessel is equipped with a 60-mL pressure-equalizing addition funnel and charged with 2,4,6-trimethylaniline (15.0 mL, 107 mmol, 1.0 equiv) (Note 5), which is added dropwise over a period of approximately 5 min resulting in a thick white slurry. Stirring is continued for an additional 15 min (Notes 6 and 7). At this time, the third port is equipped with a thermometer to monitor the internal temperature of the reaction. The addition funnel is replaced with a new 60-mL addition funnel that is charged with a freshly prepared solution of $NaNO_2$ (7.38 g, 107 mmol, 1.00 equiv) in water (20 mL) (Note 4). This solution is added dropwise so as to maintain an internal temperature of < 5 °C (Note 8). The addition funnel is washed with H_2O (5 mL) (Note 4) and stirring is continued for an additional 30 min. The addition funnel is replaced with a new addition funnel which is charged with a solution of $SnCl_2 \cdot 2H_2O$ (60.4 g, 268 mmol, 2.50 equiv) in 1:1 conc. HCl/H_2O (60 mL) (Note 9). The $SnCl_2$ solution is added dropwise over 4 h maintaining an internal reaction temperature of < 5 °C. The addition funnel and thermometer are removed. The heterogeneous orange mixture is stirred at 0 °C for 1 h before being allowed to warm to ambient temperature and vigorous stirring is maintained for 16 h. The mixture is cooled by placing the reaction vessel in the refrigerator operating at 4 °C for 2 h. The orange precipitate is collected by suction filtration (Note 10) and the reaction vessel is washed with brine (5 x 50 mL) that is poured over the collected solid with each wash. The orange solid is transferred to a 1-L round-bottom flask and Et_2O (300 mL) (Note 11) and a Teflon-coated magnetic stir bar are added. The heterogeneous mixture is placed in an ice/brine bath and the flask is equipped with a pressure-equalizing addition funnel charged with aq 10 M NaOH (200 mL) that is added over approximately 30 min. An additional 100 mL of H_2O (Note 4) is added and stirring is continued for 1 h before the biphasic mixture is allowed to warm to ambient temperature and transferred to a 1-L separatory funnel. The organic layer is separated and the aqueous phase extracted with Et_2O (2 x 200 mL). The organic fractions are combined and washed with brine (250 mL), dried over $MgSO_4$, and filtered (Note 10) into a dry 1-L, two-necked, round-bottomed flask equipped with an overhead stirrer. The other neck is filled with a rubber septum. The drying agent is washed with Et_2O (3 x 30 mL). The orange solution is placed under an atmosphere of $N_2(g)$ and immersed in an ice/brine bath. A solution of HCl (4 M in dioxane, 27.0 mL, 108 mmol, 1.0 equiv) is added via syringe over 15 min inducing a white precipitate. Stirring is maintained for 30 min and the solution is allowed to warm to ambient temperature. The overhead stirrer is removed and washed with Et$_2$O (20 mL). The solvent is evaporated under reduced pressure and MeOH (400 mL) is added. The heterogeneous mixture is heated to 50 °C until the solid is dissolved [approximately for 10 min] and the solvent is then removed to dryness using a rotary evaporator under vacuum to afford a crystalline product (Note 12). Et$_2$O (500 mL) is added and the center neck is equipped with a water condenser, while the other neck has a glass stopper. The reaction mixture is heated to reflux for 30 min and then the heterogeneous mixture is cooled to room temperature. The pale orange precipitate is collected by suction filtration (Note 10) and washed with Et$_2$O (50 mL). The crude solid is transferred to a 200-mL, round-bottomed flask and the residual volatiles are removed under high vacuum (ambient temperature, <0.75 mmHg, 12 h). A solution of 200 proof EtOH/Et$_2$O (5:1, 60 mL) is added, and the suspension is placed into a sonicating bath for 30 min (Note 13). The pale orange solid is collected by suction filtration (Note 11) and washed with a solution of 200 proof EtOH/Et$_2$O (1:1, 3 x 30 mL). The filtrate is concentrated using a rotary evaporator (35 °C, 35 mmHg) to produce an orange crystalline solid. A solution of 200 proof EtOH/Et$_2$O (1:1, 15 mL) is added, and the suspension is placed into a sonicating bath for 5 min [Note 13]. The pale orange solid is collected by suction filtration (Note 10), and washed with a solution of 200 proof EtOH/Et$_2$O (1:1, 3 x 5 mL). The collected solids are combined and transferred to a 200-mL, round-bottomed flask and the residual volatiles are removed under high vacuum (ambient temperature, <0.75 mmHg, 12 h) to afford 7.20–7.92 g (36–40% yield) of 2-mesitylhydrazinium chloride (**1**) as a pale, orange powder (Note 14).

B. *(4aR,9aS)-3-Methoxy-2,4a,9,9a-tetrahydroindeno [2,1-b][1,4]oxazine (2)*. A flame-dried, 500-mL, round-bottomed flask is equipped with a Teflon-coated stir bar (Note 15), (4a*R*,9a*S*)-4,4a,9,9a-tetrahydroindeno[2,1-*b*][1,4]oxazin-3(2*H*)-one (7.00 g, 37.0 mmol, 1.00 equiv) (Note 16), CH$_2$Cl$_2$ (185 mL) (Note 17) and trimethyloxonium tetrafluoroborate (6.57 g, 44.4 mmol, 1.20 equiv) (Note 18). The flask is equipped with a septum, placed under an atmosphere of argon through a needle inserted into the septum and stirred at ambient temperature for 16 h. The light tan solution is immersed in an ice/water bath and aq sat NaHCO$_3$ (150 mL) is added over a period of 1.5 h. Stirring is maintained for an additional 1 h, then the biphasic solution is transferred to a separatory funnel, the organic phase separated, and the aqueous phase extracted with CH$_2$Cl$_2$ (4 x 100 mL). The combined

364

organic fractions are dried over MgSO$_4$, filtered, and concentrated using a rotary evaporator (25 °C, 75 mmHg). The residual volatiles are removed under high vacuum (ambient temperature, <0.75 mmHg, 12 h) to afford 6.89–7.03 g (92–93% yield) of (4aR,9aS)-3-methoxy-2,4a,9,9a-tetrahydroindeno [2,1-b][1,4]oxazine (2) as a pale brown solid which is used without further purification (Note 19).

C.   *(Z)-2-Mesityl-1-((4aR,9aS)-4,4a,9,9a-tetrahydroindeno[2,1b][1,4] oxazin-3(2H)-ylidene)hydrazinium chloride (3)*. A flame–dried, 500-mL round-bottomed flask is charged with a Teflon-coated stir bar (Note 15), 2-mesitylhydrazinium chloride (1) (6.44 g, 34.4 mmol, 1.00 equiv) and MeOH (138 mL) (Note 20), resulting in a light orange solution after stirring at ambient temperature for 5 min. To this solution is added (4aR,9aS)-3-methoxy-2,4a,9,9a-tetrahydroindeno [2,1-b][1,4]oxazine (2) (7.00 g, 34.4 mmol, 1.00 equiv) in a single portion and the mixture is stirred at ambient temperature until a red homogeneous solution forms (ca. 5 min). A catalytic amount of anhydrous HCl (4 M in dioxane, 0.86 mL, 0.10 mmol) is added to the solution. The reaction flask is equipped with a water–jacketed reflux condenser and immersed in a 60 °C silicon oil bath and the reaction mixture is stirred under an atmosphere of N$_2$(g) for 48 h. The mixture is allowed to cool to ambient temperature and the volatiles are removed using a rotary evaporator (35 °C, 30 mmHg) followed by high vacuum (ambient temperature, <0.75 mmHg) to afford a crude orange solid. The crude material is suspended in EtOAc (125 mL) (Note 21), a Teflon-coated magnetic stir bar (Note 15) is added, and the flask equipped with a water–jacketed reflux condenser. The mixture is stirred vigorously in a silicon oil bath at 90 °C under an atmosphere of N$_2$(g) for 30 min, causing a light yellow precipitate to form. The flask is removed from the oil bath, allowed to cool to ambient temperature with vigorous stirring and immersed in an ice/water bath at 0 °C with vigorous stirring. The precipitate is collected by suction filtration and washed with EtOAc (3 x 20 mL). The light yellow solid is transferred to a 20 mL vial and residual volatiles are removed under high vacuum (ambient temperature, <1 mbar, 12 h), affording 9.96–10.1 g (81–82 %) of (Z)-2-mesityl-1-((4aR,9aS)-4,4a,9,9a-tetrahydroindeno-[2,1b][1,4]-oxazin-3-(2H)ylidene)hydrazinium chloride (3) as a light yellow powder (Notes 22 and 23).

D.   *(5aS,10bR)-5a,10b-Dihydro-2-(2,4,6-trimethylphenyl)-4H,6H-indeno[2,1-b]-1,2,4-triazolo[4,3-d]-1,4-oxazinium chloride (4)*. An oven-dried, 350-mL sealed tube is charged with a Teflon-coated magnetic stir bar

(Note 15), (*Z*)-2-mesityl-1-((4a*R*,9a*S*)-4,4a,9,9a-tetrahydroindeno-[2,1b][1,4]oxazin-3(2H)-ylidene)hydrazinium chloride (**3**) (9.00 g, 25.1 mmol, 1.00 equiv), chlorobenzene (25.5 mL) (Note 24), triethylorthoformate (33.4 mL, 201 mmol, 8.0 equiv) (Note 25) and anhydrous HCl (4 M in dioxane, 6.28 mL, 25.1 mmol, 1.00 equiv) (Note 26). The vessel is purged with N_2(g), sealed, and immersed in a silicon oil bath at 120 °C. The tan heterogeneous mixture is stirred for 1 h, resulting in a brown homogeneous solution. The solution is allowed to cool to ambient temperature, transferred to a 250-mL, round-bottomed flask (Note 27), and the solvent is removed on a rotary evaporator (60 °C, 15 mmHg) followed by high vacuum (ambient temperature, <0.75 mmHg, 1.5 h), affording a brown solid. A Teflon-coated stir bar and toluene (80 mL) is added to the flask, which is equipped with a water-jacketed reflux condenser and immersed in a 120 °C silicon oil bath. The homogeneous mixture is stirred under an atmosphere of N_2(g) for approximately 5 min at which point a solid precipitates from the solution. The resulting heterogeneous mixture is allowed to cool to ambient temperature with stirring and is then immersed in an ice/water bath. The resulting white precipitate is collected by suction filtration and washed with toluene (5 x 20 mL). The filtrate is concentrated using a rotary evaporator (60 °C, 15 mmHg) followed by a high vacuum (ambient temperature, <0.75 mmHg, 1 h) affording a brown solid. Toluene is added (30 mL) and the homogeneous mixture is placed into a 60 °C water bath until heterogeneity (Note 28) is reached (approximately 3-5 min). The suspension is allowed to cool to ambient temperature with stirring. The resulting white precipitate is collected by suction filtration and washed with toluene (2 x 20 mL). The filtrate is concentrated using a rotary evaporator (60 °C, 15 mmHg) followed by high vacuum (ambient temperature, <0.75 mmHg, 1 h) affording a brown solid. Toluene is added (10 mL) and the homogeneous mixture is placed into a 60 °C water bath until heterogeneity (Note 28) is reached again. The suspension is allowed to cool to ambient temperature with stirring. The resulting white precipitate is collected by suction filtration and washed with toluene (2 x 20 mL). The collected white solids are transferred into a 20 mL vial and residual volatiles are removed under high vacuum (ambient temperature, <1 mbar, >24 h) affording 5.54–5.89 g (60–64%) of *(5aS,10bR)-5a,10b-dihydro-2-(2,4,6-trimethylphenyl)-4H,6H-indeno[2,1-b]-1,2,4-triazolo[4,3-d]-1,4-oxazinium chloride* (**4**) as a white powder (Note 29).

2. Notes

1.   The checkers found that it was difficult to maintain stirring without the use of an overhead stirrer.

2.   The authors used an Eyela Low-Temp Pairstirrer PSL-1400 and a MeOH bath operating at –10 °C, while the checkers used a Neslab CC-100 to keep an acetone bath operating at –10 °C.

3.   Hydrogen chloride solution (ACS certified, 37%) was purchased from Fisher Scientific.

4.   Deionized water was used.

5.   2,4,6-Trimethylaniline (97%, Alfa Aesar) was fractionally distilled under reduced pressure (80 °C, 0.75 mmHg) from zinc metal (granular) prior to use.

6.   *CAUTION*: Due to the slight exotherm from the addition, HCl(g) evolves from the flask.

7.   An additional 10 mL of deionized water was used to wash the wall of flask.

8.   Sodium nitrite (Fisher) was used as received. The addition of the aq $NaNO_2$ solution was carried out over a period of 45 min with an internal temperature range of –5–3 °C and resulted in a homogeneous yellow solution.

9.   Tin(II) chloride dihyrdrate (Fisher) was used as received. The dissolution of stannous chloride in 1:1 conc. HCl/H_2O is incomplete resulting in an iridescent, slightly heterogeneous solution.

10.   A 150-mL medium-porosity Büchner funnel lined with filter paper (Whatman Grade No. 3) was used. Alternatively, a similar filter funnel of fine-porosity can be used.

11.   Diethyl ether (99.9%, EMD) was used as received.

12.   The checkers found that the initial precipitate is very fine and a long time is required for the filtration. Dissolving in MeOH and re-concentration is helpful to convert the fine powder into a more crystalline material that can be easily filtered.

13.   VWR B 2500A-MTH ultrasonic cleaner with 85 W was used and ice is added occasionally to maintain the temperature below 30 °C

14.   The product had the following physiochemical properties: Mp 195–197 °C (dec.); IR (neat) v 3296, 3002, 2964, 2911, 2691, 1564, 1515, 1479, 849, 830, 757 cm^{-1}; ^1H NMR (400 MHz, DMSO-d$_6$) δ: 2.20 (s, 3 H), 2.35 (s, 6 H), 6.60 (bs, 1 H), 6.86 (s, 2 H), 9.76 (variable bs, 3 H); ^{13}C

NMR (100 MHz, DMSO-d$_6$) δ: 17.8, 20.4, 129.0, 134.9, 136.1, 137.9; HRMS (ESI) m/z calcd. for C$_9$H$_{15}$N$_2$ (M$^+$) 151.1223; found 151.1228; Anal. calcd. for C$_9$H$_{15}$ClN$_2$: C, 57.90; H, 8.10; N, 15.01; found: C, 57.82; H, 8.19; N, 15.04.

15. A 5/8 inch x 1 ¼ inch egg-shaped stir bar was used.

16. (4aR,9aS)-4,4a,9,9a-tetrahydroindeno[2,1-b][1,4]oxazin-3(2H)-one was prepared from (1R,2S)-(-)-cis-1-amino-2-indanol, following the preceding *Organic Syntheses* procedure reported by Rovis and co-workers (*Org. Synth.* **2010**, *87*, 350-361).

17. Methylene chloride (99.9%) was purchased from Fischer Scientific and purified by pressure filtration under argon through activated alumina.

18. Trimethyloxonium tetrafluoroborate (98%, Aldrich) was used as received. Direct contact of skin with trimethyloxonium tetrafluoroborate must be avoided because of its caustic nature and alkylating properties.

19. The purity of the crude product was estimated to be about 95% based on the crude ^{1}H NMR. An analytically pure sample can be obtained via flash column chromatography (Sorbent Silica Gel 60 (230–400 Mesh), pentane/ethyl ether 7:3) and has the following physiochemical properties: R$_f$ = 0.42 (EMD precoated plates (silica gel 60 F254, Art 5715, 0.25 mm, pentane/ethyl ether 7:3); Mp 73–75 °C (uncorrected); [α]$_D^{20}$ –15.7 (c 1.26, EtOH); IR (neat) ν 1683, 1443, 1383, 1238, 1100 738 cm^{-1}; ^{1}H NMR (600 MHz, CDCl$_3$) δ: 3.02 (d, 1 H, J = 16.8 Hz), 3.20 (dd, 1 H, J = 16.2, 4.8 Hz), 3.79 (s, 3 H), 3.95 (d, 1 H, J = 15.8 Hz), 4.01 (dd, 1 H, J = 15.7, 1.9 Hz), 4.29 (t, 1 H, J = 4.8 Hz), 4.89 (d, 1 H, J = 3.8 Hz), 7.21–7.26 (m, 3 H), 7.49 (d, 1 H, J = 7.8 Hz); ^{13}C NMR (150 MHz, CDCl$_3$) δ: 37.6, 52.2, 61.4, 62.8, 75.1, 124.6, 124.9, 126.8, 127.4, 139.1, 143.2, 161.7; HRMS (APCI) m/z calcd. for C$_{12}$H$_{14}$NO$_2$ ([M+H]$^+$) 204.1025, found 204.1019; Anal. calcd. for C$_{12}$H$_{13}$NO$_2$: C, 70.92; H, 6.45; N, 6.89; found: C, 70.84; H, 6.48; N, 6.82.

20. Freshly-opened MeOH (anhydrous, EMD Chemicals) was used.

21. Freshly-opened EtOAc (anhydrous, EMD Chemicals) was used. The submitters reported that EtOAc was dried by passage through an alumina drying column under an atmosphere of Argon. The submitters have observed that if anhydrous EtOAc is not used a lower yield is obtained due to hydrolysis of the amidrazone intermediate.

22. The product was contaminated with 2-mesitylhydrazinium chloride (**1**), which could not be removed. The checkers and submitters typically observed ~5–10% contamination based on ^{1}H NMR integration of the purified amidrazone hydrochloride **3**.

23. The product had the following physiochemical properties: Mp 200–201°C (dec.); $[\alpha]_D^{20}$ 66.7 (c 1.42, EtOH); IR (neat) v 3294, 3119, 2997, 2966, 2951, 2916, 2731, 2692, 1674, 1514, 1482, 1330, 845, 739 cm^{-1}; ^{1}H NMR (400 MHz, DMSO-d$_6$) δ: 2.20 (s, 3 H), 2.22 (s, 6 H), 2.97 (d, 1 H, J = 16.9 Hz), 3.29 (dd, 1 H, J = 16.9 Hz, 4.7 Hz), 4.55 (d, 1 H, J = 16.7 Hz), 4.62 (d, 1 H, J = 16.7 Hz), 4.72 (t, 1 H, J = 4.4 Hz), 5.00 (t, 1 H, J = 3.6 Hz), 6.86 (s, 2 H), 7.13 (s, 1 H), 7.27–7.32 (m, 3 H), 7.69–7.73 (m, 1 H), 11.08 (d, 1 H, J = 3.2 Hz), 11.46 (s, 1 H); ^{13}C NMR (100 MHz, DMSO-d$_6$) δ: 17.9, 20.2, 37.0, 56.1, 60.1, 76.8, 124.75, 124.83, 126.6, 128.1, 129.3, 131.2, 133.7, 138.2, 139.7, 140.2, 159.2; HRMS (ESI) m/z calcd for $C_{20}H_{24}N_3O$ (M$^+$) 322.1914; found 322.1913.

24. Chlorobenzene (99%, Aldrich) was dried over molecular sieves (4Å) prior to use.

25. Triethylorthoformate (98%, Acros) was used without further purification.

26. Anhydrous hydrogen chloride (4 M dioxane) was purchased from Aldrich Chemical Co. and used as received.

27. CH$_2$Cl$_2$ (3 x 20 mL) was used to rinse the sealed tube and transferred into the 250-mL, round-bottomed flask.

28. As soon as toluene is added, the solid dissolves and a brown homogeneous solution is formed. Within 3-5 min of heating at 60 °C, a precipitate is formed rapidly.

29. The product had the following physiochemical properties: Mp 217–219 °C; $[\alpha]_D^{20}$ –133.5 (c 1.00, EtOH); IR (neat) v 3435, 3482, 2904, 2853, 1580, 1466, 1222, 1097, 1083, 847, 749, 729, 662 cm^{-1}; ^{1}H NMR (400 MHz, DMSO-d$_6$) δ: 2.12 (s, 6 H), 2.37 (s, 3 H), 3.16 (d, 1 H, J = 17.0 Hz), 3.50 (dd, 1 H, J = 16.9, 4.8 Hz), 4.99 (t, 1 H, J = 4.4 Hz), 5.08 (d, 1 H, J = 16.0 Hz), 5.26 (d, 1 H, J = 16.0 Hz), 6.12 (d, 1 H, J = 4.0 Hz), 7.21 (s, 2 H), 7.33–7.45 (m, 3 H), 7.65 (d, 1 H, J = 7.2 Hz), 11.34 (s, 1 H); ^{13}C NMR (100 MHz, DMSO-d$_6$) δ: 16.9, 20.6, 37.0, 59.7, 61.1, 76.8, 124.0, 125.3, 127.1, 129.2, 129.3, 131.2, 134.8, 136.1, 140.6, 141.3, 144.7, 150.0; HRMS (ESI) m/z calcd. for $C_{21}H_{22}N_3O$ (M$^+$) 332.1757; found 332.1756; Anal. calcd. for $C_{21}H_{24}ClN_3O_2$ (M+H$_2$O): C, 65.36; H, 6.27; N, 10.89; found: C, 65.43; H, 6.24; N, 10.88.

Safety and Waste Disposal Information

All hazardous materials should be handled and disposed of in accordance with "Prudent Practices in the Laboratory"; National Academy Press; Washington, DC, 1995.

3. Discussion

In 1997, Leeper reported the design and synthesis of a new class of chiral triazolium salts derived from chiral 1,2-amino alcohols.[2] While the initial application of these precatalysts to enantioselective benzoin reactions provided only moderate selectivities, subsequent work by Rovis and Enders led to the identification of broadly useful aminoalcohol-derived catalysts that have proven useful for highly enantioselective intermolecular homobenzoin reactions,[3] intramolecular Stetter reactions,[4] and intramolecular benzoin cyclizations.[5] Of particular utility, in terms of reaction yield and enantioselectivities, is Rovis' aminoindanol-derivated triazolium precatalysts bearing *N*-phenyl, *N*-*para*-methoxyphenyl, or *N*-pentafluorophenyl substituents.[4c]

Scheme 1. Reactive pathways promoted by *N*-mesityl carbene **5** (Mes = 2,4,6-trimethylphenyl).

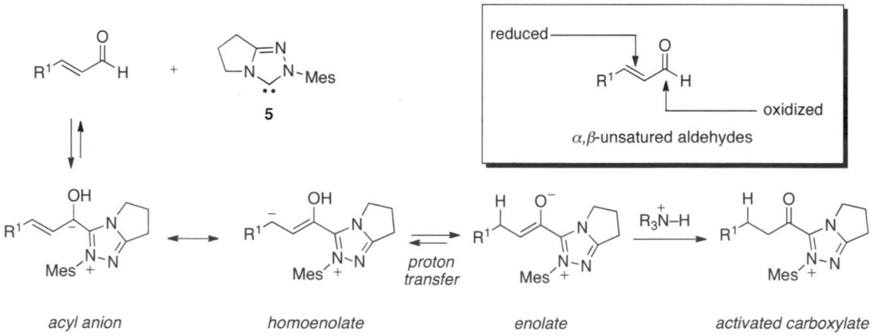

In our own studies, we have developed redox reactions of alpha-functionalized aldehydes for the catalytic generation of activated carboxylates, homoenolates, and enolates (Scheme 1). As part of these efforts, we have documented a profound, product-determining role for both the catalyst type (imadazolium vs. triazolium)[6] and the nature of the *N*-

370

substituent.[7] A series of investigations have identified a critical need for *N*-mesityl substituted triazolium precatalysts for effective conversions, a finding that has been mirrored in subsequent studies by other groups on new reactions catalyzed by chiral *N*-heterocyclic carbenes.[8] For example, in our own work we have found the *N*-mesityl-substituted aminoindanol-derived precatalyst **4** to be particularly effective in controlling the enantioselectivity of a wide variety of novel annulation reactions.[9] A selection of processes employing *N*-mesityl substituted triazolium salts reported by our group is shown in Figure 1.

Figure 1. Selected stereoselective annulations promoted by *N*-mesityl substituted aminoindanol-derived NHC precatalyst **4**.

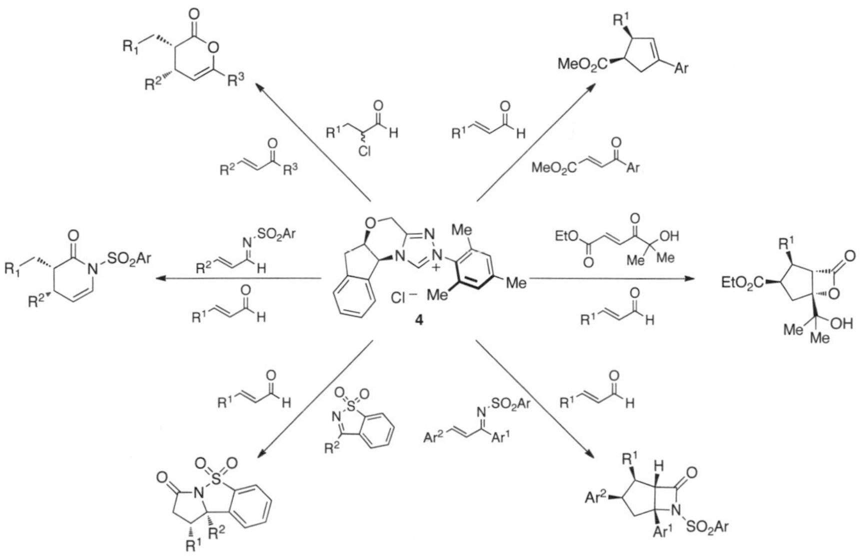

Rovis has described a concise, high-yielding synthesis of a number of chiral aminoindanol-derived triazolium salts.[10] While we have found these protocols to be very effective for the synthesis of triazolium salts bearing simple aromatic groups, we obtained capricious outcomes when attempting to apply these procedures to the preparation of the *N*-mesityl substituted variant. In our efforts to adapt Rovis' procedures to a preparative scale synthesis of *N*-mesityl substituted triazolium salts, we traced much of our difficulties to the low purity and poor stability of 2-mesitylhydrazine.[11] To circumvent this, we adopted the protocol originally reported by Knight and

Leeper,[2] in which the neutral iminoether was allowed to react with the arylhydrazine hydrochloride.

This approach required a reliable, scale synthesis of 2-mesitylhydrazinium hydrochloride (**1**). To achieve this we refined a Sandmeyer approach from the corresponding aniline.[12] Critical to the success of this reaction on scale is vigilant monitoring of the internal reaction temperature during diazonium formation and reduction steps by maintaining a cold (–10 °C) external bath and by portion-wise additions of sodium nitrite and stannous chloride. Although the isolated yield of this particular hydrazinium is moderate, this procedure is robust and reproducible on a preparative scale.

The condensation of 2-mesitylhydrazinium hydrochloride and the iminoether was carried out under the conditions reported by Leeper with the exception that a catalytic amount of anhydrous HCl was found to be beneficial for the reaction outcome.[13] The most challenging aspect of the synthesis of the *N*-mesityl substituted triazolium salts was the ring-closing reaction with triethylorthoformate. After considerable experimentation, we found two important factors that led to clean, high-yielding reactions for the production of *N*-mesityl substituted triazolium salts. First, it was essential to add an equivalent of anhydrous HCl to the reaction mixture. Second, and most critically, we found that extended reaction times were detrimental to the isolation of the desired triazolium salts. In contrast to prior protocols that employed longer reactions times (>12 hours), reaction periods of 1–2 hours were generally preferred. Qualitatively, the progress of the reaction could be followed by the solubilization of the starting amidrazone hydrochloride; when a clear solution had formed the reaction was generally complete.

This procedure is not limited to the synthesis of the aminoindanol-derived triazolium salt **4**. We have also found it to be directly applicable to other classes of *N*-mesityl substituted triazolium NHC precatalysts (Figure 2). We reported the first *N*-mesityl-substituted triazolium salt (**5**) in 2005 for the redox esterification of enals,[7] which we have conveniently named RMesCl. Further, we have demonstrated that this procedure was amenable to *N*-mesityl substituted triazolium salts based on bicyclic scaffolds of increasing ring size (**6**[14]–**7**), the chiral bicyclic scaffold popularized by Rovis (**8–10**),[10,15] and the achiral (**11**) and chiral (**12–13**)[16] morpholinone-derived bicyclic scaffold. We anticipate that the synthesis of novel *N*-mesityl substituted triazolium NHCs by our method will facilitate the discovery of

372

new NHC-promoted processes and offer the possibility to render such processes enantioselective.

Figure 2. Examples of *N*-mesityl substituted triazolium salts prepared by our reported method (Mes = 2,4,6-trimethylphenyl).

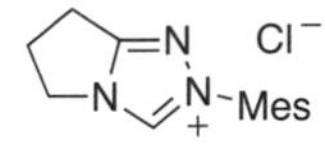

5
(RMesCl)

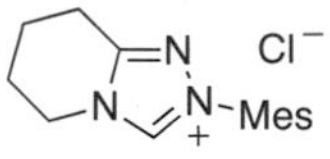

6

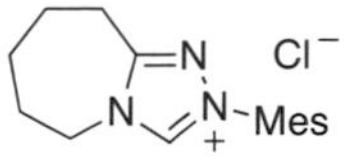

7

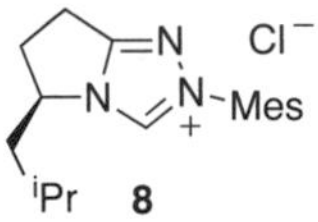

8

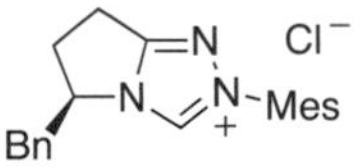

9

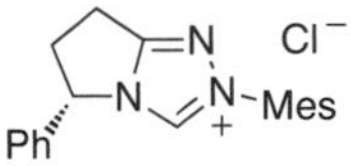

10

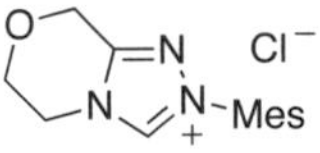

11

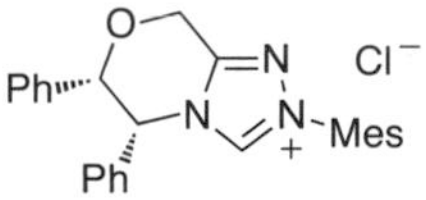

12

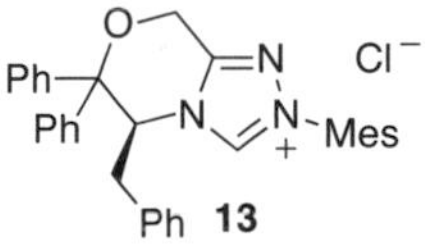

13

1. Roy and Diana Vagelos Laboratories, Department of Chemistry, University of Pennsylvania, Philadelphia, PA 19104. Current address: Laboratorium für Organische Chemie, ETH-Zürich, CH-8093, Switzerland. Email: bode@org.chem.ethz.ch
2. Knight, R. L.; Leeper, F. J. *J. Chem. Soc., Perkin Trans. 1* **1998**, 1891–1893.
3. Enders, D.; Kallfass, U. *Angew. Chem., Int. Ed.* **2002**, *41*, 1743–1745.
4. Selected examples: (a) Kerr, M. S.; de Alaniz, J. R.; Rovis, T. *J. Am. Chem. Soc.* **2002**, *124*, 10298–10299. (b) Kerr, M. S.; Rovis, T. *J. Am. Chem. Soc.* **2004**, *126*, 8876–8877. (c) de Alaniz, J. R.; Rovis, T. *J. Am. Chem. Soc.* **2005**, *127*, 6284–6289. (d) Liu, Q.; Rovis, T. *J. Am. Chem. Soc.* **2006**, *128*, 2552–2553. (e) Cullen, S. C.; Rovis, T. *Org. Lett.* **2008**, *10*, 3141–3144. (f) de Alaniz, J. R.; Kerr, M. S.; Moore, J. L.; Rovis, T. *J. Org. Chem.* **2008**, *73*, 2033–2040.

5. (a) Takikawa, H.; Hachisu, Y.; Bode, J. W.; Suzuki, K. *Angew. Chem., Int. Ed.* **2006**, *45*, 3492–3494. (b) Takikawa, H.; Suzuki, K. *Org. Lett.* **2007**, *9*, 2713–2716.

6. Struble, J. R.; Kaeobamrung, J.; Bode, J. W. *Org. Lett.* **2008**, *10*, 957–960.

7. Sohn, S. S.; Bode, J. W. *Org. Lett.* **2005**, *7*, 3873–3876.

8. (a) Phillips, E. M.; Wadamoto, M.; Chan, A.; Scheidt, K. A. *Angew. Chem., Int. Ed.* **2007**, *46*, 3107–3110. (b) Seayad, J.; Patra, P. K.; Zhang, Y. G.; Ying, J. Y. *Org. Lett.* **2008**, *10*, 953–956.

9. (a) He, M.; Struble, J. R.; Bode, J. W. *J. Am. Chem. Soc.* **2006**, *128*, 8418–8420. (b) He, M.; Uc, G. J.; Bode, J. W. *J. Am. Chem. Soc.* **2006**, *128*, 15088–15089. (c) Chiang, P. C.; Kaeobamrung, J.; Bode, J. W. *J. Am. Chem. Soc.* **2007**, *129*, 3520–3521. (d) He, M.; Bode, J. W. *J. Am. Chem. Soc.* **2008**, *130*, 418–419. (e) He, M.; Beahm, B. J.; Bode, J. W. *Org. Lett.* **2008**, *10*, 3817–3820. (f) Rommel, M.; Fukuzumi, T.; Bode, J. W. *J. Amer. Chem. Soc.* **2008**, *130*, 17266–17267. (g) Kaeobamrung, J.; Bode, J. W. *Org. Lett.* **2009**, *11*, 677–680.

10. Kerr, M. S.; de Alaniz, J. R.; Rovis, T. *J. Org. Chem.* **2005**, *70*, 5725–5728.

11. The instability of 2-mesitylhydrazine has been described previously: Carlin, R. B.; Moores, M. S. *J. Am. Chem. Soc.* **1962**, *84*, 4107–4112.

12. Hunsberger, I. M.; Shaw, E. R.; Fugger, J.; Ketcham, R.; Lednicer, D. *J. Org. Chem.* **1956**, *21*, 394–399.

13. Empirically we find the addition of catalytic anhydrous HCl and extended reaction times (48 h) to only be necessary for the morpholinone derivatives.

14. An *N*-phenyl triazolium salt similar to **7** has been reported: Chen, D. D.; Hou, X. L.; Dai, L. X. *J. Org. Chem.* **2008**, *73*, 5578–5581.

15. Chiang, P.-C.; Rommel, M.; Bode, J. W. *J. Am. Chem. Soc.* **2009**, *131*, 8714–8718.

16. Originally reported as the BF_4^- salts by (a) Phillips, E. M.; Wadamoto, M.; Chan, A.; Scheidt, K. A. *Angew. Chem., Int. Ed.* **2007**, *46*, 3107–3110. (b) Phillips, E. M.; Reynolds, T. E.; Scheidt, K. A. *J. Am. Chem. Soc.* **2008**, *130*, 2416–2417.

Appendix
Chemical Abstracts Nomenclature; (Registry Number)

2,4,6-Trimethylaniline; (88-05-1)

Hydrogen chloride, concentrated aqueous solution 37%; (7647-01-1)

Sodium nitrite; (7632-00-0)

Tin(II) chloride dihydrate; (10025-69-1)

Trimethyloxonium tetrafluoroborate; (420-37-1)

Hydrogen chloride, dioxane solution (4 M); (7647-01-1)

2-Mesitylhydrazinium chloride; (76195-82-9)

Chlorobenzene; (108-90-7)

Triethyl orthoformate; (122-51-0)

(5a*R*,10b*S*)-5a,10b-Dihydro-2-(2,4,6-trimethylphenyl)-4H,6H-indeno[2,1-*b*]-1,2,4-triazolo[4,3-*d*]-1,4-oxazinium chloride; (903571-02-8)

Jeffrey W. Bode was born in California in 1974 and studied chemistry and philosophy at Trinity University in San Antonio, Texas. He received his Dok. Nat. Sci. from the Eidgenössicsche Technische Hochschule (ETH) in Zürich, Switzerland with Prof. Erick M. Carreira in 2001. Following a JSPS Postdoctoral Fellowship with Prof. Keisuke Suzuki at the Tokyo Institution of Technology, he joined the faculty of the University of California, Santa Barbara as an Assistant Professor in 2003. In 2007, he joined the University of Pennsylvania in Philadelphia, Pennsylvania as an Associate Professor of Chemistry. His research interests include the development of new synthetic methods, catalysis, peptide synthesis, and bioorganic chemistry.

Dr. Justin R. Struble was born in Toledo, OH in 1981. An Ohio native, he received his B.A. in Chemistry in 2003 from Kenyon College. In 2004, he joined Dr. Jeffrey Bode at University of California Santa Barbara where he developed N-Heterocyclic Carbene (NHC) catalysts for their application in stereoselective transformations. In 2007, along with Dr. Bode, he moved to the University of Pennsylvania where he received his Ph.D. in 2009. Currently he is a postdoctoral research associate for Dr. Martin Burke at the University of Illinois Urbana-Champaign where he focuses on expanding the capabilities of MIDA boronates in metal-catalyzed cross-coupling reactions as well as working on the total synthesis of Amphotericin B and its derivatives.

 Yajing Lian graduated with a B.S. degree from Xiamen University, P.R. China in August 2003 with chemistry major. After graduation, he joined Dr. Robert J. Hinkle's group at the College of William and Mary in Virginia for his Masters study, investigating the bismuth catalyzed tandem cyclization reactions and the iodonium(III) chemistry. He received his masters degree in 2005. In 2006, Yajing Joined Dr. Huw Davies' group at S.U.N.Y. at Buffalo for his Ph. D study and then he moved to Emory University in 2008 together with the whole group. Currently he is developing new reactions of rhodium carbenoid and applying these to the total synthesis of natural products.

ERRATA (modification to procedure originally published in *Org. Synth.* **2003**, *80*, 38; Coll. Vol. 11, **2009**, 404.)

PREPARATION OF *O*-ALLYL-*N*-(9-ANTHRACENYL-METHYL)-CINCHONIDINIUM BROMIDE AS A PHASE TRANSFER CATALYST

Submitted by E. J. Corey and Mark C. Noe

The NMR data in Note 8 of this article was reported incorrectly. Note 8 should read as indicated below.

8. *O*-Allyl-*N*-(9-anthracenylmethyl)cinchonidinium bromide has the following properties: $[\alpha]_D23$ −320 (*c* 0.45, CHCl₃); mp 194-197 °C; FTIR (film) cm⁻¹: 3504, 3082, 2950, 2907, 2884, 1646, 1641, 1625, 1588, 1509, 1450, 1067, 996; ¹H NMR (400 MHz, CD₃OD) δ: 1.60 (m, 2 H), 1.96 (d, 1 H, *J* = 2.9), 2.17 (m, 1 H), 2.48 (m, 2 H), 2.90 (dt, *J* = 10.8, 5.9 Hz, 1 H), 3.24 (app. t, *J* = 12.2 Hz, 1 H), 3.78 (m, 1 H), 4.39-4.56 (m, 4 H), 4.96 (d, *J* = 6.9 Hz, 1 H), 5.00 (d, *J* = 13.5 Hz, 1 H), 5.54 (dd, *J* = 10.5, 12 Hz, 1 H), 5.60-5.64 (m, 2 H), 5.88 (d, *J* = 13.9 Hz, 1 H), 6.37-6.42 (m, 2 H), 6.95 (bs, 1 H), 7.61-7.65 (m, 2 H), 7.77-7.79 (m, 2 H), 7.92-7.95 (m, 3 H), 8.19-8.25 (m, 3 H), 8.44 (d, 1 H, *J* = 9.0), 8.57 (m, 1H), 8.76 (d, 1 H, *J* = 9.0), 8.89 (s, 1 H), 9.02 (d, 1 H, *J* = 4.6); ¹³C NMR (100 MHz, CD₃OD) δ: 23.4, 26.2,

27.3, 39.5, 49.9, 53.6, 57.4, 63.4, 69.9, 71.4, 117.8, 119.0, 121.8, 125.0 (2C), 125.7 (2C), 126.5, 126.6, 127.1, 129.2, 129.5, 130.5, 131.1, 131.3, 131.5, 133.0, 133.1, 133.8, 133.9, 134.6, 134.7, 134.8, 138.6, 143.0, 149.3, 151.1; FABMS: 525 [M-Br]$^-$; HRMS calcd for [C$_{37}$H$_{37}$N$_2$OBr-Br]$^-$: 525.2906, found: 525.2930. Anal. Calcd for C$_{37}$H$_{37}$BrN$_2$O: C, 73.38; H, 6.16; Br, 13.19; N, 4.63. Found: C, 73.40; H, 6.12; Br, 13.19; N 4.47.

CUMULATIVE AUTHOR INDEX FOR VOLUMES 85-87

This index comprises the names of contributors to Volumes **85**, **86**, and **87**. For authors of previous volumes, see either indices in Collective Volumes I through XI, or the single volume entitled *Organic Syntheses, Collective Volumes I-VIII, Cumulative Indices,* edited by J. P. Freeman.

Campbell, L., **85**, 15
Carreira, E. M., **87**, 88
César, V., **85**, 34
Chalker, J. M., **87**, 288
Chaloin, O., **85**, 147
Chandrasekaran, P., **86**, 333
Chang, S., **85**, 131
Chang, Y., **87**, 245
Charette, A. B., **87**, 115, 170
Chen, Q.-Y., **87**, 126
Chiong, H. A., **86**, 105
Cho, S. H., **85**, 131
Chouai, A., **86**, 141, 151
Clososki, G. C., **86**, 374
Coates, G. W., **86**, 287
Coste, A., **87**, 231
Couty, F., **87**, 231

Dai, P., **86**, 236
Daugulis, O., **86**, 105, **87**, 184
Davies, S. G., **87**, 143
Davis, B. G., **87**, 288
de Alaniz, J. R., **87**, 350
DeBerardinis, A. M., **87**, 68
Delaude, L., **87**, 77
Deng, X., **85**, 179
Denmark, S. E., **86**, 274
Ding, K., **87**, 126
Do, H.-Q., **87**, 184
Do, N., **85**, 138
Donahue, J. P., **86**, 333
Drago, C., **86**, 121
Du, H., **86**, 315, **87**, 263
Duchêne, A., **85**, 231
Dudley, M. E., **86**, 172

Ebner, D. C., **86**, 161
Ekoue-Kovi, K., **87**, 1
Ellman, J. A., **86**, 360
Endo, K., **86**, 325
Erkkilä, A., **87**, 201
Evano, G., **87**, 231

Kappe, C. O., **86**, 252
Kerr, M. S., **87**, 350
Kim, S., **86**, 225
Kirai, N., **87**, 53
Kitazawa, K., **87**, 209
Kitching, M. O., **85**, 72
Knauber, T., **85**, 196
Knochel, P., **86**, 374
Kochi, T., **87**, 209
Kocienski, P. J., **85**, 45
Koenig, S. G., **87**, 275
Kong, J., **87**, 137
Kozmin, S. A., **87**, 253
Kramer, J. W., **86**, 287
Krause, H., **85**, 34, **86**, 298
Krout, M. R., **86**, 181, 194
Kuethe, J. T., **86**, 92
Kurth, M. J., **87**, 339
Kwon, O., **86**, 212

La Vecchia, L., **85**, 295
Landais, Y., **86**, 1
Langenhan, J. M., **87**, 192
Langle, S., **85**, 231
Larock, R. C., **87**, 95
Lathrop, S. P., **87**, 350
Lautens, M., **85**, 172, **86**, 36
Lazareva, A., **86**, 105
Lebel, H., **86**, 59, 113
Lebeuf, R., **86**, 1
Lee, H., **87**, 245
Leogane, O., **86**, 113
Ley, S. V., **85**, 72
Li, B., **87**, 16
Linder, C., **85**, 196
List, B., **86**, 11
Longbottom, D. A., **85**, 72
Lou, S., **86**, 236; **87**, 299, 310, 317, 330
Lu, C. -D., **85**, 158
Lu, K., **86**, 212

Mani, N. S., **85**, 179
Mans, D. J., **85**, 238, 248
Marcoux, D., **87**, 115

382

Marin, J., **85**, 147
Marshall, A.-L., **87**, 192
Matsunaga, S., **85**, 118
Maw, G., **85**, 219
McAllister, G. D., **85**, 15
McDermott, R. E., **85**, 138
McNaughton, B. R., **85**, 27
Meletis, P., **86**, 47
Meng, T., **87**, 137
Meyer, H., **85**, 287, 295
Miller, B. L., **85**, 27
Mohr, J. T., **86**, 181, 194
Montchamp, J.-L., **85**, 96
Moore, D. A., **85**, 10
Morra, N. A., **85**, 53
Morshed, M. M., **86**, 172
Mosa, F., **85**, 219
Mousseau, J. J., **87**, 170
Movassaghi, M., **85**, 88
Mudryk, B., **85**, 64
Müller-Hartwieg, J. C. D., **85**, 295
Murakami, K., **87**, 178

Nakamura, E., **86**, 325
Nakamura, M., **86**, 325
Ngi, S. I.,
Nicewicz, D. A., **85**, 278
Nixon, T. D., **86**, 28

Oelke, A. J., **85**, 72
Ohshima, T., **85**, 118
Okano, K., **86**, 130
Olofsson, B., **86**, 308
Ortiz-Marciales, M., **87**, 26, 36
Osajima, H., **86**, 130
Oshima, K., **87**, 178

Pagenkopf, B. L., **85**, 53
Pan, S. C., **86**, 11
Pihko, P. M., **87**, 201
Pu, L., **87**, 59, 68

Rafferty, R. J., **86**, 262
Ragan, J. A., **85**, 138

RajanBabu, T. V., **85**, 238, 248
Reynolds, N. T., **87**, 350
Riebel, P., **85**, 64
Robert, F., **86**, 1
Roberts, P. M., **87**, 143
Rohbogner, C. J., **86**, 374
Romea, P., **86**, 70, 81
Rovis, T., **87**, 350

Schaus, S. E., **86**, 236
Schiffers, I., **85**, 106
Schramm, M. P., **87**, 253
Schwekendiek, K., **85**, 267
Scott, M. E., **85**, 172
Sebesta, R., **85**, 287
Sedelmeier, G., **85**, 72
Seebach, D., **85**, 287, 295
Shi, F., **87**, 95
Shi, Y., **86**, 263, 315
Shibasaki, M., **85**, 118
Shubinets, V., **87**, 253
Simanek, E. E., **86**, 141, 151
Singh, S. P., **87**, 275
Slatford, P. A., **86**, 28
Smith, C. R., **85**, 238, 248
Snaddon, T. N., **85**, 45
Solano, D. M., **87**, 339
Sperotto, E., **85**, 209
Stepanenko, V., **87**, 26
Stevens, K. L., **86**, 18
Stoltz, B. M., **86**, 161, 181, 194
Storgaard, M., **86**, 360
Struble, J. R., **87**, 362

Takita, R., **85**, 118
Tambar, U. K., **86**, 161
Taylor, R. J. K., **85**, 15
Thibonnet, J., **85**, 231
Thirsk, C., **85**, 219
Thompson, A. L., **87**, 288
Tian, W.-S., **87**, 126
Ting, P., **87**, 137
Tokuyama, H., **86**, 130
Treitler, D. S., **86**, 287

384

Truc, V., **85**, 64
Tseng, N. -W., **85**, 172
Turlington, M., **87**, 59, 68

Urpí, F., **86**, 70, 81

van Klink, G., P.M. **85**, 209
van Koten, G., **85**, 209
Vandenbossche, C. P., **87**, 275
Vaultier, M., **85**, 219
Venditto, V. J., **86**, 141, 151
Vinogradov, A., **87**, 104
Vora, H. U., **87**, 350

Wagner, A. J., **86**, 374
Walker, E.-J., **86**, 121
Wang, Y., **87**, 126
Waser, J., **87**, 88
Watson, I. D. G., **87**, 161
Whiting, A., **85**, 219
Whittlesey, M. K., **86**, 28
Williams, J. M. J., **86**, 28
Williams, R. M., **86**, 262
Wolf, C., **87**, 1
Wong, J., **87**, 137
Woodward, S., **87**, 104

Yamamoto, Y., **87**, 53
Yang, J. W., **86**, 11
Yorimitsu, H., **87**, 178
Yu, J., **85**, 64
Yudin, A. K., **87**, 161

Zakarian, A., **85**, 158
Zaragoza, F., **87**, 226
Zhang, A., **85**, 248
Zhang, H., **85**, 147
Zhang, Z., **87**, 16
Zhao, B., **86**, 263, 315
Zhao, H., **87**, 275
Zhong, Yong-L. **87**, 8
Zimmermann, B., **85**, 196

CUMULATIVE SUBJECT INDEX FOR VOLUMES 85-87

This index comprises subject matter for Volumes **85, 86,** and **87**. For subjects in previous volumes, see either the indices in Collective Volumes I through XI or the single volume entitled *Organic Syntheses, Collective Volumes I-VIII, Cumulative Indices,* edited by J. P. Freeman.

The index lists the names of compounds in two forms. The first is the name used commonly in procedures. The second is the systematic name according to Chemical Abstracts nomenclature, accompanied by its registry number in parentheses. Also included are general terms for classes of compounds, types of reactions, special apparatus, and unfamiliar methods.

Most chemicals used in the procedure will appear in the index as written in the text. There generally will be entries for all starting materials, reagents, intermediates, important by-products, and final products, which are indicated by the use of italics.

Benzyl bromide: Benzene, (bromomethyl)-; (100-39-0) **87**, 36, 137, 170

N-Benzyl-2-bromo-*N*-phenylbutanamide: Butanamide, 2-bromo-*N*-phenyl-*N*-
(phenylmethyl)-; (851073-30-8) **87**, 330

Benzyl carbamate: Carbamic acid, phenylmethyl ester; (621-84-1) **85**, 287

Benzyl chloromethyl ether: Benzene, [(chloromethoxy)methyl]-; (3587-60-8) **85**, 45

1-Benzyl-3-(4-chloro-phenyl)-5-p-tolyl-1H-pyrazole (908329-95-3) **85**, 179

(S)-N-Benzyl-7-cyano-2-ethyl-n-phenylheptanamide: Heptanamide, 7-cyano-2-ethyl-N-
phenyl-N-(phenylmethyl)-, (2S)-; (851073-44-4) **87**, 330

Benzylhydrazine dihydrochloride; (20570-96-1) **85**, 179

Benzyl hydroxymethyl carbamate: Carbamic acid, (hydroxymethyl)-, phenylmethyl ester;
(31037-42-0) **85**, 287

(E)-N-Benzylidene-4-methylbenzenesulfonamide: Benzenesulfonamide, 4-methyl-*N*-
(phenylmethylene)-, [*N(E)*]-: (51608-60-7) **86**, 212

Benzyl isopropoxymethyl carbamate **85**, 287

1 (2R,4S) [2-Benzyl-3-(4-isopropyl-2-oxo-5,5-diphenyl-3-oxazolidinyl)-3-
oxopropyl]carbamic acid benzyl ester (218800-56-7) **85**, 295

(R)-2-(Benzyloxycarbonylaminomethyl)-3-phenylpropanoic acid: Benzenepropanoic
acid, α-[[[(phenylmethoxy)carbonyl]amino]methyl]-, (aR)-; (132696-47-0) **85**,
295

Benzyloxymethoxy-1-hexyne: Benzene, [[(1-hexyn-1-yloxy)methoxy]methyl]-; (162552-
11-6) **85**, 45

Benzyloxymethoxy-2,2,2-trifluoromethyl ether: Benzene, [[(2,2,2-
trifluoroethoxy)methoxy]methyl]-: (153959-88-7) **85**, 45

(R)-*N*-Benzyl-*N*-(α-methylbenzyl)amine; *(R)*-*N*-benzyl-α-phenylethylamine; *(R)*-*N*-
benzyl-1-phenylethylamine (38235-77-7) **87**, 143

Benzyl propagyl ether; (4039-82-1) **86**, 225

(R)-(+)-1,1'-Bi(2-naphthol); (18531-94-7) **85**, 238

(R)-BINOL; (18531-94-7) **85**, 238

(S)-BINOL: [1,1'-Binaphthalene]-2,2'-diol, (1*S*)-: (18531-99-2) **85**, 118

(R)-(1,1'-Binaphthalene-2,2'-dioxy)chlorophosphine: (R)-Binol-P-Cl; (155613-52-8) **85**,
238

[1,1'-Binaphthalene]-2,2'-diol, (1*S*)-: (18531-99-2) **85**, 118

(R)-2,2-Binaphthoyl-(S,S)-di(1-phenylethyl)aminoylphosphine (415918-91-1) **85**, 238

(2-Biphenyl)dicyclohexylphosphine; (247940-06-3) **86**, 344

Bis(4-tert-butylphenyl)iodonium triflate; (84563-54-2) **86**, 308

Bis[1,2:5,6-η-(1,5-cyclooctadiene)]nickel: [bis(1,5-cyclooctadiene)nickel (0)]; (1295-35-
8) **85**, 248

[*N,N'*-Bis(3,5-di-*tert*-butylsalicylidene)-1,2-phenylenediamino-chromium-di-
tetrahydrofuran]tetracarbonylcobaltate (1); (909553-60-2) **86**, 287

4,5-Bis(Diphenylphosphino)-9,9-dimethylxanthene (Xantphos); (161265-03-8) **86**, 28

Bis(diphenylphosphino)methane; (2071-20-7) **85**, 196

Bis(4-Fluorophenyl)Difluoromethane (339-27-5) **87**, 245

Bis(4-fluorophenyl) ketone: 4,4'-Difluorobenzophenone; (345-92-6) **87**, 245

Bis-(Hydroxymethyl)-cyclopropane; (2345-75-7) **85**, 15

(S)-3,3'-Bis-morpholinomethyl-5,5',6,6',7,7',8,8'-octahydro-1,1'-bi-2-naphthol: [1,1'-
Binaphthalene]-2,2'-diol, 5,5',6,6',7,7',8,8'-octahydro-3,3'-bis(4-
morpholinylmethyl)-, (1S)-; (758698-16-7) **87**, 59, 68

(-)-Bis[(*S*)-1-phenylethyl]amine (56210-72-1) **85**, 238

(-)-Bis[(*S*)-1-phenylethyl]amine hydrochloride (40648-92-8) **85**, 238

Bis(Pyridine)Iodonium(I) tetrafluoroborate: Iodine(1+), bis(pyridine)-,
    tetrafluoroborate(1-) (1:1); (15656-28-7) **87**, 288

Bis[rhodium(α,α,α',α'-tetramethyl-1,3-benzenedipropionic acid)]; (819050-89-0) **87**, 115

3,5-Bis(trifluoromethyl)bromobenzene; (328-70-1) **85**, 248

Boc$_2$O: Dicarbonic acid, *C,C'*-bis(1,1-dimethylethyl) ester; (244424-99-5) **86**, 113, 374

9-Borabicyclo[3.3.1]nonane: 9-BBN; (280-64-8) **87**, 299

(5-(9-Borabicyclo[3.3.1]nonan-9-yl)pentyloxy)triethylsilane: 9-Borabicyclo[3.3.1]-
    nonane, 9-[5-[(triethylsilyl)oxy]pentyl]-; (157123-09-6) **87**, 299

Borane tetrahydrofuran complex solution, 1.0 M in tetrahydrofuran; (14044-65-6) **87**, 36

Boron **87**, 26, 299

Boron trifluoride etherate: BF$_3$·OEt$_2$ (109-63-7) **86**, 81, 212

Boronic acid, phenyl-; (98-80-6) **87**, 53

Bromination, **85**, 53

Bromine; (7726-95-6) **85**, 231; **87**, 126

2-[3-(3-Bromophenyl)-2H-azirin-2-yl]-5-(trifluoromethyl)pyridine **86**, 18

3'-Bromoacetophenone: 1-acetyl-3-bromobenzene; (2142-63-4) **86**, 18

Bromobenzene; (108-86-1) **87**, 26, 178

2-Bromobenzoyl chloride: Benzoyl chloride, 2-bromo-; (7154-66-7) **86**, 181

1-Bromo-8-chlorooctane: Octane, 1-bromo-8-chloro-; (28598-82-5) **87**, 299

6-Bromohexanenitrile: Hexanenitrile, 6-bromo-; (6621-59-6) **87**, 330

8-Bromo-1-octanol: 1-Octanol, 8-bromo-; (50816-19-8) **87**, 299

4-Bromophenylboronic acid; (5467-74-3) **86**, 344

4-Bromophenylboronic MIDA ester **86**, 344

p-Bromophenyl methyl sulfide: Benzene, 1-bromo-4-(methylthio)-; (104-95-0)
    86, 121

(S)-(–)-p-Bromophenyl methyl sulfoxide **86**, 121

(*S*)-(–)-*p*-Bromophenyl methyl sulfoxide: Benzene, 1-bromo-4-[(*S*)-methylsulfinyl]-;
    (145266-25-0) **86**, 121

1-(3-Bromophenyl)-2-[5-(trifluoromethyl)-2-pyridinyl]ethanone **86**, 18

2-(3-Bromophenyl)-6-(trifluoromethyl)pyrazolo[1,5-a]pyridine **86**, 18

(1Z)-1-(3-Bromophenyl)-2-[5-(trifluoromethyl)-2-pyridinyl]-ethanone oxime **86**, 18

2-Bromopropene; (557-93-7) **85**, 1, 172

N-Bromosuccimide: NBS; (128-08-5) **85**, 53, 267; **86**, 225; **87**, 16

4-Bromo-5-(thiophen-2-yl)oxazole: Oxazole, 4-bromo-5-(2-thienyl)-; (959977-82-3) **87**,
    16

4-Bromotoluene; (106-38-7) **85**, 196

2-Butanone; (78-93-3) **86**, 333

2-tert-Butoxycarbonylamino-4-(2,2-dimethyl-4,6-dioxo-[1,3]dioxan-5-yl)-4-oxo-butyric
    acid tert-butyl ester; (10950-77-9) **85**, 147

(2S)-2-[(tert-Butoxycarbonyl)amino]-2-phenylethyl methanesulfonate (110143-62-9) **85**,
    219

2-[3,3'-Di-*tert*-butoxycarbonyl)-aminodipropylamine]-4,6-dichloro-1,3,5-triazine; 12-
    Oxa-2,6,10-triazatetradecanoic acid, 6-(4,6-dichloro-1,3,5-triazin-2-yl)-13,13-
    dimethyl-11-oxo-, 1,1-dimethylethyl ester; (947602-03-1) **86**, 141, 151

N-α-*tert*-Butoxycarbonyl-L-aspartic acid α-*tert*-butyl ester (Boc-L-Asp-O*t*-Bu); (34582-
    32-6) **85**, 147

1-tert-Butoxycarbonyl-2,3-dihydropyrrole: 1H-Pyrrole-1-carboxylic acid, 2,3-dihydro-,
    1,1-dimethylethyl ester; (73286-71-2) **85**, 64

2-(*tert*-Butoxycarbonyloxyimino)-2-phenylacetonitrile; (58632-95-4) **86**, 141

N-(*tert*-Butoxycarbonyl)-piperazine; (57260-71-6) **86**, 141

Carbamic acid, N-[phenyl(phenylsulfonyl)methyl]-, 2-propen-1-yl ester; (921767-12-6)
86, 236
Carbamic acid, 2-propen-1-yl ester; (2114-11-6) **86**, 236
(Carbethoxymethylene)triphenylphosphorane; (1099-45-2) **85**, 15
Carbon disulfide; (75-15-0) **86**, 70
Carbon monoxide; (630-08-0) **86**, 287
Carbon tetrabromide: Methane, tetrabromo-; (558-13-4) **87**, 231
Carbonyl(dihydrido)tris(triphenylphosphine)ruthenium (II); (25360-32-1) **86**, 28
1,1'-Carbonyldiimidazole: Methanone, di-1*H*-imidazol-1-yl-; (530-62-1) **86**, 58
Cbz-L-proline: 1,2-Pyrrolidinedicarboxylic acid, 1-(phenylmethyl) ester, (2*S*)-; (1148-11-4) **85**, 72
Cesium carbonate: Carbonic acid, cesium salt (1:2); (534-17-8) **86**, 181; **87**, 231
Cesium fluoride; (13400-13-0) **86**, 161
Chloroacetonitrile (107-14-2) **86**, 1
4-Chlorobenzaldehyde; (104-88-1) **85**, 179
Chlorobenzene; (108-90-7) **86**, 105; **87**, 362
(3-Chlorobutyl)Benzene; (4830-94-8) **87**, 88
Chloro(chloromethyl)dimethylsilane; (1719-57-9) **87**, 178
2-Chloro-6,7-dimethoxy-1,2,3,4-tetrahydroisoquinoline **87**, 8
(Chloromethyl)dimethylphenylsilane; (1833-51-8) **87**, 178
2-Chloro-5-(3-methylphenyl)-thiophene; (1078144-58-7) **87**, 178, 184
4-Chlorophenylboronic acid: Boronic acid, B-(4-chlorophenyl)-; (1679-18-1) **86**, 360
2-Chloropyridine; (109-09-1) **85**, 88
5-Chlorosalicylaldehyde: Benzaldehyde, 5-chloro-2-hydroxy-; (635-93-8) **86**, 172
2-Chlorothiophene; (96-43-5) **87**, 178, 184
(13-Chlorotridecyloxy)triethylsilane: Silane, [(13-chlorotridecyl)oxy]triethyl-; (374754-99-1) **87**, 299
Chlorotriethylsilane: Silane, chlorotriethyl-; (994-30-9) **87**, 299
2-Chloro-5-(trifluoromethyl)pyridine; (52334-81-3) **86**, 18
m-Chloroperbenzoic acid; Peroxybenzoic acid, *m*-chloro- (8); Benzocarboperoxoic acid, 3-chloro- (9); (937-14-4) **86**, 308
Chlorotrimethylsilane: Silane, Chlorotrimethyl-; (75-77-4) **86**, 252
Cinnamyl alcohol: 3-Phenyl-2-propen-1-ol; (104-54-1) **85**, 96
Cinnamyl-H-phosphinic acid: [(2E)-3-phenyl-2-propenyl]-Phosphinic acid; (911128-46-6) **85**, 96
(±)-Citronellal: ((±)-3,7-dimethyl-6-octenal); (106-23-0) **87**, 201
Cobalt(II) tetrafluoroborate hexahydrate; (15684-35-2) **87**, 88
Condensation **85**, 27, 34, 179, 248, 267; **86**, 11, 18, 92, 121, 212, 252, 262; **87**, 36, 59, 77, 88, 115, 143, 192, 201, 218, 275, 310, 339, 362
Copper(I) bromide; (7787-70-4) **85**, 196
Copper-catalyzed reactions **87**, 53, 126, 184, 231
Copper chloride: Cuprous chloride; (7758-89-6) **85**, 209 ;
Copper Cyanide; (544-92-3) **85**, 131
Copper iodide; (1335-23-5) **86**, 181
Copper(I) iodide: Cuprous iodide; (7681-65-4) **86**, 225; **87**, 126, 178, 184, 231
Coupling **85**, 158, 196; **86**, 105, 225, 274; **87**, 184, 299, 317, 330
Cuprous chloride; (7758-89-6) **85**, 209
(S)-2-Cyano-pyrrolidine-1-carboxylic acid benzyl ester: (63808-36-6) **85**, 72
Cyanuric chloride: 2,4,6-Trichloro-1,3,5-triazine; (108-77-0) **85**, 72; **86**, 141
Cyclen: 1,4,7,10-Tetraazacyclododecane; (294-90-6) **85**, 10

392

6,7-Dimethoxy-1,2,3,4-tetrahydroisoquinoline hydrochloride: Isoquinoline, 1,2,3,4-
  tetrahydro-6,7-dimethoxy-, hydrochloride (1:1); (2328-12-3) **87**, 8
(2R,3S,4E)-N,3-Dimethoxy-N,2,4-trimethyl-5-phenyl-4-pentenamide
(3,5-Dimethoxy-1-phenyl-cyclohexa-2,5-dienyl)-acetonitrile **86**, 1
Dimethyl acetamide; (127-19-5) **86**, 298
Dimethylamine; (124-40-3) **86**, 298
4-(Dimethylamino)benzoic acid: (619-84-1) **87**, 201
N-(3-Dimethylaminopropyl)-*N*'-ethylcarbodiimide hydrochoride (EDC·HCl); (25952-53-
  8) **85**, 147
4-Dimethylaminopyridine: 4-Pyridinamine, *N,N*-dimethyl-; (1122-58-3) **85**, 64; **86**, 81
9,9-Dimethyl-4,5-bis(diphenylphosphino)xanthene: Xantphos; (161265-03-8) **85**, 96
Dimethyl-bis-phenylethynyl silane: Benzene, 1,1'-[(dimethylsilylene)di-2,1-
  ethynediyl]bis-; (2170-08-3) **85**, 53
(2R,3R)-2,3-Dimethylbutane-1,4-diol: (2R,3R) 2,3-Dimethyl-1,4-butanediol; (127253-15-
  0) **85**, 158
2,2-Dimethyl-1,3-dioxane-4,6-dione (Meldrum's acid); (2033-24-1) **85**, 147
4,5-Dimethyl-1,3-dithiol-2-one; (49675-88-9) **86**, 333
N,N'-Dimethylethylenediamine: 1,2-Ethanediamine, *N*1,*N*2-dimethyl-; (110-70-3) **86**,
  181; **87**, 231
N,O-Dimethylhydroxylamine hydrochloride: Methanamine, *N*-methoxy-, hydrochloride;
  (6638-79-5) **86**, 81
Dimethyl malonate; (108-59-8) **87, 115**
(4S)-N-[(2R,3S,4E)-2,4-Dimethyl-3-methoxy-5-phenyl-4-pentenoyl]-4-isopropyl-1,3-
  thiazolidine-2-thione **86**, 81
3,7-Dimethyl-2-methylene-6-octenal; (22418-66-2) **87**, 201
*4,4-Dimethyl-3-oxo-2-benzylpentanenitrile (*875628-78-7) **86**, 28
4,4-Dimethyl-3-oxopentanenitrile; (59997-51-2) **86**, 28
Dimethyl 2-phenylcyclopropane-1,1-dicarboxylate (3709-20-4) **87, 115**
7,7-Dimethyl-3-phenyl-4-p-tolyl-6,7,8,9-tetrahydro-1H-pyrazolo[3,4-b]-quinolin-5(4H)-
  one: 5H-Pyrazolo[3,4-b]quinolin-5-one, 1,4,6,7,8,9-hexahydro-7,7-dimethyl-4-(4-
  methylphenyl)-3-phenyl-; (904812-68-6) **86**, 252
N, N-Dimethyl-4-pyridinamine: (1122-58-3) **85**, 64
(2R,3R)-2,3-Dimethylsuccinic acid; (5866-39-7) **85**, 158
Dimethyl sulfoxide: Methyl sulfoxide; Methane, sulfinybis-; (67-68-5) **85**, 189
(–)-(*S*)-1-(1,3,2-*Dioxaborolan-2-yloxy)-3-methyl-1,1-diphenylbutan-2-amine*; (879981-
  94-9) **87**, 26, 36
2-(1,3-Dioxolan-2-yl)ethyl bromide: 1,3-Dioxolane, 2-(2-bromoethyl)-; (18742-02-4) **87**,
  317
2-[2-(1,3-Dioxolan-2-yl)ethyl]zinc bromide: Zinc, bromo[2-(1,3-dioxolan-2-yl)ethyl]-;
  (864501-59-7) **87**, 317
1,3-Diphenylacetone *p*-tosylhydrazone: Benzenesulfonic acid, 4-methyl-, [2-phenyl-1-
  (phenylmethyl)ethylidene]hydrazide; (19816-88-7) **85**, 45
Diphenyldiazomethane (883-40-9) **85**, 189
(4-((4R,5R)-4,5-Diphenyl-1,3-dioxolan-2-yl)phenoxy)(tert-butyl)dimethylsilane **86**, 130
(1R,2R)-1,2-Diphenylethane-1,2-diol **86**, 130
α,α-Diphenylglycine: Benzeneacetic acid, α-amino-α-phenyl-; (3060-50-2) **87**, 88
Diphenylphosphine: Phosphine, diphenyl-; (829-85-6) **86**, 181
(S)-(–)-1,3-Diphenyl-2-propyn-1-ol: Benzenemethanol, α-(2-phenylethynyl)-, (αS)-;
  (132350-96-0) **85**, 118

(*S*)-4-Isopropyl-*N*-propanoyl-1,3-thiazolidine-2-thione: 2-Thiazolidinethione, 4-(1-methylethyl)-3-(1-oxopropyl)-, (4*S*)-; (102831-92-5) **86**, 70

(4S)-Isopropyl-3-propionyl-2-oxazolidinone: (4S)-4-(1-Methylethyl)-3-(1-oxopropyl)-2-oxazolidinone; (77877-19-1) **85**, 158

(S)-4-Isopropyl-1,3-thiazolidine-2-thione **86**, 70

(*S*)-4-Isopropyl-1,3-thiazolidine-2-thione: 2-Thiazolidinethione, (4*S*)-4-(1-methylethyl)-; (76186-04-4) **86**, 70

(L)-*tert*-Leucine: L-Valine, 3-methyl-; (20859-02-3) **86**, 181

(L)-(+)-*tert*-leucinol: 1-Butanol, 2-amino-3,3-dimethyl-, (2*S*)-; (112245-13-3) **86**, 121

Lithium (7439-93-2) **85**, 53; **86**, 1

Lithium aluminum hydride; (16853-85-3) **85**, 158; **87**, 310

Lithium bis(trimethylsilyl)amide; (4039-32-1) **87**, 16

Lithium *t*-butoxide; (1907-33-1) **87**, 178, 184

Lithium chloride; (7447-41-8) **86**, 47

Lithium hydroxide monohydrate; (1310-66-3) **85**, 295

Lithium triethylborohydride; (22560-16-3) **85**, 64

(*R*)-(-)-Mandelic acid; (611-71-2) **85**, 106

(*S*)-(+)-Mandelic acid; (17199-29-0) **85**, 106

(R)-Mandelic acid salt of (1S,2S)-trans-2-(N-benzyl)amino-1-cyclohexanol; (882409-00-9) **85**, 106

(S)-Mandelic acid salt of (1R,2R)-trans-2-(N-benzyl)amino-1-cyclohexanol; (882409-01-0) **85**, 106

Manganese(IV) oxide; (1313-13-9)

Magnesium; (7439-95-4) **87, 178**

Mesitylamine: Benzenamine, 2,4,6-trimethyl-; (88-05-1) **85**, 34; **87, 77**

3-(Mesitylamino)butan-2-one: 2-Butanone, 3-[(2,4,6-trimethylphenyl)-amino]-; (898552-96-0) **85**, 34

Mesitylene (108-67-8) **85**, 196

2-Mesitylhydrazinium chloride; (76195-82-9) **87**, 362

N-Mesityl-N-(3-oxobutan-2-yl)formamide: Formamide, N-(1-methyl-2-oxopropyl)-N-(2,4,6-trimethylphenyl)-; (898553-01-0) **85**, 34

Metal complexation **87**, 26, 104

Metallation, **85**, 1, 45, 209; **86**, 374; **87**, 317, 330

Methansulfonyl chloride; (124-63-0) **85**, 219; **86**, 181

trans-4-Methoxy-3-buten-2-one; (51731-17-0) **87**, 192

(4*S*)-3-[(2*R*,3*S*,4*E*)-3-Methoxy-2,4-dimethyl-1-oxo-5-phenyl-4-pentenyl]-4-(1-methylethyl)-2-thiazolidinethione; (332902-42-8) **86**, 81, 181

Methyl acetoacetate: Butanoic acid, 3-oxo-, methyl ester; (105-45-3) **86**, 161

Methyl 2-(2-acetylphenyl)acetate: Benzeneacetic acid, 2-acetyl-, methyl ester; (16535-88-9) **86**, 161

(R)-Methyl 2-allylpyrrolidine-2-carboxylate hydrochloride (112348-46-6) **86**, 262

Methyl anthranilate; (134-20-3) **87**, 226

(*R*)-α-Methylbenzylamine; (*R*)-α-phenylethylamine; (*R*)-1-phenylethylamine (3886-69-9) **87**, 143

(*S*)-(α-Methylbenzyl)hydroxylamine oxalate salt: (α*R*)-*N*-Hydroxy-α-methyl-benzenemethanamine ethanedioate salt; (78798-33-1) **87**, 218

2-Methyl-3-butyn-2-ol; (115-19-5) **85**, 118
Methyl crotonate; (623-43-8) **87**, 253
(E)-3,4-Methylenedioxy-β-nitrostyrene; (22568-48-5) **85**, 179
Methylhydrazine; (60-34-4) **85**, 179
N-Methyliminodiacetic acid; (4408-64-4) **86**, 344
Methylmagnesium bromide (75-16-1) **87**, 143, 317
Methyl 2-methyl-3-nitrobenzoate; (59382-59-1) **86**, 92
4-Methyl-2'-nitrobiphenyl; (70680-21-6) **85**, 196
trans-p-Methyl-β-nitrostyrene: Benzene, 1-methyl-4-[(1*E*)-2-nitroethenyl]-; (5153-68-4)
    85, 179
*(1S,2S)-2-Methyl-3-oxo-1-phenylpropylcarbamate***86**, 11
Methyl 2-(3-oxopropyl)benzoate; (106515-77-9) **87**, 226
4-Methyl-n-(2-phenylethynyl)-n-(phenylmethyl)benzenesulfonamide: (609769-63-3) **87**,
    231
4-Methyl-*N*-(phenylmethyl)benzenesulfonamide; (1576-37-0) Benzylamine:
    Benzenemethanamine; (100-46-9) **87**, 231
(R)-3-Methyl-3-phenylpentene: [(1R)-1-ethyl-1-methyl-2-propenyl]-benzene]; (768392-
    48-9) **85**, 248
(*S*)-(−)-4-Methyl-1-phenyl-2-pentyn-1,4-diol: (321855-44-1) **85**, 118
(*E*)-2-Methyl-3-phenylpropenal; (15174-47-7) **86**, 81
Methyl propionylacetate: Pentanoic acid, 3-oxo-, methyl ester; (30414-53-0) **85**, 27
1-Methyl-2-pyrrolidone; (872-50-4) **85**, 196
1-Methyl-2-pyrrolidinone (872-50-4) **85**, 238
Methyltriphenylphosphonium bromide; (1779-49-3) **85**, 248
Microwave **86**, 18, 252
Morpholine; (110-91-8) **87**, 59
Morpholinomethanol: 4-Morpholinemethanol; (4432-43-3) **87, 59**

Naphthalene; (91-20-3) **85**, 53
Nickel(II) chloride, dimethoxyethane adduct (NiCl₂·glyme): Nickel, dichloro[1,2-
    di(methoxy-κO)ethane]-; (29046-78-4) **87**, 317, 330
Nickel-catalyzed reactions **87**, 317, 330
Nitriles **86**, 1,28
2-Nitrobenzaldehyde; (552-89-6) **85**, 27; **86**, 36; **87**, 339
2-Nitrobenzoic acid; (552-16-9) **85**, 196
2-Nitrobenzyl bromide: Benzene, 1-(bromomethyl)-2-nitro-; (3958-60-9) **87**, 339
(*S*)-5,5',6,6',7,7',8,8'-Octahydro-1,1'-bi-2-naphthol [(*S*)-H₈BINOL]: [1,1'-
    Binaphthalene]-2,2'-diol, 5,5',6,6',7,7',8,8'-octahydro-, (1*S*)-; (65355-00-2) **87**, 59,
    263
(*S*)-1-(8,9,10,11,12,13,14,15-Octahydro-3,5-dioxa-4-phosphacyclohepta[2,1-a;3,4-
    a']dinaphthalen-4-yl)-2,2,6,6-tetramethylpiperidine **87**, 263
2-Octenoic acid, 3-phenyl-, methyl ester, (2E)-; (189890-29-7) **87,** 53
2-Octynoic acid, methyl ester; (111-12-6) **87,** 53
Organocatalysis **86**, 11; **87**, 201
Oxalyl chloride: Ethanedioyl dichloride; (79-37-8) **85**, 189
Oxidation, **85**, 15, 189, 267, 278; **86**, 1, 28, 121, 308. 315; **87**, 1, 8, 253
Oximes **86**, 18; **87**, 36, 275
Oxirane, 2-methyl-; (75-56-9) **87**, 126
Oxone® monopersulfate; (37222-66-5) **85**, 278

Palladium **86**,105; **87**, 104, 143, 226, 263, 299
Palladium(II) acetate; (3375-31-3) **85**, 96; **86**, 92, 105, 344; **87**, 299
Palladium acetylacetonate; (140024-61-4) **85**, 196
Palladium hydroxide on carbon; Pearlman's catalyst (12135-22-7) **87**, 143
Palladium tris(dibenzylideneacetone); (48243-18-1) **86**, 274
Paraformaldehyde; (30525-89-4) **87**, 59
2-Pentafluorophenyl-6,10b-dihydro-4*H*,5a*H*-5-oxa-3,10c-diaza-2-
     azoniacyclopenta[*c*]fluorene; tetrafluoroborate; (740816-14-2) **87**, 350
2-Pentafluorophenyl-6,7-dihydro-5*H*-pyrrolo[2,1-*c*][1,2,4]triazol-2-ium;
     tetrafluoroborate; (862095-91-8) **87**, 350
Pentafluorophenylhydrazine; (828-73-9) **87**, 350
4-Penten-1-ol; (821-09-0) **87**, 299
1,10-Phenanthroline; (66-71-7) **85**, 196; **86**, 92; **87**, 178, 184
(1,10-Phenanthroline)bis(triphenylphosphine)copper(I) nitrate; (33989-10-5) **85**, 196
Phenylacetylene: Benzene, ethynyl-; (536-74-3) **85**, 53, 118, 131; **86**, 325
(*S*)-Phenylalaninol: (3182-95-4) **85**, 267
N-Phenylbenzenecarboxamide (benzanilide); (93-98-1) **85**, 88
4-Phenyl-1-butene: Benzene, 3-buten-1-yl-; (768-56-69) **87**, 88
Phenylboronate: 1,3,2-Dioxaborinane, 5,5-dimethyl-2-phenyl-; (5123-13-7) **87**, 209
(*E*)-1-Phenyl-1,3-butadiene: Benzene, (1*E*)-1,3-butadienyl-: (16939-57-4) **86**, 315
2-Phenyl-1-butene; (2039-93-2) **85**, 248
5-Phenyl-1,3-dimethoxybenzene **86**, 1
N-[(1*S*)-1-*phenylethyl]-benzeneacetamide*: *2-Phenyl-N-(1-phenylethyl)acetamide;*
     (17194-90-0) **87**, 218
(*S*)-Phenylglycine: Benzeneacetic acid, α-amino-, (α*S*)-; (2935-35-5) **85**, 219
5-Phenyl-2-isobutylthiazole; (600732-10-3) **86**, 105
Phenylmagnesium chloride, 2 M in THF; (100-59-4) **87**, 26
3-Phenylpropanoyl chloride: Benzenepropanoyl chloride; (645-45-4) **85**, 295
5-Phenyl-1H-pyrazol-3-amine: 1*H*-pyrazol-3-amine, 5-phenyl-; (1571-10-7) **86**, 252
Phenylpyruvic acid: α-Oxobenzenepropanoic acid, 2-Oxo-3-phenylpropionic acid; (156-
     06-9) **87**, 218
Phenylsilane: Benzene, silyl-; (694-53-1) **87**, 88
8-*Phenyl-1-tetralone* (501374-10-3) **87**, 209
N-[1-Phenyl-3-(trimethylsilyl)-2-propyn-1-ylidene]-benzeneamine; (77123-64-9) **85**, 88
Phosphoric triamide, *N*,*N*,*N'*,*N'*,*N''*,*N''*-hexamethyl- ; (680-31-9) **87**, 126
Phosphorus trichloride (7719-12-2) **85**, 238; **87**, 263
Phosphorylation **87**, 263
2-Picoline borane: Boron, trihydro(2-methylpyridine)-, (T-4)-; (3999-38-0) **87**, 339
Pimelic acid: Heptanedioic acid; (111-16-0) **86**, 194
Pinacolone: 2-Butanone, 3,3-dimethyl-; (75-97-8) **87**, 209
Piperidinium acetate; (4540-33-4) **86**, 28
Potassium *tert*-butoxide: 2-Propanol, 2-methyl-, potassium salt (1:1); (865-47-4) **86**, 315
Potassium carbonate; (584-08-7) **85**, 287; **86**, 58
Potassium hydroxide; (1310-58-3) **85**, 196; **87**, 115
Potassium *O*-isopropylxanthate: Carbonodithioic acid, O-(1-methylethyl) ester,
     potassium salt (1:1); (140-92-1) **86**, 333
Potassium trimethylsilanolate; (10519-96-7) **86**, 274
Propanol, 2-amino-, 3-phenyl, (*S*); (3182-95-4) **85**, 267

400

Silver(I) fluoride; (7775-41-9) **86**, 225
Silver nitrate; (7761-88-8) **86**, 225
Silylation **87**, 178, 299
Sodium; (7440-23-5) **86,** 262
Sodium acetate; (127-09-3) **86**, 105, 194
Sodium acetate trihydrate; (6131-90-4) **85**, 10
Sodium amide; (7782-92-5) **86**, 298
Sodium azide; (26628-22-8) **85**, 72, 278; **86**, 113; **87**, 161
Sodium bis(trimethylsilyl)amide; (1070-89-9) **87**, 170
Sodium borohydride: Borate(1-), tetrahydro-, sodium (1:1); (16940-66-2) **85**, 147; **86**,
    181; **87**, 77
Sodium carbonate; (497-19-8) **86**, 36
Sodium cyanide; (143-33-9) **85**, 219
Sodium hydride; (7646-69-7) **85**, 45, 172; **86**, 18, 194; **87**, 36
Sodium hypochlorite; (7681-52-9) **87**, 8
Sodium nitrite; (7632-00-0) **87**, 362
Sodium tetrafluoroborate; (13755-29-8) **85**, 248
Sodium tetrakis[(3,5-trifluoromethyl)phenyl]borate; (79060-88-1) **85**, 248
Sodium thiosulfate (7772-98-7) **86**, 1
4-Spirocyclohexyloxazolidinone **86**, 58
4-Spirocyclohexyloxazolidinone: 3-Oxa-1-azaspiro[4.5]decan-2-one: (81467-34-7) **86**,58
Styrene; (100-42-5) **87**, 115
Substitution **86**, 18, 141, 151, 181, 333; **87**, 104, 126, 299, 317
Sulfonation **86**, 11; **87**, 231
Sulfur; (7704-34-9) **85**, 209; **87**, 231
Super-Hydride®: Lithium triethylborohydride; (22560-16-3) **85**, 64

1,4,7,10-Tetraazacyclododecane-1,4,7-triacetic acid, Tri-tert-butyl Ester Hydrobromide:
    1,4,7,10-Tetraazacyclododecane-1,4,7-tricarboxylic acid, 1,4,7-tris(1,1-
    dimethylethyl) ester; (175854-39-4) **85**, 10
Tetrabromomethane; (558-13-4) **86**, 36
Tetrabutylammonium bromide; (1643-19-2) **85**, 278
Tetrabutylammonium fluoride; (429-41-4) **87**, 95
Tetrahydro-1,3-dimethyl-2(1H)-pyrimidinone; (7226-23-5) **87**, 178, 184
4,4a,9,9a-Tetrahydro-1-oxa-4-azafluoren-3-one: Indeno[2,1-*b*]-1,4-oxazin-3(2*H*)-one,
    4,4a,9,9a-tetrahydro-, (4a*R*,9a*S*)-; (862095-79-2) **87**, 350
Tetrafluoroboric acid: Borate(1-), tetrafluoro-, hydrogen (1:1); (16872-11-0) **87**, 288
Tetrafluoroboric acid diethyl etherate; (67969-82-8) **86**, 172
Tetrakis(dimethylamino)allene; (42928-64-3) **86**, 298
1,1,3,3-Tetrakis(dimethylamino)propenium tetrafluoroborate; (125254-01-5) **86**, 298
Tetrakis(triphenylphosphine)palladium; (14221-01-3) **86**, 315
α-Tetralone: 1(2*H*)-Naphthalenone, 3,4-dihydro-; (529-34-0) **87**, 209, 275
2,2,6,6-Tetramethylpiperidine; (768-66-1) **86**, 374; **87**, 263
1,3,5,7-Tetramethyl-1,3,5,7-tetravinylcyclotetrasiloxane; (2554-06-5) **86**, 274
(S)-2-(1H-Tetrazol-5-yl)-pyrrolidin-1-carboxylic acid benzyl ester: 1-
    Pyrrolidinecarboxylic acid, 2-(2H-tetrazol-5-yl)-, phenylmethyl ester, (2S)-;
    (33876-20-9) **85**, 72
Tetrolic acid: 2-Butynoic acid (9); (590-93-2) **85**, 231
2-Thiophene-carboxaldehyde; (98-03-3) **86**, 360; **87**, 16

5-(Thiophen-2-yl)oxazole: Oxazole, 5-(2-thienyl)-; (70380-70-0) **87**, 16
Thiourea; (62-56-6) **87**, 115
Tin (II) chloride; (7772-99-8) **85**, 27
Tin(II) chloride dihydrate; (10025-69-1) **87**, 362
Titanium tetrachloride; (7550-45-0) **85**, 295; **86**, 81
p-Tolualdehyde: benzaldehyde, 4-Methyl-; (104-87-0) **86**, 252
p-Toluenesulfonamide: Benzenesulfonamide, 4-methyl-; (70-55-3) **86**, 212
p-Toluenesulfonic acid: Benzenesulfonic acid, 4-methyl-; (104-15-4) **85**, 287
p-Toluenesulfonic acid monohydrate: Benzenesulfonic acid, 4-methyl-, hydrate (1:1);
    (6192-52-5) **86**, 194
(*S*)-3-[*N*-(*p*-Toluenesulfonyl)amino]-4-methylpentanoic acid (936012-07-6) **87**, 143
p-Toluenesulfonyl chloride: Benzenesulfonyl chloride, 4-methyl-; (98-59-9) **86**, 58; **87**,
    88, 231
p-Tolylboronic acid; (5720-05-8) **86**, 344
4-(p-Tolyl)-phenylboronic acid **86**, 344
4-(p-Tolyl)-phenylboronic acid MIDA ester **86**, 344
Tosylmethylisocyanide: Benzene, [(isocyanomethyl)sulfonyl]-; (36635-63-9) **87**, 16
1,3,5-*Triacetylbenzene*; (779-90-8) **87**, 192
Tribasic potassium phosphate; (7778-53-2) **86**, 105
Tri-*n*-butylphosphine: Phosphine, tributyl-; (998-40-3) **86**, 212
2,2,2-Trichloro-1-ethoxyethanol; (515-83-3) **86**, 262
(3R,7aS)-3-(Trichloromethyl)tetrahydropyrrolo[1,2-c]oxazol-1(3H)-one; (97538-67-5)
    86, 262
Tricyclohexylphosphine: Phosphine, tricyclohexyl-; (2622-14-2) **87**, 299
Triethylamine: Ethanamine, *N,N*-diethyl-; (121-44-8) **85**, 131, 189, 219, 295; **86**, 58, 81,
    252, 315; **87**, 231
Triethylamine hydrochloride: Ethanamine, *N,N*-diethyl-, hydrochloride (1:1); (554-68-7)
    85, 72
Triethyl orthoformate: Ethane, 1,1',1''-[methylidynetris(oxy)]tris- ; (122-51-0) **87**, 77,
    350, 362
Triethyl(pent-4-enyloxy)silane: Silane, triethyl(4-penten-1-yloxy)-; (374755-00-7) **87**,
    299
Triethylphosphine; (554-70-1) **87**, 275
Trifluoroacetic acid (76-05-1) **87**, 143
Trifluoroacetic anhydride; (407-25-0) **85**, 64; **86**, 18
2,2,2-Trifluoroethanol (75-89-8) **85**, 45
Trifluoro, 2-(trimethylsilyl)phenyl ester; (88284-48-4) **86**, 161
Trifluoromethylation **87**, 126
Trifluoromethanesulfonic acid; HIGHLY CORROSIVE, Methanesulfonic acid, trifluoro-
    (8, 9); (1493-13-6) **86**, 308
Trifluoromethanesulfonimide; (82113-65-3) **87**, 253
Triflic anhydride: Trifluoromethanesulfonic acid anhydride; (358-23-6) **85**, 88
Triisopropyl phosphite; (116-17-6) **86**, 36
(Triisopropylsilyl)acetylene: Ethynyltriisopropylsilane; (89343-06-6) **86**, 225
Trimethylacetaldehyde: Pivaldehyde: Propanal, 2,2-dimethyl; (630-19-3) **85**, 267
Trimethylaluminum: (75-24-1) **87**, 104
2,4,6-Trimethylaniline; (88-05-1) **87**, 362
Trimethyl orthoformate; (149-73-5) **86**, 81, 130
Trimethyloxonium tetrafluoroborate; (420-37-1) **87**, 350, 362
1-(Trimethylsilyl)acetylene; (1066-54-2) **85**, 88

402   *Org. Synth.* **2010**, *87*